WRITE IT DOWN

Guidance for Preparing Effective and Compliant Documentation

Second Edition

WRITE IT DOWN

Guidance for Preparing Effective and Compliant Documentation

Second Edition

Janet Gough

CRC Press is an imprint of the
Taylor & Francis Group, an **informa** business
A TAYLOR & FRANCIS BOOK

CRC Press
Taylor & Francis Group
6000 Broken Sound Parkway NW, Suite 300
Boca Raton, FL 33487-2742

First issued in paperback 2019

ISBN-13: 978-0-8493-2171-9 (hbk)
ISBN-13: 978-0-367-39313-7 (pbk)

Library of Congress Card Number 2004057917

Library of Congress Cataloging-in-Publication Data

Gough, Janet.
Write it down : guidance for preparing effective and compliant documentation / Janet Gough.-- 2nd ed.
p. cm.
Includes bibliographical references and index.
ISBN 0-8493-2171-9 (alk. paper)
1. Pharmaceutical industry--Documentation--Standards. 2. Pharmaceutical industry--Records and correspondence. I. Title.

HD9665.6.G68 2004
615'.1'068--dc22 2004057917

Visit the Taylor & Francis Web site at
http://www.taylorandfrancis.com

and the CRC Press Web site at
http://www.crcpress.com

Table of Contents

Dedication

This book is dedicated to my husband, Gary, and my children, Erin and Christian.

Introduction

This is a book about writing. As such, it presents an overview of the regulated environments in which companies develop, manufacture, and distribute therapeutic products. It has a three-pronged focus: to help writers understand the "why" of what they must write and the current industry standards for good documentation practices; to provide effective examples of a broad spectrum of documents; and to provide in-depth explanation of grammar and punctuation conventions.

It is by no means a book of regulatory guidance. While it gives an overview of the regulations, the purpose is to place the writing task in the context of the existing laws and guidances that drive documentation from discovery to dossier and beyond. Title 21 of Code of Federal Regulations is the primary regulatory focus. The documents herein are simply examples of working documents. They include data collection forms, audit reports, standard operating procedures, laboratory methods, development reports, excerpts from quality manuals and plans, and sections of dossiers. As the regulatory environment or industry standards change over time, any of the examples could be subject to revision for compliance purposes. And, indeed, that's why good document controls are important. The book touches on document controls in the context of the documents themselves whether the systems are manual or electronic, in compliance with 21 CFR Part 11 Electronic Records; Electronic Signatures.

The main purpose of the book is to help writers of English master the art of preparing effective documents. It includes extensive information on the structure of the language, with focus on those components that are particularly troublesome for non-native writers of English. Chapters on style, grammar, verb tenses in English, punctuation, and usage provide detailed guidance for writing clearly and concisely.

Many of the examples in this book have been provided by professionals in the industry. To all new contributors to this edition, I am grateful. Many contributed examples from the *Write It Down* first edition have moved into this second edition because they represent good documentation. (See "courtesy of" beneath the examples.) Still other examples are fictitious, but representative of the broad spectrum of the type of writing that occurs every day in companies that have achieved discovery and are in various stages of development and manufacture and market presence.

About the Author

Janet Gough has extensive experience as a consultant to the pharmaceutical, biotech, and medical device industries. She designs systems for compliance with the binding regulations, prepares documentation, and conducts training. She has been a director of technical communications for a biotech company and has taught English in university graduate and undergraduate programs. As a faculty member of professional training organizations, she teaches *Technical Writing in the Pharmaceutical and Allied Industries, Writing When English is Your Second Language, Medical Writing, Writing Effective Standard Operating Procedures and Other Process Documents, Good Documentation Practices,* and *Electronic Record Keeping: Achieving and Maintaining Compliance with 21 CFR Part 11 and 45 CFR Parts 160, 162, and 164.* She is the co-author of *Electronic Record Keeping: Achieving and Maintaining Compliance with 21 CFR Part 11 and 45 CFR Parts 160, 162, and 164,* both from CRC Press. She is the author of *Hosting A Compliance Inspection* and co-author of *The Internal Quality Audit, The External Quality Audit,* and *Commercial Off-the-Shelf (COTS) Software Validation for 21 CFR Part 11 Compliance* from Davis Horwood International (DHI) and the Parenteral Drug Association (PDA). She is listed in Who's Who in Medicine and Healthcare, 2005.

Janet Gough, MA
Compliant systems, documentation, and training
973-252-3731 phone
janetgough@optonline.net
www.writeitdown.biz

Acknowledgments

A number of accomplished people and organizations contributed excellent examples to this edition of *Write It Down,* reflective of the type of writing that occurs in the industry. Without their input this book would not be as complete or expansive as it is. Their contributions attest to the good, solid command of the language the people in this industry aspire to, and are exemplary documents. In particular, I wish to express my gratitude to the following people.

Amarpal Sahay
Arthur Melnick
David Nettleton
Halim Hasan
Joan Lorenz
John Cline
Kristine Ogozalek
Lasentec, Inc.
Marilyn Brown
Michael Nolan
Monica Grimaldi
Phil Cloud
Synaptic Pharmaceutical Corporation
Tom Lang

I am also particularly grateful to those family members who provided support and encouragement as this book took shape: my husband, Gary Gough, and my sister, Astrid Reynolds. My editor Steve Zollo and my friend and colleague Kathleen Monroe also deserve thanks for their efforts to make this book a reality.

1

Writing Within the Regulated Environment

As far as most regulatory bodies are concerned, if you didn't write it down, it didn't happen. Working in the pharmaceutical, medical device, or biologic milieu is tantamount to journal keeping. In fact, "Write it down" sums up what it takes to get the job done properly. Successful operations require a working union of the day-to-day activities that keep the wheels of the business turning and the documentation that affirms those activities.

Documents show the framework of a company's varied activities. When the development process for one product is winding down, for example, and a submission is forthcoming, development processes for other products may be in various stages of development, from discovery through final testing in clinical trials and launch. Concurrently, other company products may be in various stages of scale-up or production. For each product in each phase of development or production, there must be a written history that shows control of all activities related to that product. Thus, the answer to "What did the company do in March (or on Tuesday) in the production of acetaminophen?" should be easily available in the records the company keeps.

The sheer volume of documentation that takes place makes writing well a critical skill, one that is essential for success. It is also true that writing is intimidating for many people. Perhaps because writing is so closely scrutinized, people are loath to commit their words to paper. If writing is on your list of job responsibilities, there are some avenues you can take to make the task less formidable. The first is gaining an understanding of why you are writing and how that writing works in conjunction with other documentation. The next is obtaining the tools you need to deliver clear and complete messages that are grammatically correct and consistent. Acquiring the requisite tools is what this book is all about.

Writing for Compliance with Binding Regulations

Why does writing play such an integral part in companies that develop, manufacture, and market therapeutic products? The answer lies largely with the regulatory forces that drive the healthcare industry in the United States and abroad.

The regulations state what companies must do. Their documentation tells how they do it and what the outcomes are. In a pharmaceutical company, for instance, documentation is the proof that a company's activities meet the regulatory demands of Title 21 of the Code of Federal Regulations Part 211, *Good Manufacturing Practices for Finished Pharmaceuticals*. In this environment, documents delineate such diverse activities as facility and equipment qualification, cleaning, and maintenance; control of materials, from incoming components to finished goods; validation of manufacturing processes; sampling and testing activities; nonconformance and out-of-specification (OOS) investigations; and employee training. They provide the "how-to" for auditing vendors and contractors, handling complaints and recalls, and conducting annual product reviews. In short, documents substantiate that a company has complete control of all of its activities in compliance with the regulations.

In the United States (US), 21 CFR Part 820 *Quality System Regulation* delineates the Good Manufacturing Practices (GMPs) for medical devices and Part 606 *Good Manufacturing Practice for Blood and Blood Components* delineates them for biologics. Most countries have similar regulations to ensure the safety and efficacy of products. Canada, for instance has the Canadian Health Protectorate Branch (CHPB), Ireland has the Irish Medicines Board (IMB), the United Kingdom has the Committee on Safety and Medicine (CSM), Australia has the Australian Code of GMP for Therapeutic Goods, and Japan has the Ministry of Health and Welfare (MHW). In numerous other countries, regulations come from the Ministry of Health (MOH).

While countries have their own regulatory authorities for ensuring safety and efficacy in drugs, devices, and biologics, they are also recognizing that internationally acceptable standards can help put products into world markets. Companies conducting clinical trials worldwide embrace the International Conference for Harmonisation (ICH) Guidelines, established in 1964 and revised in 1975 and 1986. These guidelines, which have their origins in the Declaration of Helsinki, present an international standard for designing, conducting, recording, and reporting clinical trials in which human subjects are participants. They were developed in consideration of the clinical practices of the European Union (EU), Japan, the United States, Australia, Canada, the Nordic countries, and the World Health Organization.

In Europe, the European Medicines Evaluation Agency represents 15 member nations. The International Organisation for Standardization (ISO) promulgates standards for quality worldwide and offers certification. And Mutual Recognition Agreements (MRAs) between nations also attest to the drive toward uniform, internationally accepted standards for therapeutic products development.

One manifestation of this sort of standardization is the Common Technical Document (CTD). Many nations now mandate the CTD as the vehicle for gaining approval to market a drug, biologic, or device. Some countries recommend and will accept the format, but until they change the regulations in place, they do not make it a formal requirement. Until FDA, for instance, makes a revision to 21 CFR Part 314, *NDA* it won't be mandatory to submit a New Drug Application (NDA) in the CTD format. The outline for the CTD allows companies to devise one document that they can submit to many countries for product approval. The variable piece is the section on Quality Assurance. (See Chapter Eight for more information about the CTD.)

Few, if any, companies are driven by just one set of regulations. A typical US pharmaceutical company may be subject to the regulatory guidelines set forth by the Food and Drug Administration (FDA) in Title 21 CFR Parts 210 and 211 as well as those set forth by the Environmental Protection Agency (EPA), Drug Enforcement Administration (DEA), Occupational Safety and Health Administration (OSHA), the Department of Transportation (DOT), and various other state and federal organizations. If it seeks to manufacture or market a product outside the US, it must look to the binding regulations in targeted countries.

Once in place, regulations don't change all that rapidly. However, it takes about five years for industry standards to develop. Once a regulatory agency issues a regulation, dialog among companies and FDA takes place, and industry "best practices" develop. Consider, for example, 21 CFR Part 11, *Electronic Records; Electronic Signatures.* This regulation, vague in nature, has generated tremendous discussion, and FDA has issued many guidances as to how the law is to be interpreted. Industry standards are now in place for this regulation, although discussion continues. More recently, the US Department of Public Welfare and Human Services issued 45 CFR Parts 160, 162, and 164 for electronic record keeping of patient records. This regulation bridges into FDA-regulated industries, since developers of therapeutics engaged in clinical trials manage patient data. These regulations, part of the Health Insurance Portability and Accountability Act (HIPAA), are new, and industry standards are developing. However, companies seeking to comply will do well to understand industry standards for Part 11, since the regulations are parallel, even though issued by separate agencies.

Further, companies must also understand that new regulations don't supersede existing ones. Predicate rules, those already in place, still apply. Thus, companies complying with 21 CFR Part 11 must continue to follow the regulations that drive their operations. If a medical device manufacturer goes to electronic record keeping, the guidelines for the records themselves reside in 21 CFR Part 820. This is the way most regulations work.

Managing the regulatory maze is not easy. Yet keeping abreast of the regulations and remaining compliant makes good business sense. Monthly publications for industry, available by subscription, detail issues in the industry and governmental rulings. Industry forums, conferences, and courses offered by professional training organizations offer opportunities to remain current. It's not enough to adhere to the regulations. Companies need to understand the direction in which compliance is moving and keep in step. The last thing a company needs is a routine investigation that discovers nonstandard practices. The result will be a discrepancy observation, citing what needs to be fixed. It's always better to have everything in place, with documented proof that it is, rather than to scramble to fix what the company has been cited for—this slows productivity and makes poor business sense. One thing is abundantly clear: Documentation will continue to be critical to every facet of doing business within this highly regulated environment.

Regulatory Evolution in the United States

Laws governing therapeutic product development and marketing have evolved over time with specific laws marking milestones over a period of 100-plus years. The first US federal regulation dates back to 1884 when American soldiers died after ingesting adulterated quinine. As a result of these deaths, the government passed the Drug Importation Act, which required customs inspections on drugs coming from overseas. Then in 1901, The Biologic Control Act became law after 13 children died from a contaminated antitoxin for diphtheria. This act gave the government regulatory power over antitoxin and vaccine development. Shortly after, in 1906, the government passed the Food and Drugs Act to authorize the government to monitor food purity and safety of medicines.

In 1931, the Food and Drugs Act was renamed the Food and Drug Administration. Several other events were significant in developing binding regulations designed to protect humans and animals. The 1932 Tuskegee Study of Untreated Syphilis in the Negro Male, conducted under the auspices of the US Public Health Service, deprived infected men of effective treatment so as not to interrupt the project. Then in 1937, 107 people died after taking "elixir of sulfanilamide," which turned out to be an antifreeze solution. FDA removed the product from the market, not because it caused fatalities, but because it was mislabeled. In 1938, the government passed the Food, Drug, and Cosmetics Act. This Act expanded the role of FDA to control of cosmetics and devices.

It was during World War II, however, that experiments were done in large scale on unconsenting humans. The Nuremberg War Crime Trials brought these atrocities to light, and the result was the Nuremberg Code, which cited ten standards for ethical human research.

A wake-up call for even better monitoring came in 1962, when thousands of babies were born with defects, the result of their mothers taking thalidomide while pregnant. The drug had never been approved for marketing in the US, but was undergoing research in American women. Of these women, nine gave birth to defective infants. This event induced FDA to require notification of investigational use of drugs, which up until this time, had not been required. The result was the Kefauver-Harris Amendment to the Food, Drug, and Cosmetic Act.

At about the same time, President John F. Kennedy announced the Consumer Bill of Rights in a message to Congress. This Bill of Rights said that people have the right to safety, the right to be informed, the right to choose, and the right to be heard. In the same period, in 1964, the World Medical Association issued the Declaration of Helsinki, and physicians were tasked with embracing this statement: "The health of my patients will be my first consideration." The declaration has been amended four times, and the Code of Federal Regulations (CFR) has incorporated the basic elements.

In 1972, the National Institutes of Health transferred the regulation of biologics to FDA. This was followed by the National Research Act, which created the National Commission for the Protection of Human Subjects of Biomedical and Behavioral Research. Additional legislation has continued to promote ethical treatment of healthcare recipients. In 1978, FDA published the current Good Manufacturing Practices, 21 CFR Parts 210 and 211.

In 1988, the FDA became an agency of the Department of Health and Human Services. Since that time, the ICH has been formed. A significant ICH goal is to maintain safeguards on quality, safety, efficacy, and regulatory obligation for the protection of the public. The 1997 Food and Drug Administration Modernization Act reauthorized the Prescription Drug User Fee Act of 1992 and instituted reforms in agency practices. In 1996, medical devices became subject to Quality System Regulation (QSR) 21 CFR Part 820. In 1996, as well, the Department of Health and Human Services enacted HIPAA into law. This provided the forward momentum for broad changes in the healthcare industry, but the specifics of the regulation were still being written. Shortly thereafter, in 1997, 21 CFR Part 11 *Electronic Records; Electronic Signatures* was enacted.

The government does not issue laws without forethought. The Office of the Federal Register issues the Federal Register (FR), a weekly disclosure publication that informs citizens of their rights and obligations by providing access to the official text of approved regulations and descriptions of federal organizations, programs, and activities. It also publishes texts of proposed regulations and changes to existing regulations. This gives industry the opportunity to react and share dialog with the government agency that has ownership of the proposal. Reviewers can comment on content and wording, the date the regulation goes into effect, and the penalties for non-compliance. Comments are reviewed in a government forum, and the final text becomes the "final rule."

Once enacted, laws are published in the CFR, issued annually on April 1. Laws are enforceable by the respective divisions within the Department of Health and Human Services. It's important to note, however, that once a final rule appears in the FR, companies are responsible for instituting compliance. Thus, keeping abreast of the regulations requires constant vigilance.

The CFR contains regulations of specific government departments and agencies. The CFR has 50 "Titles," each assigned to a different unit of government. Title 21, Food and Drugs, contains regulations mandated by FDA. Title 45, Public Welfare, falls under the auspices of the National Institutes of Health (NIH). Each title of the CFR is then divided into chapters, and each chapter is divided into parts and subparts.

Remember, too, that as new regulations are enacted, they do not supersede existing regulations unless the government has rescinded them. New regulations in essence become adjuncts to the ones already in place. Companies must adhere to predicate rules and remain vigilant about industry best practices for compliance.

Document, Document, Document

Documents work with each other either concurrently or in tandem. Documents tell how things happen on a regular basis and present a "big picture" of a company's operations, usually in standard operating procedures (SOPs), quality manuals, plans, and other such documents. Documents such as protocols and proposals tell what the company plans to do. Ongoing assessment and data recording occurs as activities progress. Process reports give the results of projects. Finally, summary reports bring it all together — what is the outcome of a significant set of activities?

A single product's history may start with source data for the concept, usually a laboratory finding. After the initial discovery recorded in the laboratory notebook comes testing to see if the concept is viable. Countless studies, performed in accordance with binding regulations, such as Good Laboratory Practices (GLPs), result in decisions to pursue the product development or to abandon it. When a company determines to develop a product, preclinical testing helps to confirm the potential product's value and determine whether the company should file for approval to test the product in humans in controlled clinical trials.

If the product moves through clinical trials, the company files for approval to manufacture and market the product. At every step in this process, which can easily take a decade, documentation captures what happens. Once a product is in the marketplace, record keeping continues, and stability studies confirm continuing efficacy through the expiry date on the product. Careful and continual monitoring must occur, and if a product loses its efficacy or has other problems, the company may issue a field alert or recall. The company keeps the product records at least two years beyond the shelf life of the product itself.

> Companies need to keep extensive records of everything they do, whether it becomes part of the final product development or not. They need to know what didn't work so they don't go down the same road twice.
>
> **John Cline, Ph.D.**

Companies must show their control of systems, processes, and products in documentation. This means self-monitoring and assessment, as well as change management. There is no "magic formula" for documentation for all companies, but the common denominator is this: Companies must have controls in place, and they must have records of what they do, have done, and plan to do.

Critical to successful operations are trained employees, so training must also be a documented part of operations. In addition to the orientation employees receive about their jobs when they are hired, they must receive regular "refresher" training, as well as retraining as procedures change or as they move from position to position within a company. Similarly, companies must verify that consultants have the appropriate training for the roles they fulfill and that vendors and contractors meet the criteria for the quality standards set by the company.

In sum, compliance with the binding regulations requires extensive documentation, all of which reflects the activities a company carries out daily. As companies find better and better ways to do things, gain new technologies, and decide to manufacture products that require different formulas and procedures, they must both continue to meet current standards and verify, through records, their adherence.

Keeping the House in Order

Compliance with the binding regulations and clear and complete documentation should be the goals for all companies operating within regulatory statutes. Who will determine if they are being met? Companies can expect audits and inspections from many sources. FDA, for instance, sends investigators on site for two primary reasons: a general GMP inspection or a new product inspection. In a general GMP inspection, FDA is present to assess overall operations and determine a company's adherence to GMPs. In this type of inspection, investigators may ask to observe production — but not developmental — processes. A preapproval, postapproval, or scale-up inspection, on the other hand, focuses primarily on facilities and processes relative to a new product. Preapproval, postapproval, and scale-up inspections comprise much of the FDA's focus.

Companies that successfully undergo inspections know that it's difficult to anticipate the direction they may take; thus it's always wise to have everything in place and running effectively. If, for instance, a company employs electronic signatures, how the company has achieved compliance and maintains it will surely be a focus. Written records attest to what the company has done and what it continues to do.

What an inspector does or doesn't find marks the caliber of the company. Violations are not to be taken lightly. FDA considers a violation of cGMPs during an inspection an "incident." If, upon reinspection, the same violation is present, FDA considers the violation a "practice," and the product subsequently adulterated. These are serious issues, ones that good documentation reflecting good practices can very often prevent.

Companies must thus understand what controlled documentation they must have in place and accessible. For instance, to undergo a successful approval inspection, companies manufacturing drug products usually make the following core records available; other documents may accompany these, depending on the product and processes.

1. Manufacturing and controls segments of the application
2. Master formula
3. History section of the application
4. Development data, including product characteristics and physical properties, manufacturing procedures, finished product test results, dissolution profiles, and results of pilot and preliminary production-size batches that confirm formula ranges, specifications, in-process variables, and stability testing.
5. Materials analyses
6. Laboratory data
7. Equipment qualification and cleaning validation

8. Standard Operating Procedures, including those for change control, QA/QC investigations, field alerts, and validation
9. Finished product test results
10. Stability studies

Note that a visit by FDA or another government agency is not the only time a company receives an inspection. A company may be the subject of an audit by a firm seeking contract services or a joint venture. Such an audit will likely be every bit as strenuous as other inspections, and, once more, having documented practices in place translates to doing good business.

Document Control

While this is not a book about document control, it's important that writers understand that companies must control their documents and that writers must conform to the process in their companies. The systems vary from company to company, but effective companies know which documentation is drafted, written, under review, beginning revision, or moving into obsolescence. The degree of sophistication that characterizes the system is relative to the degree of sophistication of the company itself. A company with many sites needs systems that are more complex than those required by smaller companies.

While there are many excellent systems, most share common ground. There are fixed procedures for introducing and approving the concept for a document, and drafting, reviewing it, and giving it final approval. Documents generally have other controls and are searchable by number, title, author, and key words. In addition, documents have revision histories, so a review of the document tells the life of the document from conception to retirement. Finally, who signs what type of document needs to be spelled out. Usually companies develop a minimum required signature list that tells who has authority to sign what type of document and how many signatures are needed to approve the document. Companies typically detail how their systems work in an SOP on document management. They may also have instructions for writing specific documents such as study reports, audit reports, and submission documents.

Effective document management systems ensure that documents maintain their integrity. For instance, hard copies of documents — such as those in SOP manuals — are controlled, and when new documents are issued, previous versions are accounted for and destroyed. Approved, official copies of documents must reside in controlled environments with limited access — in a limited access area for manual systems or in software system vaults for electronic systems.

Companies must all define how their systems work. Documents in a manual system review process, for instance, may route through the system in colored folders, so reviewers know at a glance that the document in a yellow folder is a qualification, a document in a red folder an SOP, and a document in a purple folder a laboratory method. Other systems may send documents as pdf files as attachments to e-mails for review with a scheduled concurrence meeting.

Electronic record keeping (ERK) is now mandated for patient records. Many small companies rely on manual systems, while others have implemented electronic document management systems. Electronic systems are necessary for electronic submissions to many regulating agencies, so the impetus is to go electronic. But more importantly, ERK provides more efficient document management overall and cuts down on the amount of paper companies must manage in their archives.

ERK systems require extensive controls. They must be validated for the intended use of the system. There must be controls in place to ensure security, user accountability, and audit trails. Many companies have put Computer Software Validation (CSV) teams in place to ensure that validation of software-driven systems happens effectively. Once a system has gone live, it undergoes audits, and when major changes occur, it undergoes full or partial revalidation.

The bottom line for document management is this: Companies have to determine how their document management system works and then document it. Further, anyone working with documentation within the system needs to understand how the system works. That means system users must have training in the system and not deviate from it.

Standard Formats

Standardized formats also make documents easier to write and process for most companies. These formats can guide writers through the tasks of drafting and revising; they can guard against zealous rewrites by reviewers and can facilitate the approval process. Many companies have stylebooks that specify the presentation of certain information: These guides may call for a serial comma or not or direct certain SOP phraseology in delivering information common to many procedures, such as securing QA approval and signature. The extent to which companies control the details of documentation depends on each company's resources. (See Chapter Nine for more information about style.)

Document control staff should be able to identify the location of a document in a system at any given time. Staff may also write documents relative to their area or serve in the review process. They may have license to make mechanical, but not content, changes before final approval. Once a document receives final approval, through either a series of review cycles or a concurrence meeting, document control staff should issue the document with no further change. The group should also retrieve previous versions of documents, if any, and provide a history of the document's development. Document control involves exhaustive attention to detail but does not infringe on the integrity of the documents.

The Writing Task

Writing is hard work, and it is high on the list of what people hate to do most. For many, it's an intimidating task. In regulated industries this can be especially true: You may find yourself in the position of having to document what has happened, what happens regularly, what will happen. Regardless of the focus, writing always requires accuracy, attention to detail, and clarity.

In this industry, few people write in solitary. You may be called upon to prepare a report, write a technical memo, review any number of documents, draft a report that requires the participation of several people, or compile information from many sources as a basis of study. How you tackle these tasks requires some foresight. Understanding the writing project you are about to undertake is the place to begin. You may be the primary author of an SOP, a collaborative author of a dossier, or one of several authors involved in a project such as a facilities validation. You may be the primary author of an activity, such as an audit, that requires a bevy of writing to reach a conclusion. (See the text box *Put It in Writing.*)

Put It in Writing

A good audit report results from good planning. Each audit should have a record of activities, from the decision to audit through the audit review. Each step of the process requires writing it down. Here's a sequence that helps ensure each audit a company conducts gives optimal results.

1. Determine what the customer wants
   - Internal audit to determine GMP compliance
   - Focused audit of manufacturing process to determine compliance gaps or reason for a nonconformance
   - Supplier audit to determine suitability
   - Manufacturing record audit for errors, omissions, deviations

- Stability data for a product

2. Determine the audit scope
   - An entire company
   - One department within the company
   - Manufacturing records of a defined time or product
   - All product packaging operations for one week
3. Determine the type of audit
   - Planned inspection
   - Unannounced inspection
   - Document desk audit
4. Determine the governing documents
   - FDA regulations
   - ISO standards
   - Corporate procedures
   - OSHA standards
   - Departmental procedures and required documentation
   - Process maps and diagrams
5. Determine who to interview
   - Employees conducting the process
   - Newly hired employees
   - Department managers
   - All nightshift analysts conducting stability testing
6. Determine a statistical sample size
   - How many lots are manufactured in one week, month, year
   - How many complaint files in the past three months
   - How many employees in the company
7. Determine the audit duration if not predetermined
8. Conduct the audit
   - Know what should happen
   - Observe what is happening
   - Verify what happened through documentation

9. Meet with audited groups to confirm deficiencies and observations to eliminate misunderstandings and auditing errors

10. Write the report

11. Report the findings to the original customers and auditees

12. Review corrective and preventive action plans

13. Follow-up on the effectiveness of corrective action plans after implementation

Courtesy of Monica Grimaldi, Certified Quality Engineer

The good news is that for many types of writing there are clear guidelines. For writing documents such as SOPs, you need to look to the company standards; the same holds true for validation documents. For other types of writing, you can look to the regulations, industry practices, and government-issued guidances. Consider for instance, preparing a Chemistry, Manufacturing, and Controls (CMC) section of a submission for approval to market a solid-dose drug product. How will what you are to write fit into the big picture? The guideline for CMC breaks down the components into manageable groupings of information including (1) the drug substance, (2) the drug product, (3) methods validation, and (4) environmental assessment. Within each of these groupings are subgroupings. You can thus prepare components of each and assemble them accordingly. Of course, you'll have to do your homework first. Make sure you fully understand what it is that you have to say.

The preliminary work can be tedious, to be sure, but starting the actual writing is usually the toughest part. Many people complain of "writer's block," or the inability to get words down on paper. If you suffer from bouts of writer's block, there are some steps you can take to overcome these down periods. See the following.

Arnold Melnick, author of *Melnick on Writing,* a column in the *AMWA Journal,* the publication of the American Medical Writers Association, offers five questions to help writers understand their writing patterns.

Only You Can Solve Your Writer's Block

Writer's block is the "temporary inability by a writer to put words on a page." It's a common experience for writers, but there are things you can do about it. Answering five simple questions accurately and intelligently can provide an answer to this affliction.

1. Do you struggle vainly to "write" something instead of communicating information or ideas to the reader?

   According to Joel Saltzman, author of *If You Can Talk, You Can Write,* write anything as though you are talking to a friend. Write whatever words might be associated with your document, without pausing to criticize or edit. Intersperse it with whatever random thoughts come to your mind. Then, edit and edit carefully. Good writing is good editing. Very few writers can get their desired effect in the first draft. For writers, how they edit determines whether the writing is good or not.

2. Do you know your own patterns of creativity?

   What are the most favorable work conditions for you? Do you write best early in the morning, late in the day, or at night? Do you do better work with a dish of candy next to your computer or while abstaining from sweets? Do you work better alone or with people nearby? To get the most out of your writing, observe and respect your own personal idiosyncrasies. They guide creativity — or at least they don't block it.

3. Do you work best with notes or without notes?

   Writers work in different patterns. Some do better with copious notes, others with outlines, others with sketchy notes, and still others without any notes at all. In some cases, writers do better using notes for factual documents (as in reporting data) and without notes for less concrete material (as in light correspondence) — or vice versa. People have different patterns of behavior for different types of writing.

4. Is your problem ideas or words?

   If your difficulty is in ideas, it means that you have no concept of how to get where you want to go. In such a case, here are two recommendations. First, just scribble some notes or words about your idea and about your concepts. Later on, you can flesh out these notes. Second, handwrite some of your thoughts because in the extra time it takes to write out concepts you will probably be able to fill in some of the creative thoughts you had in the first place.

   If your difficulty is in words, it means that you know what you want to say, but can't quite say it. One of the better ways to approach this difficulty is to determine which section of the document you are most sure of, then write it first, even if it is out of order. Everything does not have to be written in sequence.

   A second approach is to write down a few of the key words of your document and then expand them by word association. For example, if you are writing a report on a meeting, you might jot down "meeting," "election," "conference room," and "Tuesday morning." You can then add other words to each of those original words until you have sketched an outline that will permit you to start writing. Next, add material to it. Remember, you can edit out all the extraneous material.

5. Are you a procrastinator?

   Procrastination may well be a genetic thing: some people are procrastinators, some are not, and some swing back and forth. What is important is that each writer recognize personal patterns of procrastination. When given a task, do you attack it immediately, or almost immediately, regardless of when the deadline is or what the import? Or, no matter how serious the job, do you put it off until almost the last minute? Look at how you shop. Look at how you pay bills. Look at how you study for examinations. Good examples, all. Examine your behavior in writing situations and determine whether or not — or how much — you are a procrastinator.

   Here's a recommendation to help procrastinators: sit and sit and sit. Station yourself in front of your computer and do not yield to the temptation to get up and walk around or do anything else. Stay seated for a reasonable period until your thoughts start to flow. Others recommend two other approaches. Interestingly, they are opposites. Some experts say start with the most difficult task and get it out of the way, noting that the rest will then be easier. Others recommend the reverse: start with the easiest things because they can be done quickly, and then gradually work your way up to the most difficult task. Meanwhile, you will already have written much of the work. Still another recommendation is to "take five." Walk around the building, take a short coffee break, do some deep breathing. But, if you "get away" like this, try not to substitute something you enjoy, such as eating ice cream. In essence, don't reward behavior that you shouldn't encourage.

   It's also wise to get a sounding board if you have difficulty organizing your thoughts or words. Use a dictating machine, or find a colleague who will act as a sounding board so you can tell what you want to say. You will then probably have created your own first rough draft. This process ties in with natural law that you can talk or dictate about ten times as fast as you can write, so when an idea strikes your brain you can record it in a shorter period of time by speaking, losing far less of the thought. Then transcribe what you've said.

   In essence, to get rid of writer's block, or at least reduce it, study your own style of writing and your own personality. Once you understand yourself, you will be well on the way. Stay with who you are, and you will be rewarded.

Excerpted from *KYOS—Five Easy Questions to Erase Your Writer's Block,* the AMWA Journal, Vol. 17, No. 1, 2002.

Courtesy of Arnold Melnick

Writing and Revising

The best motivation for writing is a deadline.

Kristine Ogozalek, Regulatory Manager

Just about anyone can write *something* — it's what happens to it after the first draft that makes it good. In short, pretty much everything that's written can use some skillful editing and revision. Unless you are a genius, good writing doesn't just happen. It's the result of drafting, revising, reassessing, and revising again. Further, the more eyes that see a piece of writing, the better it usually is. This is especially true of the highly technical writing that's the norm in regulated industries.

Most writers have had the experience of proofreading their own words and giving the copy an okay, only to discover too late that they overlooked glaring errors because they did what humans tend to do: They saw what they expected to see and not what was there. On the other hand, a writer may spend hours developing an idea or researching a detail, then notice that to include it would confuse or mislead. In such circumstances, the only recourse is to cut the passage. Developing your own writing is no easy task, for you are dealing with yourself as a writer. You may know exactly what you mean, whereas your readers may not. That's why review and revision play a strong role.

If you write simple memos, e-mails, faxes, and letters that no one but you sees before distribution, it's best to draft the document and let it rest, if you can. Come back to it and look at it again. Read it out loud if you have the luxury of time. (This helps you "hear" as well as "see.") If the piece is important, ask someone else to read it through, and be open to suggestions. Be appreciative when a typo or misspelling comes to light, so you can make changes to improve your writing for the better. Do the best fine-tuning you can; draft and revise until the document is as good as you can make it.

If you are writing a document for a formal review, remember that the better the quality is prior to the review process, the quicker the approval will be forthcoming. Your reviewers, in particular, will thank you for your diligence, because their task will be easier. And in the long run, you may spend less time trying to get the text through final approval.

Collaborative Writing

Collaborative writing means that two or more people conjointly contribute information to the draft and completion of a single document. For example, work that runs continuously, such as pilot plant operations, requires systematic record keeping across shifts. Those records may ultimately feed into reports, with several people preparing sections. Certainly, equipment installation and operation protocols and qualification reports require the expertise of all who work on a specific project. Clinical trial reports may have more than one writer, and certainly dossiers headed to regulatory agencies have a host of authors who have provided input.

> Common sense is not so common.
>
> **Voltaire**

Writers working collaboratively on documents must offer information that ultimately serves one purpose, and although that can be difficult, it's common. What's needed when people embark upon a joint writing venture is a clear understanding up front and a sense of document ownership. Many a collaborative writing project has gone awry because none of the writers assumed ownership, and the end product became a document with no clear purpose, simply a compilation of information without unity.

Writers need to agree on the main purpose and supporting points for the document. Often each writer can clarify the others' thoughts because all have a solid — but somewhat differing — vision of the main idea. Discussion helps clarify the purpose of the report, and this discussion is best done up front before the writing process begins. The next thing to do is to decide who is going to write what. If you write collaboratively, work with your coauthors to define the process that's easiest for all involved. The two approaches that follow are equally workable, and both require some negotiation skills.

The First Approach

The first approach calls for a designated person to draft the document and for the others to add and amend. That's not to say the first person shouldn't review and be permitted adjustments to the text before submission of the finished product. The strength of this system is that the person with the strongest language skills does the "cleaning up," while the writers with the strongest technical expertise have their say. Alterations in the text are with the approval of all writers. You'll find this approach to be particularly efficient in the composition of short documents.

The Second Approach

The second approach requires more planning than the first approach. In this approach, the writers assess needs of the document and assume ownership of specific portions. All writers need to understand the components of the planned document and what needs to reside where. They then agree on the formatting conventions and the time for text completion.

Writers then meet to combine the elements and polish the document, with each reading and making comments on the entire text. Revision and refinement should come through tactful commentary and with the consent of the writer responsible for each individual section. This approach is usually the most effective in the composition of reports or other documents of length.

Reaching Agreement

Trust in other people's expertise and a willingness to accept their judgment are crucial to collaborative writing. Remember also that two or more people will have distinct writing styles, and that those styles may vary dramatically; yet sometimes the style distinctions will be barely discernible. Try not to make arbitrary alterations in your coauthors' work; similarly, be tolerant of any minor changes a coauthor may make in your writing, and reach agreement as to the clarity and completeness of the message. And remember, nothing does as much for a common goal as conversation. If you feel a change is necessary, discuss it. Chances are greater that your collaborators will agree after they've heard your explanation. Similarly, you'll feel better about text adjustments after you've had the opportunity to hear why your coauthors feel they should be made. Discussion, after all, is the bond that makes collaboration workable in the first place.

Finally, when writing collaboratively, make every effort to present a document that's cohesive, clear, and grammatical. Getting a document to this point may take many readings, discussions, and revisions. Your collective goal should be the end result: a quality document ready for either immediate distribution or formal review that will be well received.

It doesn't matter what kind of writing you are collaborating on. The following is an abstract written by three people: a vice president of development, a regulatory manager, and a consultant. In preliminary discussions, the three determined to submit an abstract to present at an industry conference. The requirement called for a maximum of 300 words. They had several ideas, then narrowed them down to defining how a company can make the transition from discovery to a compliant development operation. The regulatory manager tackled the task of getting the idea down on paper. Here is the first pass. Notice that the manager asks a few questions of her coauthors, and that this draft is far from complete. It has 114 words

The final word count is 300, and the message is succinctly delivered. Here's a happy note: The abstract was accepted and the authors presented at the conference.

First Draft/Idea Stage

ABSTRACT TITLE:

From discovery to development

SUMMARY:
Session focuses building a development organization from the ground up. It includes how to build project management, regulatory, and document functions to take a product from discovery to market.

LEARNING OBJECTIVES:
Basics for managing development activities with project management, regulatory expertise, and documentation functions.

ABSTRACT:
Newer companies are entering a new arena – development. Different skill sets are required for development than are required for research and discovery. Contract organizations and consultants are often used to acquire the expertise that the company itself doesn't have. When consultants are used, companies may not have the knowledge they need.

In-house vs. farmed out?
Documentation?
Main force is all three?

The consultant then reviewed the text and added some information to address the manager's queries and to refine the writing.

2nd Draft: Expanded Text

~~FIRST DRAFT/IDEA STAGE~~

ABSTRACT TITLE:

~~From discovery to development~~
Discovery! Now What?

SUMMARY:
~~Session~~This session focuses on building a development organization from the ground up. It includes how to build project management, regulatory, and document functions to take a product from discovery to market.

LEARNING OBJECTIVES:
~~Basics~~Understanding the basics for managing development activities with effective project management, regulatory ~~expertise~~awareness, and compliant documentation~~ functions~~.

ABSTRACT:
Newer companies are entering a new arena -- development.
~~Different~~Development requires different skill sets ~~are required for development~~ than ~~are required for~~do research and discovery. ~~Contract organizations and consultants are often used to acquire the expertise that the company itself doesn't have. When consultants are used, companies may not have the knowledge they need~~The transition to development thus presents a learning curve. Newer companies often turn to contract organizations and consultants to acquire the expertise that they lack. Companies relying solely on contractors or consultants, however, risk not having control of their own products; worse, they may be going down the wrong path and not know it. So even with outside expertise, companies must equip themselves with certain essentials so they can manage the development process effectively.

~~In-house vs. farmed out?~~
~~Documentation?~~
~~Main force is all three?~~
The first factor in building a solid core for development is assessment. Namely, who will start and oversee the development activities? What can occur in house and who can handle it? How will contractors interface with the core company teams? Companies must understand the activities that must occur before they build the infrastructure. A key function is therefore project management. As essential is the regulatory role – the driver between the company and the agency that will ultimately approve the product moving into development. The last component is a documentation system that captures development activities, from standard operating procedures to data gathering and reporting. Good records provide the "proof" that the company is compliant; without records, everything else amounts to nil.
With the right functions, built on a critical understanding of binding regulations and good business practices, companies can avoid the fits and starts that are inevitable without effective controls.

The revision has 298 words. When the Vice President had a look, she had a few changes, notably that the scope was too broad. In one short presentation, the trio could not discuss discovery to marketplace. Thus the scope is limited to development, and the abstract discusses only that bridge. Further, she focuses on "young" rather than "newer" companies. She also opted to say "research and development" as a single entity; thus the verb is singular. The result is a better abstract. (Her edits are underlined.)

Final Draft

ABSTRACT TITLE:

Discovery! Now What?

SUMMARY:
This session focuses on creating the essential building blocks for a sound development organization from the ground up. ~~It includes how to build~~ These essential building blocks include project management, regulatory, and document functions necessary to take a product from discovery to ~~market.~~development.

LEARNING OBJECTIVES:
Understanding ~~the~~of basics for managing early development activities with effective project management, regulatory awareness, and compliant documentation.

ABSTRACT:
~~Newer~~Many young companies are entering a new arena -- development. Development requires different skill sets than ~~do~~does research and discovery. The transition to development thus presents a learning curve. ~~Newer~~Most young companies ~~often~~ turn to contract organizations and consultants to acquire the expertise that they lack. Companies relying solely on ~~contractors or~~ consultants, however, risk not having control of their own products; worse, they may be going down the wrong path and not know it. So even with ~~outside~~contracted expertise, companies must equip themselves with certain essentials so they can manage the development process effectively.

The first factor in building a solid core for development is assessment. Namely, who will start and oversee the development activities? What can occur in house and who can handle it? How will contractors interface with ~~the~~ core company teams? Companies must understand the activities that must occur before they build the infrastructure. A key function is therefore project management. As essential is the regulatory role – the driver between the company and the agency that will ultimately approve the product moving into development. ~~The last component~~Next is a documentation system that captures development activities, from standard operating procedures to data gathering and reporting. Good records provide the "proof" that the company is compliant; without records, everything ~~else~~ amounts to nil.

With the right functions, built on ~~a~~ critical understanding of binding regulations and good business practices, companies can avoid the fits and starts that are inevitable without effective controls.

The final word count is 300, and the message is succinctly delivered. Here's a happy note: The abstract was accepted and the authors presented at the conference.

English: A Living Language

American English, like other living languages, is in constant transition. It is ever adjusting to reflect the changes in culture and technology. The alterations in the language can be controversial and have drawn criticism from many who decry the changes. Others, however, say the ability of this language to embrace change, particularly in the acquisition of new words, is its very strength. Here are some facts about this vital language.

- English belongs to the Germanic group of languages, which are part of the Indo-European system of languages. Germanic languages include German, Dutch, Afrikaans, Swedish, Danish, Norwegian, and Icelandic.
- Mandarin Chinese is spoken by more people, but English is more widely spoken around the globe and has wider dispersion than any other language. English is the official language of England, Ireland, the US, Canada, Australia, and New Zealand. It is also the official language of Ghana, Liberia, Nigeria, Uganda, and Zimbabwe in Africa; Jamaica, the Bahamas, the Dominican Republic, and Barbados in the Caribbean; Vanuatu, Fiji, and the Solomon Islands in the Pacific; and a dozen other nations and territories. In more than 20 nations, English shares official status with another language. Some of these nations are Singapore, the Philippines, India, and Pakistan. In still other nations, English holds no official status but is widely spoken, particularly in the business sector. English is also the official language of the United Nations.
- English is divided into three periods: old English (about 449 to 1100 AD); middle English (about 1100 to 1500); and modern English, from 1500 on.
- English is widely used in science and other technical arenas. More Nobel Prizes in literature have been awarded to more writers using English than any other language.
- The word is the basic element of the sentence, and therefore of writing itself. English contains more than one million words. Of these three-quarters are technical. Only about 20,000 words are currently in common use. Of these one-fifth are Anglo-Saxon; three-fifths come from French, Latin, and Greek. The rest come from languages around the world.

- The average person has three vocabularies: reading, speaking, and writing. The three are interconnected, but the reading vocabulary is by far the largest. Since we speak more than we write, the speaking language is the second largest. Writing, thus, is the smallest vocabulary. This accounts in part for our difficulty in committing information to paper.
- In speaking, words carry less than 10 percent of our messages. The rest is conveyed by facial expressions, tone of voice, gestures, and posture. In writing, however, words must carry more than 90 percent of the message. Punctuation and graphics convey the rest.
- The subject of syntax is word order, or sentence patterns. Syntax is about the relationships between words. Word order changes meaning. Consider these two sentences:
- Have you left anything?
- Have you anything left?
- A group of words is a sentence if it contains a subject, verb, and complete idea. If it does not contain a complete idea, it is a clause.

Document Formal Review

Many documents are subject to a formal review process. You may participate as either author or reviewer, and both roles can be daunting. An understanding of the process itself and of the revision that's inherent as a result will make your task as writer or reviewer easier.

Serving as a Reviewer

As reviewer, the first thing you must understand in practicing your skills is this: Although reviewing in itself is far from child's play, altering a piece of writing is easier than creating the original. Your task is difficult because many people are sensitive about what they've written, and they have a right to be. The writer who has a sense of ownership in what he has written communicates pride in the work, and that spirit is conducive to good business.

Before you put pen to paper, first understand that writing always reflects the writer; it's a portrait of the person who created it, and a writer will usually defend what she's created, even if it doesn't "measure up" to your criteria. Often, too, a writer's defense of her work stems from an insecurity about her own writing ability. You must, therefore, exercise some human resource skills in addition to your language skills. Make your comments with respect.

First, understand what's basic to human nature: The very act of writing is a thinking process; and words, your own and others', trigger new ideas and, often, alternate ways of saying the same thing. Thus, it's extra easy to have your own sense of language stimulated by what you're reading, and it's something to consider with care.

Your job when reviewing anyone else's writing is to assist in creating clear, readable documents. Here's a bit of common sense: There's always more than one right way to say just about anything. Another person's way of saying something may not be like yours, but that doesn't mean it's not equally good.

The Writer's Voice

The words a writer uses and the structures of his sentences are as uniquely his as his fingerprints. These elements flavor what he says. If you're reviewing a document you can change *ambivalent* to *ambiguous* or *between* to *among* when it's necessary, and you can let a writer know when more or less background or explanation is needed to make meaning clear. You can suggest concrete, specific, everyday words for an audience that will be unimpressed — or worse, confused — by inflated, generalized, uncommon ones. But you should steer clear of changes that don't materially affect the way the message is understood, or the way the intended audience will receive it.

Making Comments

Reviewing is different from collaborative writing. You can see what happens when an editor's own writing style is imposed on someone else's writing. A supervisor submitted an update on a laboratory project that included the following line:

Three unknowns are inherent.

His manager crossed it out and wrote the following:

Three unknowns are innate.

Such correction does nothing to improve communication, and it delivers its own unspoken message: "You can't say it right, but I can." This sort of supercilious, authoritarian revision serves no good purpose; in fact, it hinders the business of getting the job done. So if you are reviewing someone else's writing, remember, if a piece works, leave it alone. The practical effectiveness of the final product is your only concern. And a good final product depends not only on your sense of the factually, grammatically correct but also on your respect for the writer's judgment. To change without purpose only confuses and often angers. Surely poor judgment lies behind a comment such as the following that appeared on an SOP as it moved through review:

"Confusing. Rewrite!!"

A dictate such as this fuels resentment, and in fact, it's an attack. Would it not be better to say, "I'm not sure whether you mean A or B. Can you clarify?" This critique places the responsibility for a reader's understanding on the writer's shoulders.

> Asking a working writer what he thinks about critics is like asking a lamppost what it feels about dogs.
>
> **John Osborne, Writer**

Most documents require numerous reviewers, and people need to think about content — what works and what doesn't. Epiphany may not come until a reviewer has looked at a document for a second or third time. When a writer drafts a document like an SOP, it usually goes into a mandatory review cycle, so the norm is reviewers' comments and subsequent revisions based on those comments. The same is true for documents of great length — reviewers may need to see a document in its various phases of development and in its final stages more than one time.

One final note: Writers are most effective if they feel relaxed and encouraged, with a person to consult when they're blocked or unsure about mechanics. Your job as a reviewer will be easier and your results better if you approach it looking for the positive and downplaying the negative. Good reviewers can inspire people to build their strengths; as that happens, the negatives evaporate! Focusing on negative performance, on the other hand, does not eliminate weaknesses, but rather diminishes the strengths. To get the best documentation, first identify the writer's best: Then build from there. When the document completes a review cycle, the writer can readily tell what is needed to create a complete, working vehicle.

Guidelines for Reviewing Documents

How many times have you picked up a pen or pencil, ready to annotate before you've read a word? Sometimes, initial restraint facilitates the task. Keep the following guidelines in mind when you must review the writing of others.

1. Determine the purpose/objective of the document. Make sure you fully understand the job the document has to do.
2. Read the document through in its entirety for substance. Do not hold a pen in your hand. The temptation to make comments can be overwhelming; yet an emendation to a document on page one may prove unnecessary if the author covers the same ground, perhaps even more effectively, in a later section.
3. Read the document through a second time if you have the luxury; then reflect on it before coming back to it.

4. If the document covers a process, reread the sequential steps again. If you can, allow some "resting" time before you make any changes.
5. Be positive, not negative, in your comments. Suggest, don't insist.
6. Annotate in order of importance for the following:
   - Content, completeness, and logic — Is the document comprehensive and understandable?
   - Consistency — Do all parts of the document work together?
   - Language — Are there typos and spelling errors? Is the grammar correct? Are word choices appropriate?
7. And remember, if the document works, leave it alone. Never "fix" what doesn't need it.

Serving as Author During Document Review

Those who review your writing may have valid comments about the content. Someone else's perspective may be just what your document needs to make it do its job. If, too, a reviewer is less than constructive in his or her criticism, step back emotionally. Don't take negative, unproductive notations personally. Such commentary often reflects any number of agendas coincidental to the task at hand. Often posing a diplomatic question to a reviewer yields not only conciliation but also positive results in the document.

It's a secure writer who questions reviewers' comments. Often the outcome is a discussion that yields an entirely different — and infinitely better — result. Be open to different perspectives for getting the job done well. And remember, the call-outs a reviewer finds may make the difference between a document that goes the distance or one that comes back for revision again.

Considering the Effects of Diversity

Not all writers of American English are native born. What that means, of course, is that the writing that gets done in English may reflect the native tongue of the writer. Leaving out articles, for instance, is a deviation common to writers whose native language doesn't have them, such as Asian and Eastern European languages. These are not serious omissions and usually don't create context problems. Reviewers can readily insert them.

Prepositions are pesky, too, for many writers. The Germanic languages have them in abundance, and English is a Germanic language. The Romance languages — French, Spanish, Italian, Portuguese, and Romanian — have them, but they don't translate perfectly. Still other languages, such as Chinese, don't have them at all. Most foreign-born writers struggle with some elements of the language, and the difficulty of knowing what and how to document is often compounded by insecurity about the working structure of English.

If you are such a writer or work with the documentation of such writers, here's the important thing: The errors typical in the writings of those who have learned English as adults don't reflect lack of intelligence; they simply reflect a struggle for mastery of the language. Accept the reality and give or take suggestions with grace.

Setting Priorities for Writing

Ideally, to produce a clear, easily understood document, whether you have written all or part of it or are reviewing it, you should adhere to commonsense guidelines for writing and reviewing. To have the time to allow a document in various stages of development sit while you think about the content, perhaps as you work on another job or while you attend a meeting or go to lunch, is a luxury, to be sure, but it offers advantages. When you come back to the document, read it again; then make your adjustments. This may seem impossible, given the demands the working day places on you, but important documentation especially needs all the effort you can give it. Understanding the strictures under which you work will help. The most omnipresent is allotted time available: It is often simply not enough. All you can do then, of course, is the best possible job in the time you have. Intelligent assessment of any task will always reveal that the amount of time spent on a job will directly affect its outcome. Be realistic. For some projects there is very little time; for others, there is more.

To allot the maximum time to the tasks at hand for every document, set priorities. If, for instance, several "rush" jobs need to be finished by the end of the workday, take some initial time and evaluate your given tasks. Arrange them in order of importance, most urgent to least, and address them in that order. That way, you'll address the most important task first, and, with luck, you'll be able to let it "rest" while you tackle another document and come back to it with a fresh eye.

This approach may not always work. There's no accounting for the last-minute emergency, the job demanding you let everything else drop and attend to it, but many times setting priorities will give you the upper hand and let you control your work and prevent your work from controlling you.

2

Connecting Writer and Reader

People who assess it overwhelmingly point to one problem in writing: a lack of understanding of and perception about the audience. Knowing whom your documents address and what response you want is a key to successful technical writing. This kind of writing informs people of past activities, findings, and decisions; it presents data and makes recommendations; it provides records of ongoing projects; it tells people how and why to take certain actions; it tells what kind of outcome is likely from intended actions; and it always has an audience — sometimes immediate and well known, other times projected. Often documents directed to a future audience go to immediate readers who need the information for interim work processes.

> When you write to FDA, the entire industry is a potential audience.
>
> **Monica Grimaldi, Certified Quality Engineer**

You can create grammatically correct and efficient passages that offer the information you need to convey. But overlooking the effects of your words on your readers is all too easy. That's why, when you write, you need to step back and evaluate the readers. Who are they? How will they respond to the information you're giving? Most people run the risk of not being objective about their readers, because when they write they are much more involved with what they have to say than they are with how readers will receive it.

This holds particularly true in the technical disciplines, where an initial reader may have knowledge of a project equal to that of the writer, but the intended audience may not be privy to the underlying details of a project. Other documents, such as technical notes, have no direct or immediate audience, and the intent is to supply a record of a problem resolved or record an issue. Nevertheless, while you may not know *now* who will seek a future record, you can assume someone will, so you will need to include enough background information to make the message understandable to the possible future reader.

The laboratory notebook is a good example of writing that's done for an immediate audience: those persons coordinating projects and overseeing routine activities who are in positions to make decisions. Should the data in the notebook lead to significant discovery, then the data in the notebook become the foundation for development. So future audiences may include anyone who seeks a record of a project, including a regulatory body such as the Food and Drug Administration (FDA). Thus, information must be clear, complete, and comprehensible to all pertinent, possible readers, not just the immediate ones.

> The palest ink is better than the best memory.
>
> **Chinese proverb**

Similarly, you may write a report to inform a scientific area of the company of a project's progress, but a projected audience may well be a regulatory body who will look at the report at some point in the future. Thus, it's smart to include information that the projected audience may require for complete understanding — but which is, nevertheless, not crucial for your immediate audience. Consider the case of a Quality Assurance (QA) Director who served as the project lead in the validation of a renovated plant facility. He included far more background and information than he had to for the simple reason that the facility would eventually be subject to a compliance inspection. He deemed it best to capture the complete history of the renovated facility, because he did not want to be in the position of having to recreate it at a future point.

A dossier that goes to a regulatory agency for review and product approval usually includes a large number of supporting documents, such as those prepared as the development progressed often over a period of ten years or more. When information is summarized, such as within individual portions of a New Drug Application (NDA) or Common Technical Document (CTD), writers really won't know who the precise readers will be for the sections or their supporting documents. There may be many readers who will read various sections of a submission. For instance, a reviewer at FDA for a Chemistry, Manufacturing, and Controls (CMC) section of a submission will most probably not be the same reviewer as the one who looks at clinical data. And neither reviewer will have seen the supporting study documents when they were actually written. Those will have had a more immediate audience at the time the actual study reached completion. The common link here is usually the project manager at the agency who has been working with the company's regulatory people and providing insight and suggestions for compiling the submission. And that dialog should set the tone for the writing.

Questions to ask before putting pen to paper — or booting up the computer — are the following:

- Who are the immediate readers?
- Who may read this document in the future?
- How much do the immediate readers already know about the topic?
- What do future readers need to know to understand the topic?
- What response do you want?

Consider this passage from a Chemical Hygiene Plan. It is incomplete and vague. What is missing that the readers need to know? How can it be more explicit and direct?

> Where there is no immediate danger to the skin from contact with a hazardous chemical, it is still prudent to select clothing to minimize exposed skin surfaces. A laboratory coat should be worn over street clothes and be laundered regularly. A laboratory coat is intended to prevent contact with direct, chemical dusts, and minor chemical splashes of spills. If it becomes contaminated, it should be removed immediately and affected skin surface washed thoroughly. Employees should wear long legged clothing and avoid short trousers or skirts to cover areas that lab coats do not. Shoes should be worn in the laboratory at all times. Sandals and perforated shoes are not permitted in the laboratory. In addition long hair should be confined.

The above passage falls short because it does not directly address the readers and does not follow through in its directive. Further, the passage does not include related information the readers need. Who launders the lab coats? In the following rewrite, there is no question. The rewrite also gets rid of the conditional ("should be"), speaks directly to the reader, and tells how to follow through.

> Where there is no immediate danger to the skin from contact with a hazardous chemical, it is still prudent to select clothing to minimize the risk of exposure. Wear clothing with long legs to cover areas that lab coats do not. Avoid short pants or skirts to minimize exposure risk. Wear socks and closed shoes at all times; do not wear sandals and perforated shoes. In addition, confine long hair for your safety and to prevent compromising the integrity of the lab work. In addition, do not wear jewelry that can catch in laboratory equipment. Most important, always wear a laboratory coat over street clothes. Laboratory coats prevent contact with dirt, chemical dusts, and minor chemical splashes or spills. Always wear clean laboratory coats and dispatch soiled ones to the hamper bins in the change room.
>
> Note: If you suspect your lab coat has become contaminated, remove it immediately, bag it, and wash all exposed skin surfaces thoroughly. Refer to the MSDS sheet for any additional measures. Call the Chemical Hygiene Officer to collect the contaminated lab coat.

Readers' Language Skills

Whether you are writing for internal distribution or external, you must consider the language skills of your readers. When writing to a regulatory body, you can be confident that readers will have a degree of sophistication to enable them to understand fairly complex data. These people have read similar documentation repeatedly.

If you prepare a manuscript, you can also be confident that if you write according to the style set forth by the publication, the readers will readily comprehend what you say. These readers are your colleagues and will have comparable language skill sets. The same holds true for posters and slide sets. The audience that reviews posters in an exhibit hall does so because of interest in what others in the industry are doing; they attend presentations for the same reason.

Not all readers possess the same skills. Some documents must be comprehensible to a much lower reading level, and it is critical that they be. Work instructions for a manufacturing environment are best in clear, direct language that speaks to a tenth grade or lower reading level. Informed consent for potential participants in clinical trials is also best written to an eighth grade or lower reading level as well, as is patient literature.

Informed Consent

What Is Informed Consent?

Informed consent is not a signature on a piece of paper. It is a process undertaken to assure that people are fully informed about the choices available for their health care. Informed consent is based on the legal and ethical rights of individuals to direct their own care and the ethical duty of physicians to involve their patients in their own care. Individuals have the right to refuse information about procedures or surgeries, but they must initiate the request, not be offered it. After it is determined that the person understands the proposed intervention and has reached a decision, his/her signature on the informed consent form denotes acceptance of the proposed intervention

When Is Informed Consent Needed?

Informed consents are required for invasive interventions (for example, procedures, anesthesia, and surgeries); before being given certain medications (for example, flu shots) or receiving certain tests (for example, HIV testing); and before participating in research projects.

What needs to be included in an informed consent?

Informed consent needs to include:

- The rationale for the intervention
- The nature of the proposed intervention
- Reasonable and medically appropriate alternatives to the proposed intervention
- The risks and benefits of the proposed and all alternative interventions
- The anticipated outcome of the proposed intervention.

Who Can Give Informed Consent?

Conscious, competent individuals can give informed consent. An incompetent or incapacitated individual needs a surrogate decision maker to give informed consent.

Parents give informed consent for their dependent children.

In most states, emancipated minors can give informed consent. Most states also allow minors to seek treatment for sexually transmitted diseases, pregnancy, and drug or alcohol abuse on their own.

Tips for Writing Informed Consent

When writing informed consent forms, be sensitive to a person's ability to digest complicated or bad news, and be sensitive to personal (cultural or religious) beliefs of the regional population.

Rely on the following when writing informed consent forms:

- Provide alternative language (Spanish, for example) informed consent forms for individuals whose first language is not English.
- Explain the proposed intervention in consumer friendly language, not technical jargon.
- Define all terms and write out phonetic pronunciation of all unfamiliar terms.
- List risks in as nonthreatening a way as possible.
- Put risks in perspective. For example, about 1 in 50,000 healthy people die from general anesthesia; you have about double that chance to die in an automobile accident.

Courtesy of Joan Lorenz

Another consideration is whether readers are native born speakers and writers of English, or if they are foreign born. In most pharmaceutical, device, and biologic environments, the mixture of employees is multinational. The same is true of patient populations for clinical trials. The result is that, although your readers may have sufficient language skills to comprehend most communication related to activities with which they are familiar, they may not understand a term you consider critical, or they may have difficulty decoding a complex passage. If you suspect that's a possibility, it's easy to give a quick explanation or include a brief glossary if the document is long, or to rewrite to simplify presentation. Remember that it's a rare reader who will confess to "not understanding" if he thinks his lack of comprehension stems from his own inabilities. Provide the meanings you intend your readers to have so that there will be no misunderstanding.

Further, if you are a writer for whom English is a second language, you may be bringing certain conceptions about writing, unfounded in standard American conventions, to your workplace. In some cultures for instance, when a writer sums up information for a reader, it is tantamount to saying, "You probably won't understand the data, so I'll have to explain it." Such a message can seem insulting. In written English, however, the standard is to explain the significance of data. Always tell your readers what you mean and what the significances of certain facts are. Never leave the message open to interpretation.

The average person has three vocabularies:

- Reading is the largest.
- Speaking is the next largest.
- Writing is the smallest.

In writing, words must carry 90 percent of our messages. Punctuation and graphics convey the rest.

Writing Directly to the Reader

Many documents work best when they speak directly to the readers. When you seek an immediate response from specific readers, the best way to make sure the response is what you want is to involve them. You may write directly to fellow employees where you work; you may also write directly to readers external to the company. Sometimes you will write to one person specifically; just as often, your immediate readers may be two or more.

Many pieces of writing address readers directly. Letters, memos, and e-mails certainly do. So does informed consent for clinical trials. Other documents that speak directly to readers, such as methods, instructions, procedures, and operator manuals, usually work best in the imperative voice, which simply means the "you" in the writing is understood. The imperative voice uses the present tense and is the only instance in English where you can compose a complete sentence without actually writing out the subject. For example, "Write it down" is a directive, and the "you" is understood to be the reader. Understood as well is that the audience for these documents has been trained in the processes. Typical steps in a procedure, such as these that tell how to measure flow rate, have an implied "you":

6.1 Make sure the clamp securing the funnel to the vibrator is properly affixed.

6.2 Open the air valve and adjust the regulator to 16 psi.

(See Chapter Five for more information about procedure writing.)

A Preventive Maintenance Memo

Consider this internal memo about preventive maintenance to a project engineer. Here the author writes directly to the engineer but does not include him in the communication at all.

Original

JORSTAD LABORATORIES Internal Memorandum

DATE: August 16, 2005
TO: Matthew Zarelli
FROM: John Lopez
RE: Preventive Maintenance

At the weekly meeting on August 14, 2005, it was decided to review aspects of the preventive maintenance program to assess its sufficiency. The preventive maintenance for sophisticated components of the fluid bed dryer has been contracted out to Glatt and Co. The in-house work orders for the past month have all been reviewed, and none are related to preventive maintenance.

During the weekly staff meeting, the question of possible "downtime" in production because of simple maintenance needs was not discussed. A meeting should be arranged that includes the maintenance manager. Regular preventive maintenance must be addressed.

While the paragraphs in this memo are grammatically reasonable and present key information, they're not doing the best job they can. The writer has focused on the information that precipitated the memo. Seriously lacking is what the writer expects from the recipient. How does the information relate to the reader? What role has the reader played in the development of events? And who should arrange the meeting that will address preventive maintenance? Will the reader be involved in making the decision about preventive maintenance?

A rewrite that addresses the engineer directly, using "we," "our," "you," and "your" produces a much clearer message that's more likely to get results. In the first version, the writer refers to himself once, but not at all to the reader. In the second version, the writer includes the reader ten times and directs the engineer to action. The rewritten memo is more effective because the reader knows what action he must take, and he knows it up front.

Rewrite

JORSTAD LABORATORIES Internal Memorandum

DATE: August 16, 2005
TO: Matthew Zarelli
FROM: John Lopez
RE: Preventive Maintenance

Please arrange a meeting and include the maintenance manager so that we can discuss preventive maintenance and devise a plan for action. As we discussed at our weekly meeting on August 16, 2005, we are going to review the work orders for your area to see if aspects of the current regular preventive maintenance program are sufficient.

As you pointed out then, the preventive maintenance for "sophisticated components of the fluid bed dryer" was contracted out to Glatt and Co., and thus that equipment is regularly serviced. However, maintenance of equipment not under contract remains an issue. Since our meeting, I have reviewed all the work orders for the past month, and you are right: Preventive maintenance has not been a priority. In addition, during the weekly staff meetings, we did not discuss whether any equipment has suffered "downtime" because of maintenance needs. I firmly agree that we must address these issues, and our meeting will be the place to start.

> When we give serious consideration to who will read what we've written, what they already know, and what they need to know, we are much more likely to hit the mark.
>
> **Arlene Johnson, Author**

Announcing an Inspection

The following memo effectively addresses several key people who need the information contained therein to prepare for an FDA inspection. Notice how the memo tells people directly what they need to do in preparation for the FDA visit. Further, while it refers recipients to the established protocol, it reiterates what the writer feels is essential. In addition, it refers to the recipients 15 times, with the words "you" and "your" and "we" and "our," as well as with the use of the imperative voice of "you" understood. This technique clearly makes the readers part of the message. This memo is effective.

LEON LABS, INC. **Memorandum**

Date: November 6, 2005

To: Frank Grimaldi
Linda Zwagerman
Lamar Brown

From: Beatrice Solomon

Re: Upcoming FDA Inspection

We anticipate FDA will be on the premises sometime next month. To prepare our facility for the visit, we will be holding a mock inspection on Friday, the thirteenth. Please inform your respective staff. You may want to review the company protocol for handling inspections, as well.

We plan to provide two escorts per inspector. In addition, I'd like to reiterate that our position is to do the following:

- Provide requested documentation to the company inspection leader as quickly as possible.
- Avoid being confrontational.
- Answer only the questions you are asked.
- Clarify all observations.
- Agree to implement corrective action, if any, as soon as possible.

We will evaluate our mock inspection on Monday to identify any potential problems. If you have any questions, please contact me before week's end. Thanks.

Announcing a GMP Audit

The same common-sense approach works when you are writing to readers outside the company: Include them in the message. The following notice to a contractor advises of an impending current Good Manufacturing Practice (cGMP) audit. Notice how it addresses the reader directly and calls for action.

Ronway Laboratories

2323 Ronway Drive Jackson, Wyoming 83001
June 5, 2005

Dr. Calin Cionca
CEO and General Manager
Micro Systems Management, SA
737525 Hundingsland Boulevard
Chelsea, West Virginia 24608

Dear Dr. Cionca,

We at Ronway are pleased with the range of services Micro Systems is prepared to offer us. Our dramatic increase in manufacturing puts us in a position to use contract services, and we believe Micro Systems will augment our operations satisfactorily. As I mentioned to you during our phone conversation, all that remains is a routine GMP audit of your facilities.

We plan to send two auditors to your facilities on Tuesday, July 14th, to perform an evaluation. They will assess your analytical testing to verify your Certificates of Analyses, your manufacturing site, and those processes that concern our manufacturing requirements. Please provide them with the appropriate documentation as well as guided access to your facilities.

Thanks for your cooperation. We at Ronway look forward to a productive working relationship with Micro Systems.

Sincerely,

RONWAY LABORATORIES

Christian Horton
Director of Quality Control

Giving Product Information

The following letter provides information about a product. It speaks directly to the recipient and includes both reader and writer, referring to previous conversations.

Lasentec

15224 NE 95th Street Redmond, WA 98052

August 28, 2005
Benjamin Smith
Professor of Chemical Engineering
University of Maine
6982 Jenness Hall
Orono, ME 04469

Dear Professor Smith,

I look forward to continuing our discussion of your experiences in crystallization research. I have enclosed the latest information available describing the application of Lasentec® technologies in the monitoring and control of batch crystallizers.

The two instrumentation systems we discussed, FBRM© and PVM©, were developed by Lasentec for real-time *in-process* measurement of Particle Count, Size, Shape, and Imaging. Both systems are probe-based for easy installation and application. The probes are inserted directly into your process vessel or pipeline. They do not require material extraction and sample preparation.

LASENTEC FBRM© (**F**ocused **B**eam **R**eflectance **M**easurement) provides a continuous, high-speed measurement of the Particle Population, enabling you to track the rate and degree of change both on a particle count and particle dimension basis. Even at high solids concentration, FBRM© can provide a *sensitive* and *precise* measurement of

- Population in *independent* size ranges
- Rate and degree of nucleation
- Rate and degree of growth, reduction and agglomeration
- Process endpoint
- Batch to batch consistency
- Continuous process stability

LASENTEC PVM© (**P**article **V**ision and **M**easurement) provides a continuous stream of high-resolution microscope images of your particles and particle structures as they naturally exist *in process.*

PVM© images will provide an in-depth knowledge of your particle-to-particle interaction, and gives you the ability to visually track dynamic process changes in real-time. It is a powerful R&D tool for imaging particle, droplet, and bubble systems at full process concentrations.

Lasentec-2

I believe that you will find the enclosed papers relate directly to your planned research project. After you have had a chance to review this material and are prepared to further discuss our collaboration, I would be pleased to carry out a feasibility test to demonstrate the effectiveness of our instruments in your particular application.

Please call me at your convenience, at 1-800 LASENTEC (1-800-527-3683).

Best regards,

Terry P. Redman
Application Engineer
TerryR@Lasentec.com

Enclosures

Courtesy of Lasentec, Inc.

(See Chapter Four for more examples of correspondence.)

Focusing on the Information

Writing directly to readers, when you know who they are and want a specific response, is usually the most effective way to deliver information. However, much of the time, your purpose in writing will be to present information that's part of a larger picture. Such documents as analytical reports, summary reports, and project analyses frequently make little or no mention of the reader, nor do they need to, because their purpose is to provide a history or other information.

Documents that don't call for reader action directly must still present what readers need to know to understand what has happened, what is happening, or what will be happening, as well as the conclusions and recommendations resulting from specific findings. Often, documents are but links in a chain of documents, and getting the overall picture of a project may require understanding of much more than what the document at hand presents.

Even if immediate reviewers and approvers comprehend what's in a report, filling in the gaps can be useful, even if it means repeating some information the initial readers already know. In the same vein, controlling your delivery so that it is clear and direct also makes sense. Using internal jargon, for instance, in a report that may be subject to future outside scrutiny is foolish, for doing so invites questions that demand answers. How much better to be lucid and complete in the first place!

It's better to err on the side of giving more information than your immediate audience needs than to be sparse in data or detail, which you may be asked to produce at some future point. There's a fine line, here, however. Your written histories should provide the information relevant to certain activities; they should not trigger unrelated questions that arise from the inclusion of inappropriate details. Worse is to make statements that are unsupported; blanket statements without support are sure triggers for queries from readers, and when readers start to question, you may ultimately wind up giving more information than you intend or need to. Good writing gives readers what they need to know. Thus, a writer may deliver the same information in several versions depending on the readers' need to know and level of technical comprehension.

Summarizing an Investigation

Many documents focus purely on events. Investigations, for example, tend to be directed toward activities leading up to a conclusion. Readers play no role other than that of understanding the information and perhaps responding to it routinely. For instance, a manufacturing deviation report will present the findings of an investigation; the result of those findings may provide the basis for deciding whether to rework or discard a batch.

The following technical note does not call for reader action directly because the purpose is to provide a history of an occurrence. Addressed to the Director of Quality Control, the memo summarizes the result of an investigation.

JORSTAD LABORATORIES **Internal Memorandum**

DATE: February 18, 2005

TO: Ellen Measday

FROM: Hal McCornac

RE: 098C3-Q1234RT-18M and 098C3-Q1235RT-18M test failures

February 15 retesting of 098C3-Q1234RT 18M and 098C3-Q1235RT-18M failed to confirm the test results obtained from the dissolution tests performed on January 29 and 30, 200- The February 15 testing showed a higher potency than the earlier testing did. An investigation revealed that the low potency was a result of the vials being previously opened, and consequently the samples had absorbed moisture and increased in weight. Thus, while the weights in the tests were identical, the ratio of active ingredient to total weight was not, since the first test samples had proportionately higher water content, and therefore proportionately smaller active ingredient.

Retesting using a new standard from an unopened vial verified a potency of 947.67 mcg/mg and a standard factor of 0.0671 mcg/ml. Assays LIC-99-0039 and LIC-99-0040 substantiate these results.

Summarizing a Complaint Investigation

The following memo focuses on the events and does not call for action from the reader; its purpose is to inform the recipient, a Regulatory Director, of the results of a complaint investigation. The Director, in turn, will compose a complaint response.

KIM LABORATORIES, S. A. Memorandum Page 1 of 1

DATE: October 11, 2005

TO: Rui Li

FROM: Eugenia Cline

RE: 2 mg/5 mL complaint

Quality Assurance has reviewed the following 0.5 mg/1 mL product complaint and found radioactivity in the excess precipatated material and low final activity in the product.

Site	Lot	Isotope Manufacturer	Isotope Lot
Wellman Pharmacy	66B2	Radiostuff Inc.	456

The isotope from Radiostuff Inc. has periodically become bound in the excess precipitate formed during the labeling of the product. The isotope caught up in the precipitate is a result of isotope colloids forming during the preparation of the isotope-product complex. The formation of isotope colloids is dependent on the lot of isotope. In addition to this product complaint, three additional complaints the same day proved to be related to the labeling of the products with isotope lot 66B2. In addition to isotope lot–related issues, there are several other interrelated variables, including the importance of strictly following the labeling procedure on the package insert. Quality Assurance completed a report on the causes of isotope colloid formation during the labeling procedure in January 2005. It is available in the QA Archives.

A review of the production records found them to be acceptable. The complaint incident does not represent a serious or unexpected drug experience, so a report of the incident to FDA is unnecessary. However, a review of isotope-related complaints from last year is now underway to determine if further clarification of this recurring situation is possible.

Courtesy of Michael Nolan

Explaining Trade Dress Revisions

The thrust of the following memo is to announce a trade dress revision. It does not call for any action on the part of the readers and simply presents information about a change in market image of a product. Ultimately it will become part of a product development outline. It is direct and to the point and uses the corporate "we."

Lily Labs Interoffice Memo

To: Dr. Maureen Conlan
Dr. Kathleen Monroe
Dr. John Tessman

From: Seth Porter

Location: New Product Development

Date: Nov. 22, 2005

Re: Trade Dress Revisions

We have made the following changes to the market image of Nadolol tablets, 300 mg. Labetalol HCI Tablets, 300 mg: The tablet will be coated with an Opadry to match blue 552 in the Opadry color kit. These tablets will be film coated and have no bisect. The imprint "LL" will appear on one side of the tablet.

The Words You Choose

Much has been said about objectivity and subjectivity. Remaining objective is generally the goal for writing, unless we are preparing a piece of correspondence that contains a personal message. What does being "objective" rather than "subjective" mean? Being objective really means focusing on the best, most direct delivery of information for the reader's understanding without interjection from the writer. Subjectivity allows the writer's opinion to come through.

However, many people who teach and assess writing have come to understand that, while some passages can be more objective than others, no writer can rid himself of subjectivity completely. The words writers choose are the results of their thinking patterns and their personalities; so are their language constructions. Thus, the writer's voice really reflects the individual who first composes a document. Perhaps that's why so many eyes see most documents generated within companies developing or manufacturing therapeutic products, at contract research organizations, and at clinical sites. In an effort to be objective, rather than subjective, most technical people focus on the activity rather than the doer of the activity when recording information. However, whereas this approach is logical, it's not law. People, after all, comprise companies. People communicate what they know to others. They present their findings and ideas so that other people can examine them, either immediately or in the future.

Readers generally receive information better if it includes a doer of the action, an active "we did this" rather than "this was done" approach. Studies show that when reading the passive voice the audience tends to turn the information around anyway before digesting it. Thus, it often makes sense to say "we" when you mean your department or area. Indeed, the corporate "we" is a handy device when you speak on behalf of a unit. You can also refer to your department or area directly: "New Product Development (NPD) has completed the formulation." (Chapters 9, 10, and 11 give more information about the passive voice.)

Don't shun using "I" either. Many documents that speak directly to people appropriately incorporate "I," the writer. "I'm sending you the dissolution data you requested yesterday" is direct and to the point in a brief memo. Even in longer reports, it's okay to include yourself. "During July, I completed the validation of the in-house method" delivers information straightforwardly in a periodic report. Howard Kanare, author of *Writing the Laboratory Notebook,* points out that a notebook entry that says "I saw the mixture turn blue after 10 min.," is preferable to what scientists often write: "It was observed that the mixture changed color after a few minutes had elapsed."

A word of caution is probably wise here, however. Many technical arenas have for so long avoided the use of the first person and first person plural in correspondence and documentation that it feels unnatural, and may even seem "unprofessional." Exercise some judgment, but make your decision knowing that including the doer of the action violates no professional writing standards.

The following passage is an excerpt from a Drug Master File from a Pre-Market Approval Application (PMAA). Notice that it is objective and factual, yet the writer here does not choose to eliminate the corporate "we." The result is direct and easy to read text.

> Abrams, Inc. began development of the active pharmaceutical ingredient in 2003. Development was in three phases. An initial small scale processing produced material for toxicology studies and research. In 2003 through 2004, Abrams scaled the process to provide sufficient final intermediate to allow production by an alternate manufacturing source for the final active pharmaceutical ingredient. No manufacturing occurred before 2004.
>
> In 2004, Abrams, under our directive, undertook a major development project to improve process safety and reduce process cycle-time and process waste. The result of this final phase development effort was the drug product which is the subject of this submission.

> We provide the details of the lots produced during development and explain the changes in the processes relative to lot usage in the development studies.

Using Suitable Language

You must also make sure your memos, letters, and reports present your information in terminology your readers will understand. If you are a chemist writing to another chemist, you share expertise, and complex technical terms will probably pose few problems. If you are a chemist, however, writing to an information systems specialist, you may very well run into difficulties, just as you might in conveying information if you are a chemical engineer writing to a line supervisor. Again, awareness of the expertise of the recipient is tantamount to successful delivery and acceptance.

The following excerpt from an engineering department planning a laboratory renovation was addressed to a director of analytical development.

> Enclosed herewith is the laboratory master plan proposal which abandons the one previously proposed by engineering in favor of a proposed master plan for a facility with six additional modified workstations, and one additional door to the corridor to facilitate access and egress and interfacing between the analytical laboratory undergoing restoration and quality control.

Here the writer has forgotten that the reader of his review, a chemist, is not a facility engineer and has not been a party to all the discussions that have gone into developing the original proposal. Even though the director may understand what is necessary in the laboratory, she may not be prepared to work out the difference between a "master plan proposal" and the plan "previously proposed." She can easily become befuddled if the vocabulary impedes the message.

The phrasing "to facilitate access and egress and interfacing between the laboratory undergoing restoration and quality control," means, simply put, "to make going back and forth between the analytical development lab and quality control easier." In his choice of words, the writer betrays either ignorance of the person who will be reading his analysis or carelessness in presenting the information.

You can avoid the grief that will surely befall you if you submit writing such as this. For one thing, look for words such as "proposed" and "proposal" that each seem to be one word referring to two different things or ideas. This is called "equivocation," and here it is confusing—not richly ambiguous as it would be in a piece of literature. Second, if it isn't possible to reduce technical verbiage to concrete terms, supply brief definitions—a phrase or two—or examples to clarify things for the nontechnical reader. Third, on the assumption that as master plans, such documents are subject to review long after they have been put together and after their authors have moved on to other jobs, ask the reviewer to supply context or briefly repeat background information that other memos might include in more detail—that is, the circumstances of the suggestions under review.

At times it's appropriate to use technical terminology. Most professions have their own "shorthand"—terms that are unique to a particular subject area. Among specialists, this shorthand, or jargon, can save time and even communicate information more precisely than if it were eliminated. When you say "We're working out the 'bugs' in the system," or "The computer is down," you're communicating succinctly to people who are computer knowledgeable but perhaps unclearly to those who aren't. What you must do, then, is evaluate your audience and make sure you're using diction that will do the best job for you. When a chemist refers to "two unknowns," another chemist will immediately understand that two impurities have not been identified. A less technical person might not.

Evaluate what needs to be said and to whom. What can happen as a result of unnecessary jargon is that your reader will think you're trying to put something over on him, or worse, won't understand what's being said at all. In addition, since most companies have their own jargon, you may run the risk of including terminology that means one thing to you, but something quite different to an external reader.

Controlling Acronyms

Acronyms are specialized jargon. They're created from the first letters of a group of words, or from a combination of letters and parts of words. For example, GC is an acronym for gas chromatography; IPA is an acronym for isopropyl alcohol; and SOP is an acronym for Standard Operating Procedure. These acronyms are perfectly clear to most technical people and they probably need no explanation.

Yet acronyms are endemic in this industry—so much so that FDA includes a list of common acronyms on its website. A very real danger of acronyms wreaking havoc can occur when a company uses an acronym to mean a thing easily understood internally, but that means something entirely else to an external auditor, joint-venture manufacturer, or FDA inspector. Even cGMP, which most of the industry understands to mean current Good Manufacturing Practices, can indicate something quite different—cyclic guanosine monophosphate.

So unless you are sure your reader knows that DTA means differential thermal analysis, you'll be better off spelling the words out. It never hurts to handle acronyms the way most technical writing experts advise: Simply spell out the full term in the first citation and follow it with the acronym in parenthesis.

We hope to complete the Modified Release Facility (MRF) by June.

In subsequent references within the document, simply use the acronym.

The MRF facility will be producing solid dosage products.

To prevent confusion and ensure consistency, careful writers follow these guidelines as well.

- In most cases, do not use periods within or after the acronym, except at the end of a sentence.
- Do not use an apostrophe with a plural acronym.
  - CRFs, IRBs, INDs
- When affixing a prefix to an acronym, hyphenate the prefix and the acronym.
- pre-IND submission activities

Most companies have also realized that it's sensible to maintain a list of acronyms specific to the company. This they usually do as part of a style guide or style sheet. Such a tool makes the appropriate acronym available to writers and deflects the possibility of confusion or misunderstanding as to what company writers mean. (Chapter 14 provides a lengthy list of industry-specific acronyms.)

Connotation and Denotation

The English language is such that we have many, many words to indicate the same thing. These are synonyms. And while synonyms will appear in the dictionary as definitions for each other, there are often subtle variations. So the first definition is the primary meaning of a word, the denotation; what follows are usually the connotations of the word. A word like "smell" may be defined as "odor" or "scent," but the three words have different connotations, the associations we place on them, with "odor" being most negative, "smell" somewhat less negative, and "scent" as positive and light, and aroma also positive, but stronger. You must, therefore, choose your words carefully so that the precise understanding of the message is the result.

Defining Terms

Once you know to avoid unnecessary jargon and unidentified acronyms that may baffle and befuddle readers, concern yourself with clarifying the rest of the words essential to the message. Remember, information is useful only if it makes sense, and what makes sense to you may not make sense to your readers.

Readers with technical expertise similar to yours will most likely understand specialized data without lengthy explanation. But readers with less technical expertise than yours will generally have trouble absorbing the information. Therefore, evaluate the appropriateness of your words and select terms so as to make the information readily comprehensible and impossible to misinterpret. In a work environment, only the rare reader will consult a dictionary to read a letter or memo, let alone a long report. In technical writing, terms open to interpretation deserve definition.

For instance, if an engineer writes a report on the safe installation and operation of electrical equipment that includes as its audience new technicians, he may need to include a definition of "grounding." Similarly, an in-house manual that describes an SOP system should include a definition of key terms as they apply to the system. For instance, "Active SOP" can be defined to include the existing version of an SOP undergoing revision. When the revised version becomes "active," the previous version is "retired"; however, a "withdrawn" SOP can denote a document no longer in use in any version. The terms "retired" and "withdrawn" can be easily interchanged unless they are clearly defined. Consider these terms: "current document," "obsolete document," "inactive document," "disabled document," and "enabled document." These designations are common to document control systems and will vary from company to company. Hence, they demand definition.

There are several effective ways to define terms. These include parenthetical, restatement, classification, operation, etymology, background, and negation. Often a definition will employ more than one technique. In essence, here's how they work.

Parenthetical Definition

Parenthetical definitions include an explanation in parenthesis after the word.

> We discarded the effervescent (bubbling) mixture.
>
> Identify the scientific (taxonomic) name of a phylum, class, order, family, or genus.

Restatement Definition

Restatement definitions offer an appositive word or phrase that restates the term. Note that this type of definition calls for two commas when it appears within a sentence to set off the restatement, unless it comes at the beginning or end of a sentence. Sometimes a restatement definition is a separate sentence.

> A polygraph, an instrument used in lie detection, records changes in pulse, blood pressure, and respiration.
>
> The trees outside the new facility are deciduous; that is, they shed their leaves by the end of October.

Classification Definition

Classification tells what family a word belongs to.

> A dog is a member of the canine family.
>
> A tumor is a neoplasm.
>
> Bluetongue II virus is a member of the Reoviridae family.

Operation Definition

An operation definition tells how something works or happens.

> A disorder of the pituitary gland or pancreas causes diabetes, a metabolic disease characterized by excessive urination, persistent thirst, and often, an inability to metabolize sugar.
>
> Air-to-air solar heating circulates cool air from inside the facility, across a collector plate, which is heated by sunlight on the roof, and then back into the facility.

Etymology Definition

Etymology looks back in time to the roots of words. Approximately two thirds of the words in the English language have their origins in Latin and Greek.

> "Biology" comes from the root "bio," meaning "life," and "ology," meaning "study of."
>
> "Chromatograph" comes from the root "chroma," meaning "color," and "graph," meaning "write."

Background Definition

A background definition gives some history.

> Gasohol, a mixture of 90% unleaded gasoline and 10% ethyl alcohol (ethanol), has gained some acceptance as a fuel, since it is comparable in performance to 100% unleaded gasoline with the added benefit of having superior antiknock properties.

Negation Definition

Definition by negation means telling what something is not.

> Adsorption is not a misspelling of absorption. Adsorption causes liquid to adhere to a surface, like dew on a leaf, while absorption pulls liquid in, much like a sponge drinks up water.

Nondiscriminatory Language

Cautious writers take care to avoid using language that sounds discriminatory. Racism and sexism have worked their way insidiously into the American language. It's probably impossible to rid writing of every metaphor that holds meanings for male and female or ethnic affiliation, but you can certainly act to be sure you use the language to promote social equality rather than hinder it. Choose terms and expressions that don't refer to people in ways that can be considered negative or discriminatory.

One way to avoid discriminatory statements is to choose qualifiers carefully. To say "a woman manager" is discriminatory. To refer to "the Hispanic technician on the six o'clock shift" is as well. And to write that "her innovative ideas belie her age," even with the intent to compliment, is foolhardy. Unnecessarily drawing attention to differences without cause is discriminatory.

> The geographical distribution of the Germanic languages, of which English is one, has been more extensive than that of any other group of languages.

Be careful, too, not to use "he," "him," or "his" exclusively when referring to colleagues. One way to avoid offending is to use "he or she" or "him or her," for instance. But, as you can see, that can be awkward. Many writers, for that reason, use the plural, or interchange gender pronouns. It's also a good idea to choose the sexually neutral word over the traditional equivalent. For instance, "chair" or "chairperson" works just as well as the commonly used "chairman."

Living Language

American English, like other living languages, is in constant transition. It is ever adjusting to reflect the changes in society. The alterations in the language can be controversial and have drawn some criticism from purists who wish to retain the form they learned initially. Yet the reality is that English borrows words from other languages incessantly, sometimes with the foreign pronunciation and sometimes without, adhering more to the standards for American English pronunciation. Immigrants bring terminology that quickly gets absorbed into the vernacular; new ideas receive new labels; and words are regularly shortened and combined with other words. And so language alters itself.

Unlike many other nations, the United States has no official organization to prevent the language from changing. Indeed, the language has been called a polyglot; that is, it has a vocabulary stemming from myriad languages, a vocabulary that is ever-evolving. It's easy to let this phenomenon bog you down—even overwhelm you. Try not to let it; concentrate on the logic of what you have to say to whom and strive to write clearly and precisely. Let the language be the common denominator in communication. That means using language your audience understands.

3

Organizing and Delivering Information

> When we begin to write, we may anticipate the result, but writing has a way of taking on its own life.
>
> **Alexander Butrym, Ph.D., Professor of English**

Once you've identified to whom you are writing and assured yourself that your data are acceptable, consider how to put ideas and information together. This is often the most difficult part of the writing process. You may know what you need to say but may be stymied as to how to get it down on paper. This is normal; most people simply don't like to write. There are, however, some techniques for organizing information that can help you control your ideas as you commit them to paper. Don't be surprised, either, if you violate your initial concept of how you are going to deliver your information. The act of writing itself triggers new ideas, and this is to the good.

Any time you create a document, look at it from two angles: logic and development of ideas. The distinctions aren't always as neat and clear-cut as they are for the purposes of this book, however, and you may find many different ways of presenting information that work well. These guidelines are merely that: guidelines. They'll help you control your information when you're not sure how to present it.

You will frequently encounter problems in organization—situations where the ideas themselves are good, but not presented in a way that makes their meaning obvious or their purpose clear. In short, how can you best organize your information so that your messages work the way they should?

Categorize Your Information

To effectively organize information, it helps to categorize what you are conveying—is it good? routine? negative? persuasive? If you anticipate the reader's reaction to different categories of information, you'll know better how to present it. Surely you wouldn't respond to a request for an internal job transfer by saying, right off the bat, "You can't have the job because you lack the necessary skills." Yet, you might begin a note of recognition like this: "Congratulations, Joe. You certainly deserve this promotion." In the first instance, the reaction would surely be more negative than you desire. In the second, the reaction will likely be as positive as you hope. You gain nothing by irritating the recipient—nor do you benefit by working up to a message of goodwill. More important, withholding information or obscuring it can have deleterious effects on the day-to-day activities that make a business run.

> Writing is easy. All you do is sit staring at a blank sheet of paper until drops of blood form on your forehead.
>
> **Gene Fowler, Author**

The majority of the writing you will do in the pharmaceutical industry will fall into the category of delivering routine information. For the most part, you are not eliciting an emotional response; you are presenting information your readers need to have. Use these guidelines for logical presentation. But remember: They are not cast in stone.

Finally, writing isn't easy. Nothing says you have to "begin at the beginning and go to the end." You may be best off getting the details down and then rearranging them for readability and comprehension.

The Direct Approach

> I hate to write, but I love having written.
>
> **Dorothy Parker, Author**

Use this pattern for presenting most of the information you relay in writing. The direct approach is effective for routine information—the work-specific writing you do. When you are writing a short memo or technical note, for instance, let this pattern govern your total delivery of information. When you are writing a document of length, break it into segments, and use this pattern to help you control your overall delivery as well as your delivery in each part.

The direct approach calls for making the key point first. This way, your reader will immediately know what a document, or portion of a document, is about. Your point logically leads to information that supports it. Unfortunately, when most people write about what they've done or what they're going to do, they tend to approach documenting the past or projecting future activity sequentially, the way they have addressed or will address completion of the task itself. Writing sequentially seems logical, and indeed, it reflects inductive reasoning: A series of facts leads to a conclusion. However, the reader is best served by understanding the point first—whether it is a summary of findings, a strategic decision, or the results of an investigation.

Usually, delaying the point doesn't make a document any more effective. Your reader does not know the purpose of a document the way you did before you even began to write it. Thus, inductive presentation is not the best way for most readers to receive information. With inductive presentation the reader doesn't get the point until he reaches at the end of a document or passage in a document. Readers understand messages best when a point is followed by a series of facts that support the point.

For instance, by the time an engineer prepares a summary of an installation/operation qualification (IQ/OQ), she has been working on the project for a long time. She has already written a protocol of what she planned to do and has secured the necessary approval. Then she's gone through the testing process. Chances are, she may have had some nonconformances along the way. Once she's assured herself the equipment installation and operation meet qualification criteria for the company's needs-specific purposes, she has gone through a step-by-step process.

Yet there's no benefit in preparing a report for approval of the qualification by saying, "In July it was time to requalify the 200-gallon batching kettle with double motion counterrotating agitator" and then detailing events chronologically. What the readers need here is an immediate statement that asserts that the equipment is qualified, partially qualified, or not qualified as a result of the testing according to the protocol. An IQ/OQ summary report is easy to understand when it begins straightforwardly like this:

> The Lee Industries 200-gallon batching kettle with double motion counterrotating agitator meets all the installation/operation qualification requirements. All test functions established in Protocol Q023 are complete and reconciled. In carrying out the testing, we addressed three nonconformances. All test results are attached.

> If you have an important point to make, don't try to be subtle or clever. Use a pile driver. Hit the point once. Then come back and hit it again. Then hit it a third time—a tremendous whack.
>
> **Winston Churchill**

The summary report then can go on to explain nonconformances and their impacts and state requirements for scheduled requalification and so forth. This way, the readers know immediately the results of the qualification testing and are ready for what the report contains. Here's a basic pattern for presenting information:

The beginning:	State your purpose. What point do you want your reader to know? What, if anything, do you want your reader to do?
The supporting information:	Offer details to substantiate your purpose.
The close:	Tell your reader what he or she needs to know to be convinced. Restate the point or call for action.

Packaging Specification Change

The information in this short memo isn't as effective as it could be because it doesn't make its point up front. The reader is subject to explanatory details before learning the purpose of the communication.

JORSTAD LABORATORIES Internal Memorandum

DATE: August 23, 2005

TO: Ginger Ogden

FROM: Linda Tanabe

RE: Quinine Sulfate Packaging Specification Change

We decided to make Quinine Sulfate 200 mg capsules dose proportional to the 325 mg strength. For the Quinine Sulfate 200 mg capsules we had been using number 2 size capsules and 100 cc capacity bottles.

However, the number 2 size capsules are inadequate for the 325 mg dose. Therefore we decided to use the number 1 size capsules. Then we determined it to be too difficult to accommodate one hundred number1 size capsules in the 100 cc bottle along with the insert. We also realized that 100 number 1 size capsules and the insert would be difficult to fit into the 100 cc bottle. To accommodate the new capsules, we changed the bottle size from 100 cc to 150 cc and added rayon to fill the empty space on top of the capsules. This solution is satisfactory, and we have changed our packaging specifications accordingly.

Even though this memo is short, it's easier to understand when the key idea comes first. The point is not to explain the decision to make the capsules dose proportional to 325 mg, but to detail the consequences and subsequent decisions about packaging. In the following version, the reader reasonably looks to the information after the first statement to back up that statement, and it does.

Rewrite

JORSTAD LABORATORIES Internal Memorandum

DATE: August 23, 2005

TO: Ginger Ogden

FROM: Linda Tanabe

RE: Quinine Sulfate Packaging Specification Change

For 325 mg strength Quinine Sulfate, we are now using number 2 capsules and 150 cc size bottles with rayon to fill the empty space on top of the capsules. We have changed our packaging specifications accordingly.

For the Quinine Sulfate 200 mg capsules we had been using number 2 size capsules and 100 cc capacity bottles. When we decided to make Quinine Sulfate 200 mg dose proportional to 325 mg strength, we realized that the number 1 size capsules, rather than number 2 size capsules, were necessary, because the number 2 size capsules could not accommodate the new dosage. In turn, we realized it would be difficult to fit 100 number 1 size capsules and the insert into the 100 cc bottle. Increasing the size of both capsule and bottle has proven satisfactory.

Drug Recall Letter

The Code of Federal Regulations (CFR) defines the content of recall letters; it calls for a clear presentation of the point up front. This letter does that effectively and is a solid example of the direct approach for organizing information. The purpose is stated immediately in bold letters, before the salutation, as well as in the second sentence. The reader knows at a glance what's afoot.

Ronway Laboratories

2323 Ronway Drive Jackson, Wyoming 83001

October 16, 2005

URGENT: DRUG RECALL
Stability Assay Failure and Unknown Degradant Products
Re: 50 mg SUPPOSITORIES, USP, ALL LOTS
NDC 0168-1949-46 Ronway Laboratories

Dear Wholesaler/Retailer:

Recent tests indicate that some outstanding lots of 50 mg suppositories do not meet minimum assay requirements through their expiration dates. Because we are uncertain that all lots will meet assay requirements throughout their shelf life, we are recalling all unexpired lots of this product. These lots have potency assays ranging from 59% to 100%. In addition, we have been unable to identify a degradation product; the concentration levels are unknown. This represents a potential health hazard, but we believe that the probability of serious adverse health consequences is remote.

Please determine if you have any lots on hand. If you do, discontinue distribution of these lots and return the merchandise to our facility promptly, to my attention.
Please complete and return the enclosed response form as well. If you have any questions call 306-123-4567, extension 525.

If you have distributed any of these lots, please contact your accounts immediately; advise them of the recall; and have them return their outstanding recalled stocks to you. We are conducting this recall to the retail/dispensing level. Return these stocks as indicated above. We will reimburse you by credit memo for the returned goods and postage.

We have informed FDA of this recall. We appreciate your assistance in this matter.

Sincerely,

RONWAY LABORATORIES

Erin Andreas
Vice President, Regulatory Affairs

The Direct Approach in Report Sections

You can apply the same principles of delivery in segments of reports. Even if a report has a clean, comprehensive summary at the beginning, the information within each component of the document is easier to comprehend when it makes a point and logically supports it.

The following is a portion of a product development report. The point is to identify which batch the company will use as the standard to manufacture a product. Notice that the writer has delivered the information in the sequence in which it occurred. Where does the information the reader wants to know lie?

> Ronway Laboratories initiated the formulation development for sodium capsules in January, 2005. Our strategy has been to employ exponents similar to those used in brand formulation combined by a dry mixing method. Folles SA manufactured the available active entity.
>
> The goal was to develop 50 mg strength first and then follow up with 100 mg strengths. Appendix A includes the formulation experiments conducted during formulation development.
>
> The formulation lot numbers are five digits. The prefix 123 refers to the laboratory notebook housing the records of the development of this product, and the subsequent numbers refer to the batch.
>
> Lots 123-01 to 123-14 incorporated Lactose Anhydrous DT, and Lactose Hydrous SD Starch P21 in different ratios and combinations with Sodium Lauryl Sulfate, Magnesium Stearate, and, in the case of lot 123-11, Aerosil 200, to achieve good encapsulation properties of the blend and to match the dissolution profile of the brand capsules.
>
> We accomplished this effort with lot 123-14, and manufactured pilot batch PB987 accordingly. However, this batch developed stability problems under accelerated conditions.
>
> Lots 123-15 to 123-28 reflect the attempts to develop a second formulation employing Magnesium Stearate and Aerosil 200 in combination with Starch P21. Lot 123-28 exhibited good encapsulation properties and similar dissolution patterns to the brand product. However, concurrent to development of this batch, material became unavailable for further development.
>
> We then made and tested Lot 123-29, similar in formulation to lot 123-28, but employing Magnesium Stearate manufactured by Nogodawa Pharmaceuticals. Results indicate that this lot is comparable to brand in dissolution behavior and is stable under accelerated conditions. Thus, Lot 123-29 is the base for pilot batch PI988.

The very last sentence is what the readers want to know. All the information is here—but a simple revision, almost a cut and paste, makes this information easier to digest. Since the key information rests in the last paragraph, moving that information to the beginning both clarifies and allows for logical explanation of what follows. A simple edit to paragraph two and a brief statement of closure completes this section of the development report.

> Ronway Laboratories will use Lot 123-29 as the base for pilot batch PI988. Results indicate that this lot is comparable to brand in dissolution behavior and is stable under accelerated conditions.
>
> We initiated the formulation development for XYZ capsules in March, 2005. Our strategy has been to employ exponents similar to those used in brand formulation combined by a dry mixing method. Folles SA manufactured the available active entity.
>
> The goal was to develop 50 mg strength first and then follow up with 100 mg strengths. Appendix A includes the formulation experiments conducted during formulation development.
>
> The formulation lot numbers are five digits. The prefix 123 refers to the laboratory notebook housing the records of the development of this product, and the subsequent numbers refer to the batch.
>
> Lots 123-01 to 123-14 incorporated Lactose Anhydrous DT, and Lactose Hydrous SD Starch P21 in different ratios and combinations with Sodium Lauryl Sulfate, Magnesium Stearate, and, in the case of lot 123-11, Aerosil 200, to achieve good encapsulation properties of the blend and to match the dissolution profile of the brand capsules.
>
> We accomplished this effort with lot 123-14, and manufactured pilot batch PB987 accordingly. However, this batch developed stability problems under accelerated conditions.
>
> Lots 123-15 to 123-28 reflect the attempts to develop a second formulation employing Magnesium Stearate and Aerosil 200 in combination with Starch P21. Lot 123-28 exhibited good encapsulation properties and similar dissolution patterns to the brand product. However, concurrent to development of this batch, material became unavailable for further development.
>
> We then made and tested Lot 123-29, similar in formulation to lot 123-28, but employing Magnesium Stearate manufactured by Nogodawa Pharmaceuticals with satisfactory results.

Good News

> Reading makes a full man, conference a ready man, and writing an exact man.
>
> **Francis Bacon**

For most documentation purposes, you will deliver information most effectively using the direct approach. Sometimes, however, you may need to deliver information that is not directly linked to documentation, but serves to drive everyday activities. You will find that the direct approach works well for good news, too. People, of course, like to receive good news, and when you make your point up front, your readers are much more likely to absorb the details that reinforce the positive point. The following memorandum from a company president announces the appointment of three people to company positions. It follows this pattern:

Paragraph one:	States the good news
Paragraph two:	Gives details about Gary Williams
Paragraph three:	Gives details about Christian Matthews
Paragraph four:	Gives details about Stan Syvertsen
Paragraph five:	Closes on a friendly note

Letter of Welcome

Ronway Laboratories ***MEMORANDUM***

DATE: March 2, 2005

TO: All Employees

FROM: Maryann Sorensen

RE: New to Our Staff!

We are pleased to welcome the following people to Ronway Laboratories: Dr. Gary Williams as Vice President, Scientific Affairs; Christian Matthews as Director, Quality Assurance; and Stan Syvertsen as Manager, Methods Development. All three will assume key responsibilities within the company.

Dr. Williams comes to us from Mandal Laboratories, Inc., where he directed the scientific group in methods development, analytical, and quality control. He has over fifteen years of industry experience, and will prove to be an asset to our newly restructured laboratory.

Christian Matthews, formerly with Bergen Drugs, will head up the Quality Assurance unit. He established the Quality Unit at Bergen, and is experienced in ISO 9002. One of his first projects will be to bring ISO 9002 certification to Ronway.

Stan Syvertsen joins us as Manager of Methods Development. With 10 years of experience at Mandal Laboratories, Inc., he will work closely with Dr. Williams in building a dynamic scientific team.

Please join in welcoming the new members of our team!

You can see how starting with your main point lets you logically develop your document by offering appropriate explanatory detail in the same sequence you've established at the beginning.

Submission Approval Memo

The following simple memo posted on a company bulletin board uses the same approach to announce the approval of a submission and to thank the employees for their efforts. This direct, friendly approach is effective in goodwill messages.

JORSTAD LABORATORIES Internal Memorandum

DATE: August 23, 2005

TO: All Employees

FROM: Arnold Johannessen

RE: We Have FDA Approval!

We're pleased to announce we have gained FDA approval to manufacture Indomethacin suppositories. We expect this product to be a significant addition to our over-the-counter suppository line.

This achievement is due to the concentrated efforts of the Indomethacin development team, as well as those of you who attend to the day-to-day activities that make us successful. Thank you!

Indirect Approach

While this book advocates the direct approach as the primary pattern for delivery of information for documentation, sometimes you may find the indirect approach more appropriate. The indirect approach calls for presenting information that leads to a point.

This pattern is useful for certain information. Reports like investigations typically deliver a sequence of events, an assessment, and a plan for action. And less-than-positive information works best when the negative news is delayed a bit.

Negative News

If you are in a position to determine the suitability of contract manufacturers or suppliers, for instance, you may on occasion have to deliver negative news. Suppose you determine an outside laboratory's testing standards have deteriorated and you wish to stop employing the services, at least temporarily, until the laboratory can improve its standards. In such a case, the indirect approach may work well. You can reasonably explain your position before you state it. This provides a better opportunity for retaining goodwill should circumstances change in the future and you wish to reestablish business relations.

Delivering negative news is never easy. People naturally hesitate to come right out and state the message they know will be ill-received, particularly if they wish to retain goodwill. But at times, people do have to bear bad news. This letter to a production team provides a paradigm for delivering bad news:

Paragraph one:	Presents mutually agreeable information
Paragraph two:	Offers explanatory details that lead to the main point, states the point, and offers more information
Paragraph three:	Offers further explanation, as appropriate
Paragraph four:	Closes on a positive note

Work Cessation Memo

MEMORANDUM ***Lyngdal Laboratories***

TO: All Employees

DATE: March 21, 2005

FROM: Ingvald Landrud

RE: Third Shift Notice

We appreciate the dedicated efforts of all of you on the third shift. You have been a vital part of our production activities and an integral part of Lyngdal Laboratories.

However, we have lost two significant accounts to competitors who can offer more rapid production because of a significantly smaller product line overall, and newer facilities. That's why we must temporarily suspend our third shift activities while we modernize our packaging equipment to be competitive. You will, of course, receive unemployment compensation and other benefits to which you are entitled. Your supervisors will further explain the impact of our temporary suspension of the third shift and direct you to any additional assistance you may need during this period.

We hope to resume production as quickly as possible; we do not anticipate a prolonged work cessation. We will notify you of shift resumption at the earliest practical time.

In the meantime, please be reassured. This work suspension in no way reflects negatively on the performance of the third shift. We will make every effort to resume production as quickly as possible, and we hope to have you there on the team when it does.

Persuasion

> Whatever we well understand, we express clearly, and words flow with ease.
>
> **Nicholas Boileau**

On the occasion when you need to ask a favor of someone but anticipate a negative response, use an indirect pattern. Bear in mind that most requests are routine and the direct approach will suffice. A contractor or vendor, for instance, is generally happy to supply a response to a request. Using the indirect approach, however, will help you when you need to convince someone to take action, to respond to a request when you anticipate the initial reaction may be negative. It's human nature to be happy to be asked but reluctant to comply. Consider asking someone to speak at a professional meeting, for example. Most people hate, more than anything, to speak in front of a group. So, in this case, the reader might be flattered but reply negatively, which is not what you want.

The following letter can use revision.

Speaker Request Letter

Ronway Laboratories

2323 Ronway Drive Jackson, Wyoming 83001

March 6, 2005

Dr. Kurt Waldheimer
Environmental Systems, Inc.
842-60th Street
Casper, Wyoming 83005

Dear Kurt,

As you know, I've been pleased to serve on the program committee for our state Chamber of Commerce. Our next chamber meeting will be held on Tuesday, April 15, and we're hoping you'll consent to be our speaker.

Our topic is "Controlling Pollution in Industrial Areas." Your experiences with the Department of Environmental Protection make you an expert on this topic, and we know you'll be able to contribute enormously to the success of our meeting.

We hope you'll consider this request. We know we can all gain from your expertise.

Sincerely,

Rodney Erickson
Facilities Engineer

> Make everything as simple as possible, but not simpler.
>
> **Albert Einstein**

What's likely to happen here is this: The recipient will read as far as the second sentence, and his mind will say "no way." Even though the second paragraph recognizes the reader's expertise, he has already mentally responded to the request made in the first—very likely negatively. Reworking the request so that interest develops first makes this letter much more effective.

Rewrite

This revised version builds to the request. And by the time the request is made, the reader knows that he has something to offer that no one else has. It's hard to say "no" to that. This letter follows this pattern:

Paragraph one:	Gain your reader's interest.
Paragraph two:	Offer details and build to your request.
Paragraph three:	Close and call for action.

Ronway Laboratories

2323 Ronway Drive Jackson, Wyoming 83001

March 21, 2005

Dr. Kurt Waldheimer
Environmental Systems, Inc.
842-60th Street
Casper, Wyoming 83005

Dear Kurt,

As you know, I've been pleased to serve on the program committee for our State Chamber of Commerce. Our next Chamber of Commerce meeting is April 15, and our topic is one that we know interests you: "Controlling Pollution in Industrial Areas."

The board has been continually impressed with the expertise you've gained working with the Department of Environmental Protection. Your knowledge can effectively aid us all as we search for ways to control and eliminate pollution at our respective plant sites. That's why we're hoping you'll consent to be our keynote speaker.

I'll contact within the next week to discuss our agenda further. Again, we hope you'll share your expertise with us.

Sincerely,

Rodney Erickson
Facilities Engineer

Uniting the Direct and Indirect Approaches

Many documents, especially reports of length, combine the direct and indirect approaches for delivering information effectively. For instance, a technical report may begin with a summary that states the main idea in the first sentence and cites key points to follow. Individual sections then present information that supports those points. Some sections in the report may provide information sequentially, particularly if they discuss procedures or events leading up to a current position.

A Standard Operating Procedure (SOP), for example, uses a mixed pattern of delivery. A purpose or objective statement tells at a glance what the document is about—and the reader is prepared for the sequential steps that will lead to completion of the task.

Using an Outline

Many writers wouldn't venture into the murky waters of writing without an outline, yet outlines aren't for everybody. An outline may indeed help chart your course and serve as your guide as you develop your document, particularly if it is lengthy. But there are some cautionary notes here. While outlines can help, they can also defeat if they prevent you from exploring an innovative idea triggered by the writing process. If you prefer to use an outline, use it as a guide, not a mandate. Some writers work best with no outlines; they impose structure as the document progresses. Still others prefer a rigid outline, which will work for those who do a great deal of mental mulling before they write. Still others work best with a loosely defined outline that expands and contracts as the document takes form. Which method you adopt depends on you as a writer. There's no "must do" here—it's a matter of discovering what works best for you.

Developing Paragraphs

Think of each paragraph you write as a mini-piece of documentation. Generally, give each idea its own paragraph and provide enough information to convince. You can use the direct or indirect approach for your paragraphs as well, depending on the message you want to deliver. In most instances, each paragraph should have one sentence that states the controlling idea. The other sentences should provide proof to support that idea.

Starting with the Controlling Idea

You can present information clearly using the direct approach in individual paragraphs. When you give your readers the controlling idea first, they will be receptive to receiving what is next. The following paragraph discusses a two-tier team effort, and the first sentence sets the stage.

> Instituting a program of production equipment maintenance at Ronway Laboratories requires a two-tier team effort. A core team of Maintenance and Engineering employees trained in the aspects of TQM should form the first-tier team, with one employee serving as group leader. Each member of the core team, in turn, will assume responsibility for directing a team in developing the maintenance program for a specific piece of equipment or group of equipment. The intent is to impart "ownership" of the equipment maintenance to designated teams.

Sometimes one controlling sentence can carry more than one paragraph of explanatory information. This is a call you must make. The following paragraph is from an information report generated when a company decided to purchase a new mixer for making a topical product. While it does the job, it is lengthy, with several ideas that support the first sentence, and offers the reader no chance to pause while absorbing the information.

> The mixer we purchase must be suitable for our needs. The shape of the mixer and the rigidity of the blades, as well as the effectiveness of scrapers, are primary concerns. Since thorough and consistent scraping of the mixer walls is vital to blend uniformity, internal surfaces must be accessible to receive adequate scraping so that residual material does not become part of the batch and result in nonuniformity. The scraper blades must be rigid enough to effectively scrape the walls, but not so much so as to damage the walls. Another concern is mixing capability; inadequate mixing that permits portions of the batch to remain stationary can result in "dead spots." Overly rigid mixer blades may provide good mixing, but impede scraping. Blades with too much flexibility, on the other hand, may not effectively mix a formula with the viscosity of ours. Blades should therefore be of a firm plastic that will provide adequate mixing and permit scraping of mixer walls during processing without damaging the interior surface.

Readability improves dramatically when this lengthy passage is broken into two paragraphs, both controlled by the first sentence in the first paragraph.

> The mixer we purchase must be suitable for our needs. The shape of the mixer and the rigidity of the blades, as well as the effectiveness of scrapers, are primary concerns. Since thorough and consistent scraping of the mixer walls is vital to blend uniformity, internal surfaces must be accessible to receive adequate scraping so that residual material does not become part of the batch and result in nonuniformity. The scraper blades

> must be rigid enough to effectively scrape the walls, but not so much so as to damage the walls.
>
> Another concern is mixing capability; inadequate mixing that permits portions of the batch to remain stationary can result in "dead spots." Overly rigid mixer blades may provide good mixing, but impede scraping. Blades with too much flexibility, on the other hand, may not effectively mix a formula with the viscosity of ours. Blades should therefore be of a firm plastic that will provide adequate mixing and permit scraping of mixer walls during processing without damaging the interior surface.

Leading to a Conclusion

Often a paragraph, such as the following one, offers a series of details that lead to a conclusion. This paragraph, from an out-of-specification (OOS) report, details a series of events; it is a causal presentation of information. The result of an OOS investigation is the main point of the paragraph, and it comes last.

> As soon as the microbiologicial out-of-specification result occurred, the analyst notified the supervisor. The supervisor then authorized a second analyst to retest the original laboratory sample using a sample triple the size of the original. While the analyst retested the sample, the supervisor reviewed all documentation relative to the OOS. The retest failed to isolate the aberrant data and verified the original result. Analyst error, technical malfunction, and media contamination were ruled out, and an assignable cause proved to be inconclusive. The supervisor notified QA, and QA subsequently rejected the batch.

The following paragraph effectively cites a condition first, with an explanation of subsequent action.

> QC discovered that the in-house HPLC/2323/ASI Revision 3 analyses of the related compounds tests were flawed, and the results are invalid. The lot involved was L211; the test stations were one-month accelerated and three-month accelerated. QC corrected the problem at the nine-month test station. As a result, the 12-month test station results are valid, and the 12-month results show that all the room temperature samples are within specification.

Giving Adequate Information

Even if you organize your information appropriately, your paragraphs may somehow lack the essential ingredients for convincing your reader. This problem is often one of idea development. Ideas need to be spelled out, or enhanced in such a way that they elicit the right response from the reader.

The following paragraph from a chemical hygiene plan would benefit from more detail.

> Hot glassware can be dangerous. Be careful when handling it.

Why is hot glassware dangerous? What can laboratory personnel do to minimize the danger? What do they need to do? A good directive includes the answers to questions such as these. Consider this rewrite:

> Take care when removing glassware from heat since hot glassware can cause severe burns. Never attempt to hold hot glassware without adequate protection, because the item can fall and break, causing hot liquids to splatter. Use asbestos gloves or appropriate tongs. Use utility tongs to handle small apparatus and crucibles, and beaker tongs to hold and carry small beakers.

This rewrite has the strength of focusing on concrete specifics in an orderly way, and it provides thorough coverage of a procedure. Remember when English teachers told you that paragraphs must have five sentences or seven, or more, or less? Of course, such arbitrary strictures are ridiculous, and writers shouldn't be bound by them. However, short paragraphs can be inadequate. What they may lack most is convincing detail. They're fine for transitions, closings, and some openings, but guard against them when you need to convince.

Using Transitions

When you write, give logical signposts to move your reader through your statements as well. A reader should receive one message, and one message only, from the information you present. The following paragraph, delineating a problem with productivity, fails to make the necessary relationships clear.

> Absenteeism ran about 9 percent. People would take days off without notification. Three days' notification is required. The foreman would shut down his operation until he could either reassign employees or call someone in for overtime. Overtime costs for last year amounted to 60,000 work hours.

Watch what happens when these ideas are joined by words that indicate their relationship to each other. Here, time and cause and effect serve as connectors.

> *Last year* absenteeism ran about 9 percent *because* people would take days off without notification, *despite* the three-day notification requirement. *Consequently,* the foreman would shut down his operation until he could either call in or reassign someone for overtime. This pattern resulted in overtime costs amounting to 60,000 work hours.

Even though the original paragraph contains the essential information, the rewrite is easier to read because the transitional terms signal the relationships between the facts and guide the reader along.

Sometimes paragraphs will be purely explanatory. Then you may need words such as "because," "in addition," "further," and "finally." When a paragraph shows causal relations, you may begin with a word such as "when," followed by information connected with "as a result," "consequently," or "finally."

Omitting transitions is never disastrous to meaning, but including them usually makes a passage more fluid and facilitates reading. Further, you may occasionally write a one-sentence paragraph that serves as a transition between key points in a document of length.

Transition words can help you control your writing and assist the flow from idea to idea. First, assess the relationships between your ideas. Is it cause and effect? A phrase such as "as a result" clearly signals that something has resulted from some other action. If the document explains the causes or effects of some event, see that you signal your intentions by using such clear indicators as "because," "grows out of," or "results in." Time sequences, similarly, should be logical: Words such as "initially," "second," "then," "subsequently," and "finally" serve to indicate chronological sequence. When these and other terms—such as "in conclusion" and "thus"—are used, make sure they signal the relations they stand for and not some vague connection you only half convey. In other words, see that each paragraph is tightly coherent in the way it connects details.

Transitions of Sequence

after
after that
as
as soon as
at first
at last
at the beginning
at the onset
at the same time
before
conclude by
concurrently
continuing on
during
finally
first
following
following that
immediately
immediately after
immediately before
immediately following
in conclusion
in time to
initially
just before
last of all
lastly
next
prior to
second
sequentially
shortly after
subsequently
the next step
then
thereafter
third
to adhere to
to complete
to end
to finish
to follow
to terminate
to wind up
upon completion
when
while

Transitions of Addition

a case in point
a mirror image
additionally
also
also important
also notable
analogous to
and a correlative
and again
another such
as an example
at the same time
besides
by association
by the way
closely related
conjointly
equally important
for one thing
further
furthermore
in accordance
in addition
in addition to
in all
in fact
in like manner
in line with
in support of the point
in the same vein
in this case
indeed
it encompasses
like
likewise
more
more specifically
moreover
namely
one more example
one sample
one such
similar to
similarly
still another
such as
this too
thus
to a similar degree
to add to that
to delineate
to delineate further
to exemplify
to illuminate
too
yet another

Transitions for Explanation and Restatement

a summation of
above all
accordingly
again
all in all
another way to look at it
as noted
as stated
backing this
clearly
for example
for instance
for one thing
further
furthermore
in a nutshell
in all
in brief
in conclusion
in effect
in fact
in other words
in particular
in retrospect
in review
in short
in sum
more specifically
mostly
not only, but also
notwithstanding
on the whole
once again
once more
one more time
one perspective is
overall
simply stated
still
that is
that is to say
to clarify
to confirm
to elaborate
to elucidate
to illustrate
to look at it another way
to paraphrase
to put it differently
to recap
to reconfirm
to reiterate
to repeat
to rephrase
to restate
to retrace
to reword
to sum up
to summarize
when all is said and done

Transitions for Comparison and Contrast

a clear difference
a conflicting viewpoint
a different perspective
a strong distinction
after all
akin to
albeit
although
an opposing position
and yet
as
as compared with
as well as
at the same time
balanced with
but
by comparison
by the same token
comparatively
conforming to
consistent with
contrary to
conversely
counter to
despite
disproportionate to
disproportionately
dissimilar to
distinctly
equally important
equivalent to
even so
even though
faster than
for all that
granted that
however
in much the same way
in opposition to
in parallel
in spite of
larger
less
like
likewise
more
nevertheless
notwithstanding
of course
on the contrary
on the other hand
smaller
the antithesis of
though
to resemble
too
unfortunately
versus
while it is true that
similarly

Transitions for Cause and Effect

a by product is
a causative factor
a correlative effect
a secondary effect
a side effect is
absent that
accordingly
after
as a consequence
as a result
as opposed to
because
because of this
by means of
by reason of
comes to rest at
concludes
concluding in
consequentially
consequently
due to
thereafter
therefore
thereupon
this began
following that
for that reason
for the purpose of
for this purpose
for this reason
generated
grows out of
hence
henceforth
in conclusion
in effect
in lieu of
in opposition to
in spite of
incidentally
initiated
it follows that
it launched
it produced
it shakes out as
it winds up as
this created a movement to
this induced
this started
little by little
on account of
otherwise
precipitated
resulting directly from
since
so
terminates at
that results in
the aftermath
the by-product
the end result
the impact of
the impetus to
the long term effect
the outcome
the outgrowth
the primary effect
the spoils of
the yield
then
to the end that
to this end
with the result that

Types of Paragraphs

Paragraphs play many roles. Their purpose is to fully explain and support a controlling idea. Often they employ different techniques to get the job done. In total, a series of paragraphs comprise a document or sections of documents. Here are some examples of various types of paragraphs. Note that some longer documents can be totally descriptive, totally chronological, or totally comparative; others may require a combination of information delivery methods. Consider a clinical trial protocol: It will have elements of description, such as the study design, the patient volunteer profile, and facilities. It will also have sections identifying procedures, the "how to" of the activities that are to occur. This type of writing is chronological. The following examples show different methods for delivering information effectively.

Descriptive Paragraphs

Descriptive paragraphs create an image of an event, a person, a concept, or a thing. Paragraphs that describe facilities are descriptive. So are those that present conditions, such as what a product in development should be like, or offer narrative information about patients or animals in a toxicology study. The following paragraph gives an overview of standard operating procedures.

> Standard Operating Procedures (SOPs) of a proactive company form the backbone of its operations. SOPs delineate processes from vendor qualification to materials receipt and testing, to product manufacturing and packaging, to distribution to the marketplace, and finally, to tracking the product through its shelf life. SOPs must adapt as systems and processes change, and they must undergo rigorous controls to remain accurate and current.

The following paragraph is also descriptive/informative. It is part of a document that gives an overall explanation of the validation process.

> Actual users comprise the most effective validation teams for commercial off-the-shelf (COTS) software-driven systems. Teams may have a few or as many as twenty or more members. The number of team members is directly proportional to the scope of the project. If the validation is for software that the entire company will use, a representative from each area should be on the team. A member of the Information Technology (IT) group should also be a team member. And, one person should serve as the Validation Lead.

Extended Definition Paragraphs

Sometimes writers need an entire paragraph or even two to define a person, thing, concept, or event. This type of paragraph is closely related to descriptive paragraphs. For example, an extended definition paragraph about a system may tell what the system is, how it works, what it does, what it looks like now, and how it works with other systems. The following paragraph defines a clinical thermometer.

> A clinical thermometer is a narrow, tubular, closed glass device with one bulbous end containing mercury. It works on the principle of heat expansion. When the temperature of the bulb increases, the mercury inside expands, causing a thread to rise within the hollow stem until it stops by a marked measurement. The measurement indicates the degree of warmth.

Cause and Effect Paragraphs

Cause and effect writing usually results when someone tries to figure out what has happened and why. In essence, one thing causes something else to happen, which in turn often causes another event. Consider for example, that diphtheria causes vomiting. Vomiting in turn causes dehydration. Dehydration causes death. Note that diphtheria does not cause death.

Cause and effect writing is useful when there is a deviation or nonconformance and the need is to determine and record what happened. This type of writing is useful for exploring a probable outcome for an activity, or anticipating a result of a planned action. The following paragraph is an example of cause and effect writing. Note that the controlling idea is in the first sentence.

> Software modification can have a ripple effect. A simple change request from operations can affect multiple functions of the system. A change to the way a report form displays data on a computer screen can affect every place in the system that uses the same form. So, a seemingly minor change can be a difficult information technology assignment in a complex system.

Chronological Paragraphs

Chronological paragraphs put information into a time sequence. Chronological paragraphs appropriately tell how things happen, have happened, or will happen. Procedures, completed studies, background sections of abstracts, and protocols are examples of chronological writing. The following paragraph gives a history.

> As early as the 1980s, pharmaceutical manufacturers started using automated batch record systems. While offering the advantages of paper reduction and increased efficiency, these systems brought new issues. The paper-based cGMP regulations did not address electronic record keeping. So, in 1991, a group of pharmaceutical manufacturers met with FDA to determine how to use paperless record systems and remain compliant with cGMP requirements. This effort led to an FDA task force to look at electronic records in all areas — GMP, GLP, and GCP. The result was the publication of 21 CFR Part 11 Electronic Records; Electronic Signatures Final Rule in the Federal Register. The rule became effective on August 20, 1997.

Comparison and Contrast Paragraphs

Comparison and contrast looks at two or more things, concepts, events, or people for similarities and differences. If a company wants to purchase new equipment, such as a capping machine, a report that compares and contrasts possible equipment may be the fulcrum for a decision. Such a document might compare capacity, size, compatibility with existing equipment, price, service, and ease of validation for each piece of equipment under consideration.

Comparison and contrast typically occurs in identifying test and control articles in studies and results from study to study. It is also essential for evaluating vendors and suppliers. Here is an example of comparison and contrast in two paragraphs. Note that the transition word in the second paragraph is simply "also" and the comparative word is "but."

> Deoxyribonucleic acid (DNA), a nucleic acid, forms from a repetition of simple building blocks called nucleotides. Nucleotides consist of a phosphate (PO_4), sugar (deoxyribose), and a base that is either adenine (A), thymine (T), guanine (G), or cytosine (C). In a DNA molecule, this basic unit repeats in a double helix structure made from two chains of nucleotides which link between the bases. The links are either between A and

T or between G and C. The structure of the bases doesn't allow other kinds of links. The double helix structure resembles a twisted ladder.

Ribonucleic acid (RNA) is *also* a nucleic acid, *but* it has a single chain and the sugar is ribose rather than deoxyribose. The bases are the same as those of DNA, except that the thymine (T) which appears in DNA is replaced by another base called uracil (U), which links only to adenine (A).

Writing Headings

Headings are important, whether they are for a journal article, an abstract, or presentation for dissemination of information to the professional community at large or for documents directed to select readers. In disclosure writing, headings serve to attract readers' attention and impel them to read an article, abstract, poster, or other publication. In documents of length, headings assist in guiding readers through the text. They serve to separate elements and identify at a glance those sections of a document that may be of interest to specific readers.

Good headings give information. If, for instance, the heading is for an abstract or professional meeting poster, it's often the first, and sometimes only, thing people read. When writing such a headline, it's even more important to make every word count — and to present the main idea succinctly. Consider the following heading:

CleanBlood™ Pathogen Inactivation of Red Blood Cells

What about pathogen inactivation? A headline that makes a point grabs the reader's interest and can induce her to read the rest of the abstract.

CleanBlood™ Inactivates Protozoa, Viruses, and Bacteria in Red Blood Cells

The second example is a complete sentence, and it's packed with information about the CleanBlood product.

Writing informative headings is important in documents as well. A good document, whether a long memo or a more inclusive report, benefits from headings where every word tells. To call a section "Supporting Data" is insufficient. That phrase could apply to every report generated! Just as a newspaper wouldn't entitle a feature article "News Story," a report shouldn't label a vital section "New Information" or "Significant Data." Headings work best when they encapsulate what's in the text that follows. "Development of Method 2204" is more precise, for instance, than "Method Development Two" in a report that talks about the development of several laboratory methods.

The same caveat applies to headings for tables. The following table heading is vague and uninformative.

> Summary of the Results of Toxicity Studies Conducted with Experimental Drug S-252X

A rewrite gives more information. Further, a table is a summary, so identifying it as such is redundant. Getting rid of words that don't do their duty and replacing them with content delivers stronger information.

> Results of In Vitro and In Vivo Toxicity Studies of S-252X in Sprague Dawley Rats

Make Headings Parallel in Structure

Headings don't have to be complete sentences (although they can be), but they should be parallel in structure. That means elements within the headings themselves should be in the same form, and the headings in the entire document should parallel each other. The following report section heading lacks parallel structure.

> Using Salt Ice Mixtures and the Employment of Dry Ice for Cooling

The word "using," a verb acting as a noun, and "employment," a noun, are awkward together in that they are different word forms. Selecting one form brings the heading elements together, and, as often happens, eliminates extra words:

> Using Salt Ice Mixtures and Dry Ice for Cooling

Similarly, a report should not have several headings that conflict in structure with others contained in the report. These headings, gleaned from a report on plant maintenance, reveal inconsistencies in parallelism.

> Estimates of Electric Heating Costs Rate Determination
>
> The System Will Require Semiannual Assessment
>
> Cost Data Have Been Checked by the Supplier
>
> Projection of New Fiscal Year Cost
>
> Allowing for Rate Adjustments, Estimates of Five-Year
>
> Expenditure

These headings work better when nouns, not verbs, control them. Further, some headings are sentences; others are phrases. Making these headings consistent assists in making the report readable.

Electric Heating Costs Estimation

Rate Determination

Semiannual Assessment Requirements

Supplier's Cost Data Evaluation

New Fiscal Year Cost Projection

Five-Year Expenditure Estimation

Organize Headings by Rank

Short documents may require few headings; longer reports may have myriad sections and subsections. Headings for either should follow a pattern that allows delivery of a primary point first, with subsequent support relegated to subheadings. Often, main headings are centered on the page, and subheadings aligned left, but companies use a creative range of positioning. Whether you bold face, capitalize, or change the font size depends on the facilities you have available and the standards within your workplace. Generally, however, for sections of equal weight, use the same heading type size and face. Don't give equal standing to a major section and a subpart of that section.

Tables and Visuals

A picture is "worth a thousand words." This axiom is particularly true in documents dealing with technical or complex data. Consider a report explaining a fire drill procedure. Wouldn't a map assist in defining the designated routes? Or, think about directions for installing a piece of batching equipment. Wouldn't a diagram help "show" how to do it? If you were compiling dissolution data, wouldn't a chart or table be more comprehensive than just words? To illustrate clinical research results, for instance, a table generally works best. Trying to put complex information into words is tricky business; further, doing so can present problems for readers who must make sense of data which is often complex.

Tables are particularly useful to condense or summarize large bodies of data. When data are complex or detailed, tables are almost always the best vehicle for presentation. Tables allow readers to compare individual values and groups of data. Moreover, tables accelerate the comprehension process of readers.

The style and format of visual information will vary from company to company, but most have these elements:

- A table or figure number
- A table or figure title
- A legend, which often resides beneath the table or visual

When tables or other visuals deliver information, it's preferable not to reiterate in text what the visuals say. Write a simple statement in the text that calls the readers' attention to the visual. "Table 2 shows the results of the rat forced swim test." Then, present the table.

Principles of Table Construction

To communicate quickly and accurately, tables require readers to 1) identify how the information is organized, 2) find the information of interest, and 3) interpret the information once it is found. Experience, convention, and some research studies have identified at least five principles that should guide the construction of tables.

1. Tables should have a purpose; they should contribute to and be integrated with the rest of the text.

Data should not be reported for their own sake. Rather, they should be part of a larger effort to answer the four questions of research: "What did you do?" "Why did you do it?" "What did you find?" and "What does it mean?" Thus, tables should be used only when they can communicate information more efficiently or effectively than can be done in text or figures.

2. The purpose of the table should determine its form.

A table created to collect data is not necessarily the same table that should be used to communicate these data. A table created to organize a large amount of data so that a value can be referenced easily will not necessarily be the same as a table constructed to emphasize patterns in the data or comparisons among the patterns.

Tables may be structured for analytical or reference functions. Analytical tables are designed "from the inside out" by organizing the data field to help reveal patterns in the data. Reference tables are designed "from the outside in" by organizing the column and row heads to help readers find specific information quickly.

3. Tables should be organized and formatted to assist readers in finding, seeing, understanding, and remembering the information.

A table that contains all the necessary data but forces readers to organize the data before understanding it does everyone a disservice: it increases the time needed for readers to evaluate the data and does not assure that the author's understanding of the data will match the readers' interpretation of it.

4. Values to be compared should be placed side-by-side.

English is read from left to right and from top to bottom. Thus, at least in English language publications, placing values side-by-side is not only the easiest way to compare them, but it also encourages this comparison. In biomedical research, where a treatment group is compared to a control group, values for each group should be given in adjacent columns so that the variables in each row can be compared more easily.

5. Organize the table visually as well as functionally.

Graphic elements, including spacing, should be used to help organize the table visually. Elements such as lines, bold type, outlined cells, spacing, and shading, can help readers make within- and between-group comparisons, differentiate more important values from less important ones, highlight patterns in the data, indicate special circumstances associated with the data, and so on.

Most publishers specify their format for tables, and some may not allow the full use of design elements as recommended here.

6. Data presented in tables should not be duplicated in the text.

Describing in the text data that are also presented in a table remains a common problem, even though most style guides and journals advise against the practice. Duplicate information takes valuable space and so is to be avoided in print publications. Values, groups, or comparisons in tables can be mentioned in the text, of course, but the table should present the data.

Tables should also be kept as simple as possible. Include only the information that is relevant to the purpose of the table.

Courtesy of Tom Lang, MA, from a new chapter in the second edition of his book, *How to Report Statistics in Medicine: Annotated Guidelines for Authors, Editors, and Reviewers.*

Tables don't have to be reserved for complex data. They often present simple information precisely and directly. The following table is effective in a quality assurance SOP that calls for the inspection of packaging inserts. Inspectors can tell at a glance what to look for. They can then enter the results of the inspection accordingly on an inspection form.

Inspection	Specification	Method	Classification
Product Code	Clear, complete, and identical to printer's proof	Visual	0
Batch Number	Clear and complete	Visual	0
Label Copy	Clear, complete, and identical to printer's proof	Visual	0

The following table provides complex information that would be difficult to present verbally and equally difficult to understand were it in another format. Notice how the text of the report directs the reader to the table.

> We calculated the percent deviation of five replicate injections of the standard solution, and one injection of additional weighing of standard from the true value, as determined from the linear regression line. Table III shows the results.

TABLE III

Sample	Conc. (mg/mL)	Peak Area	Calculated Concentration	% Accuracy
1	0.16016	3837945	0.15871	99.10
2	0.16016	3876328	0.16033	99.90
3	0.16016	3875795	0.16030	99.91
4	0.16016	3898299	0.16125	99.32
5	0.16016	3881189	0.16053	98.77
6	0.16072	3953200	0.16356	98.23
AVG				99.37
SD				0.65
RSD				0.65

Illustrations, pie charts, bar graphs, scatter diagrams, and the like all work well in relaying information. Visuals often accomplish what words can't. They show patterns of data and trends and allow readers to draw comparisons more readily. They can also condense data more efficiently than text can. The following visual, from a feasibility report, visually presents a process that the text identifies.

The new system will allow us to track the dispersion of mineral oil droplets in an aqueous system with sufficient precision and sensitivity to allow the development of real-time, closed-loop control of droplet dimensions based on manipulation of the mixing rate.

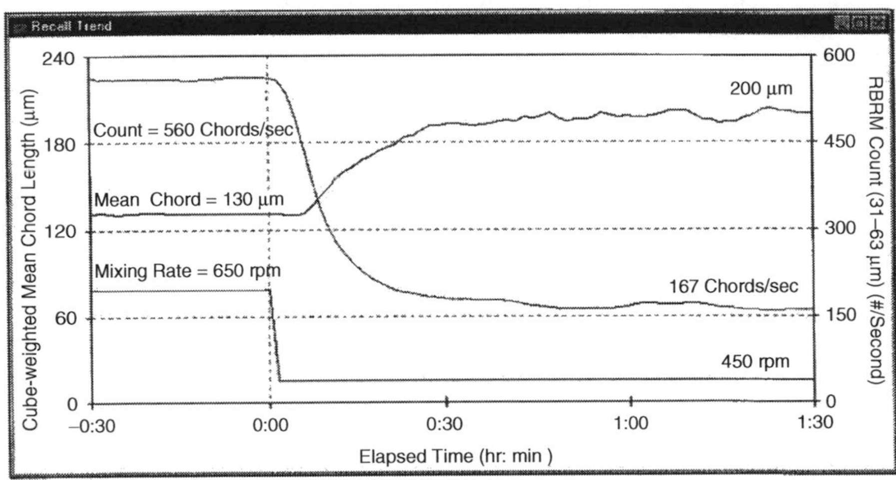

Monitoring the real-time response of a droplet dispersion to a step change in the shear rate (50% Oil/Water mixture).

Courtesy of Lasentec, Inc.

Principles of Figure Construction

1. Figures should have a purpose; they should contribute to and be integrated with the rest of the text.

As is the case with tables, data should not be reported in figures just for the sake of displaying them. Figures should be used only when they can communicate information more efficiently or effectively than can be done in text or tables.

2. The figure should be designed to assist readers in finding, seeing, understanding, and remembering the information.

When designing a figure, its purpose should be emphasized. Is the purpose to show the variability or the stability of data? To emphasize similarities or differences between groups? To show trends over time? To show linear or nonlinear relationships?

3. Figures should contain only those elements that are necessary to fulfill their purpose.

Conciseness is a value in figures, as well as in scientific writing in general. Make sure that all the lines, symbols, numbers, and words in the figure are necessary and sufficient to allow readers to interpret it. In particular, avoid three-dimensional figures unless the data are actually three-dimensional.

4. The data should be emphasized over other elements in the figure.

The advantage of figures is that they focus attention immediately on visual patterns of data. Thus, anything that distracts from this focus reduces the utility of the figure. For example, data points and lines should be larger or heavier than other graphic elements, such as the scales, the borders of the data field, or reference lines.

5. Figures should be consistent with the principles of Gestalt perceptual psychology.

Abstracting and interpreting data from a figure is a process of visual perception. Visual perception, in turn, is influenced by several principles identified by Gestalt perceptual psychology. Following these principles when designing figures should improve the utility of the figures.

- **Primacy:** *the larger arrangement ("the Gestalt") is seen before its components.*

The overall visual impression of the figure should be consistent with the actual meaning of the data. This principle can be used to manipulate readers' perceptions: the four most common examples are the "suppressed zero," the "elastic scale," the "superfluous dimension," and the "double-scale" problem (see below).

- **Proximity:** *objects near each other tend to be seen as a group.*

The characters string, ••• •••, is seen as two groups, whereas •• •• •• is seen as three. Thus, put data to be compared close to each other, and separate data that are not to be compared. This principle is especially important for placing labels with respect to the data they identify.

- **Similarity:** *similar objects tend to be seen as a group.*

The characters string | | — — is seen as two groups, rather than as four lines. Thus, display data from the same group in an obviously and uniquely consistent way, and display data from different groups in obviously and uniquely divergent ways. This principle is essential when graphing three or more variables on the same graph. Plotting marks and data lines from the same group should look alike. They should also differ enough between groups that the groups are not confused with one another.

• **Continuation:** *data arranged in an obvious pattern tend to be seen as a group.*

The character string, – – – – – –, is seen as a single group, whereas – – – _ _ _ is seen as two. So, when possible, indicate data from the same group by providing an obvious pattern, and disrupt any patterns that are coincidentally comprised of dissimilar data.

• **Closure:** *a break in a pattern is automatically "filled in" to complete the pattern.*

For example, in the sequence, _– – _, readers usually imagine the missing symbol that would complete either the pyramid: _ – ‾ – _ , or repeat the sequence: _– _ – _. So, emphasize any breaks that represent actual discontinuity in a pattern, and make the pattern clear when the data actually form a pattern (so readers do not have to "fill in" to complete the pattern).

6. Data presented in figures should not be duplicated in the text.

As is the case with tables, do not describe in the text data that are also presented in a figure. Rather, identify in the text the important aspects of the figure to help readers interpret the data.

Courtesy of Tom Lang, MA, from a new chapter in the second edition of his book, *How to Report Statistics in Medicine: Annotated Guidelines for Authors, Editors, and Reviewers.*

References, Works Cited, Works Consulted

Good writing credits sources. To use someone else's words without credit is plagiarism. Even when you paraphrase someone else's information, that is rephrase in your own words, you need to give credit where credit is due. When you are writing for a journal, the guidelines for authors will tell you how to include references. Many companies use a software program like Endnotes, or simply employ the Endnotes feature built into Word. Some companies have style guides that show how to handle references. If that's the case where you work, simply adhere to the style guide. Within many companies, however, such guidance is often lacking.

If no such guidance exists where you work, it's best to consult a style reference book. The American National Standard for Bibliographic References (ANSI) provides guidelines for referencing information. Other style guides do so as well.

4

Correspondence

When you consider the momentous effort that goes into creating a validation report or a standard operating procedure (SOP) on a complicated task, composing letters and memos may seem like child's play by comparison. But it's not. A letter may well be the first contact a party has with a company, and that first impression is important. Even a New Drug Application (NDA) submission relies on a cover letter to set the stage for review. Good letters are vital tools of doing business; so are memos. The primary difference between memos and letters is that letters generally have an external audience, whereas memos receive internal distribution.

Facsimile (fax) and e-mail (electronic mail) messages are also common. A fax is an electronically transmitted message that serves the same purpose as a letter or memo. E-mail can travel to parties internal or external to a company. Technology has made transmission of information instantaneous, and with that ability come new avenues for going awry.

All letters and memos or their electronic counterparts benefit from scrutiny. Indeed, companies are taking notice of the impressions they present. Consider a job applicant wearing a navy blue suit dotted with lint to a crucial interview. The unspoken message is lack of attention to detail and lack of pride in presentation. The visual image people project affects the responses they get. So it is with written messages: They convey, in a very accurate way, the degree of professionalism behind them. Every piece of correspondence should reflect the thought and effort that went into producing it.

When you're totally confident in your ability to say what you want lucidly and succinctly, you must also consider how your message looks on paper. Is it pleasing to the eye? Does the white space facilitate reading? Is the format consistent? These are all valid questions — ones that people often don't consider when they write for the world of work. It's all too easy to fall into careless and irregular patterns of delivery that detract from otherwise effective messages.

> The word "format" derives from the Latin word *formatus*, which means "formed." The phrase liber formatus originally described a book formed in a certain shape and size. Publishing still uses the word "format" in reference to shape, size, and layout. Webster's Dictionary, however, offers this definition: "A general plan or arrangement of something."

Conventions of Letter Writing

Depending on your position, you may write many letters or few. If you work with contract manufacturers, suppliers, outside researchers, or regulatory bodies, you may generate letters frequently. If your company has an established standard, by all means adhere to it. If, however, the company has no "set" format, it's up to you to give your letters the performance edge. Over the years, certain conventions have been accepted as the professional standards for letter writing.

Note that the companies in the United States (US) use $8^1/_2 \times 11$ inch paper. In Europe and elsewhere A4 paper is the standard. The difference is that A4 paper is somewhat narrower and longer, at 8.27×11.69 inches.

Return Address and Dateline

Company stationery carries a letterhead, which usually includes the name or logo of the company, the address, phone number, a fax number, and often other information. When such a letterhead exists, include a dateline two or more lines below the letterhead. Where it's positioned depends on the length of the message and the balance of white space. A good rule of thumb, however, is to limit the space between the letterhead and dateline to no more than six, but not less than two, lines.

XYZ Company
23 West Ellen Avenue
Coldwell, Nebraska 10923

January 16, 2005

If you're using stationery that has no letterhead, you should include your address and the date, but *not* your name, like this:

123 Main Street
Anytown, NJ 07000
March 5, 2005

Inside Address

Depending again on the letter's length, place the inside address two to six spaces below the dateline, aligned left. The inside address should show the courtesy title, addressee's name, position title, and address.

Dr. Yelena Kovalerchik
Director of Research and Development
Bien Sur Pharmaceutical Co.
600 Bergen Avenue
East Hanover, New Jersey 07936

Courtesy Titles

You can include the courtesy title even when the address includes a position title. The traditional courtesy titles are Mr., Mrs., Miss, and more recently Ms. Miss is used rarely, if at all. Even though some women object to the title Ms., just as many feel that Mrs. or Miss violates their privacy because each designates their marital status. Further, if you're unsure of the recipient's marital status, you risk offending her by using an inappropriate title. Thus, it's probably best to use Ms. for women, unless you know a specific woman's preference.

When addressing people holding advanced degrees or appointed titles, make sure you indicate their status correctly. People often get into trouble because they address the audience inappropriately. The two most troublesome titles are *Dr.* and *Esquire.* When more than one acceptable form of address exists, use only *one,* never both.

Dr., of course, can indicate medical doctor, or M.D.; doctor of dentistry, or D.D.S.; Doctor of Veterinary Medicine, D.V.M.; Doctor of Philosophy, or Ph.D.; Doctor of Education, or Ed.D.; or Doctor of Jurisprudence, or J.D. The title *Dr.* is professionally appropriate and preferable for all; however, it is permissible to use the following forms.

Hester Harte, M.D.   Georges Cheval, D.V.M.

Lawrence Pullit, D.D.S.   Victoria Norton, Ed.D.

Jane Degnan, Ph.D.

Note that for some of the above, the periods are often omitted: MD, DVM, DDS.

When you address an attorney, the preferable form is Mr. Charles Chaser, Attorney at Law. Sometimes, however, an attorney may be addressed this way:

Charles Chaser, Esq.   or Charles Chaser, Esquire.

The term *esquire* has traditionally not been used to address women attorneys, but that convention is now considered sexist. *Attorney at Law* is always appropriate.

Addresses

The company name should always conform to the official name of the company you're writing to. That means if the name includes *Incorporated,* spelled out, you should follow suit. Similarly, if the name of the company abbreviates *Incorporated,* your address should also do so. And, if *The* is a part of the official registered company name, you should include it. There's no rule for abbreviating words that stand for street direction, so it's acceptable to write both "West Mandrake Avenue" and "W. Mandrake Avenue."

When you have numbered addresses, you can treat them in several ways. Again, the key is consistency. If you write "One Federal Plaza" on your envelope, you should not write "1 Federal Plaza" in the inside address.

> A letter is like a conversation between writer and reader in which the writer does all the talking.

Always spell out city names; never abbreviate them unless the official name of the city is abbreviated, as in St. Paul. States and territories may be abbreviated, however. The following abbreviations are the ones the United States Postal Service recommends. Using these conventions helps eliminate the confusion caused by abbreviating both Massachusetts and Maine as MA, for instance. It also eases sorting. Florida, for example, can be spelled out or abbreviated: Fla., Flor., and FL; and there are probably other creative abbreviations as well. It's best to adhere to the established postal standards in all cases. Here are the acceptable abbreviations for US states and territories.

Alabama, AL
Alaska, AK
Arizona, AZ
Arkansas, AR
California, CA
Colorado, CO
Connecticut, CT
District of Columbia, DC
Florida, FL
Georgia, GA
Guam, GU
Hawaii, HI
Idaho, ID
Illinois, IL
Indiana, IN
Iowa, IA
Kansas, KS
Kentucky, KY
Louisiana, LA
Maine, ME
Maryland, MD
Massachusetts, MA
Michigan, MI
Minnesota, MN
Missouri, MO
Montana, MT
Nebraska, NB
Nevada, NV
New Hampshire, NH
New Jersey, NJ
New Mexico, NM
New York, NY
North Carolina, NC
North Dakota, ND
Ohio, OH
Oklahoma, OK
Oregon, OR
Pennsylvania, PA
Puerto Rico, PR
Rhode Island, RI
South Carolina, SC
Tennessee, TN
Texas, TX
Utah, UT
Vermont, VT
Virginia, VA
Virgin Islands, VI
Washington, WA
West Virginia, WV
Wisconsin, WI
Wyoming, WY

Zip Codes

Write zip codes out fully. The US Postal Service recommends the hyphenated nine-digit code, which looks like this:

07701-1234

Salutations

The salutation appears on the second line below the letter's inside address. The standard is to begin the salutation with *Dear* followed by the title and the last name only and a colon. If the letter is less formal and addressed to someone you know personally, begin with *Dear* followed by the person's first name and a comma.

Dear Mr. Jones:

Dear Chuck,

No satisfactory guidelines exist for addressing several people within an organization. Women rightfully dislike being addressed as *Gentlemen,* and *Ladies and Gentlemen* offers an inappropriate connotation. *Dear Sir or Madam* is dated and somewhat pompous; in addition, many women object to being addressed as *Madam.* Finally, *To Whom It May Concern* is vague, impersonal, and frankly old-fashioned.

It's common to eliminate the salutation in such cases. An attention or reference line can take the place of the salutation.

Attention Lines

You have the option of using an attention line when you are writing to a company and want a particular person whose name you don't know to receive your letter, or when you have a mixed audience. The attention line appears below the inside address and can replace the salutation.

Krumkage Company

1946 Central Circle

St. Edwardson, Minnesota 01062

Attention: Production Manager

or

Attention: Members of the Corporate Management Committee

Subject and Reference Lines

These letter parts are specialized, and you may use them when you feel your reader needs to know immediately the purpose of your correspondence. These components are gaining popularity in professional letter writing, for they facilitate quick comprehension of a message. But, to be effective, they must be succinct and encapsulate the purpose of the document. The subject or reference line should appear either beneath the salutation, beneath an attention line, or by itself in place of a salutation. Allow two lines between the subject or reference line and the preceding element.

Subject: Long-Range Planning Seminar

Re: Final Documentation for NDA

Letter Bodies

Start your letter two lines below the salutation or subject/reference line. It used to be that longer letters were single-spaced, with double-spacing between paragraphs, and that shorter letters were double-spaced, with triple spacing between paragraphs. However, double-spacing has steadily been losing popularity and is rarely seen. The standard now is to single-space. Be sure to balance the text of the letter on the page. Word processors permit insertion of white space easily, so letters can be single-spaced and balanced on the page after the fact. Most word processing programs offer a "view document" or "print preview" option that permits the writer to see how the letter lies on the page. By all means use it. It's handy.

Complimentary Closings

The closing you choose should coincide with the degree of formality reflected in the salutation and with the tone of the correspondence. It should also reflect your relationship with the addressee. Place your closing two lines below the text of the letter and punctuate it with commas. The following closings are standard and reflect, from first to last, a decreasing order of formality. Currently, *Sincerely* and *Sincerely Yours* are the most popular.

Respectfully,
Yours truly,
Sincerely yours,
Cordially yours,
Sincerely,
Cordially,
Best wishes,
Best regards,
Warmest regards,
Regards,

Signatures

The most common signature format includes the writer's full name and title on the fourth and fifth lines below the complimentary closing. The two are usually aligned with the closing. Of course, the signature goes in between.

Sincerely yours,

Juliette Boone
Associate Director, Environmental Issues

When a signature indicates legal responsibility for the message on behalf of the company, the following format is standard. It includes the company's name in capital letters two lines below the closing and the author's typed name and title four lines below that.

Sincerely yours,

RONWAY PHARMACEUTICAL COMPANY

Erin Andreas
Vice President, Regulatory Affairs

It used to be that the Food and Drug Administration (FDA) required black ink for signatures. This reflected the ability of copying machines to reproduce black ink legibly but not blue or other colors, and the use of black ink for any signatures on any documents became the norm. These days, few professionals pay inordinate attention to the color of the ink they sign their names with, and in the grand scheme of things it's probably not all that important. Recent studies, however, have shown that people respond best to blue ink: It seems to indicate a personal touch and a degree of sincerity. Also, it's always nice to know a real person took time to personally sign a letter when the signature could as easily have been reproduced.

Finally, it's not customary to include *Dr., Mr.,* or *Mrs.* below your signature. When you sign your letters, use your given and surname only.

Respectfully yours,

Michelle Miller
Director of Training and Development

Enclosure Notations

When you include other documents with your letter, an enclosure notation is appropriate. Include the notation one or two lines below the closing. The enclosure notation should indicate the number of accompanying documents and may indicate what they are specifically. Use a period, a colon, a dash, or no punctuation at all.

Enclosures 3

Enclosures: 2

Enc. 2

Enc: 2 checks, 1 statement

Enclosures—2 checks, 1 statement

Distribution Notations

When you plan to distribute copies of your correspondence to persons other than the addressee, indicate your intent by including the letters *cc:* or simply *c:* . *CC* indicates "carbon copy," and most copies are no longer carbons; *c:* indicates "copy," and is becoming the standard. Also acceptable is *pc:,* for "photocopy," or simply *copy* or *copies* spelled out.

Copies: Charla Bannett

Diane Wanat

Mary Ehret

cc: Members of the Management Committee

Occasionally, you may wish to give a copy of a letter to a second or third party, but not indicate that recipient on the primary letter. When that's the case, you can write *bc:,* which indicates "blind copy," or *bcc:* which indicates "blind carbon copy." Then on your file copy you can include the secondary recipient's name.

bc: Hengshe He

bcc: Martin Weinstein

Word Processor's Initials

> It's a rare writer who can produce an effective piece of correspondence in the first draft. Precision usually comes with revision.

The individual who operates a word processor may include his or her initials after the closing to indicate that someone other than the author had a hand in preparing the letter. The word processor's initials, in lower-case letters, usually follow the author's initials, which are in capital letters. A colon, slash, or asterisk are common punctuation marks. Occasionally the author's name is spelled out. However, more and more frequently, the author's name or initials are omitted, for they merely repeat. This entire convention will most likely soon be acceptably eliminated.

GW:bj
GW/bj
Gary Williams/bj

Postscripts

Except on sales letters and letters that add a personal note, postscripts are infrequent. Postscripts should not add, as an afterthought, an idea you neglected to include in the body of your letter. If you forget to mention a key point, it's best to rewrite.

Postscripts are very effective in sales letters, however, because they can reemphasize a *key point,* such as a particular benefit to a new client. Sales letters often include as the postscript the point that's most likely to get the recipient to react positively. Postscripts appear at the very end of letters, after the last notation.

The Page Layout

Letters should be neatly prepared and balanced with uniform margins. Usually, the bottom margin is slightly larger than the top. You may opt to align your text in the right-hand margin as well as the left. Doing so, however, may mean that you will have large spaces between long words in a line. Aligning the left margin only creates a ragged right margin for your text, but readability is usually better when the spaces between words are uniform. You should also avoid breaking words at the ends of lines with hyphens, unless they are unusually long. However, don't break two lines in a row. Visually, your message will become difficult to absorb if you do.

Correspondence of Two or More Pages

Whereas most business correspondence can neatly sit on one page, occasionally a letter of two or more pages must go out. When that's the case, begin the second page five lines from the top with a line identifying the addressee, the date, and the page number. The identifying line is usually flush left, but it can be centered, or flush right, depending on preference.

Dr. Ansel P. Angstrom, June 5, 1993, page 2

You can abbreviate the month or use numerals. Begin the text of your second page two lines below the identifying line.

Standard Letter Formats

Letter writers have options when formatting their correspondence. There is no official "rule" for laying information on a page, and companies often adopt stylized formats that become part of their signature look. Several formats, however, have become standard, and they are presented here.

Semiblock

Semiblock has enjoyed status as the most traditional format for many years. The dateline and closure appear on the right-hand side of the page, approximately two to five characters right of the center of the stationery. The inside address, salutation, and end notations align left. Paragraphs indent five spaces, or more. Some companies actually use up to fifteen spaces. Semiblock allows for some interesting and dramatic stylized correspondence.

Product Alert Letter

The following letter, a field alert, adheres to the conventions for semiblock.

Complaint Response Letter

The next letter, a response to a complaint, adheres to the conventions for modified block.

Nouveaux Diagnostics
25 Brigham Way West
Miami, Florida 33156
(305)-565-5645, fax: 305-565-5646

May 25, 2005

Acute Test Labs
4242 Wilmeet Avenue
Charles, Arkansas 16052

Dear Valued Customer:

It has come to our attention that Rubella IgM Test Kit bearing Lot Number XYZ may, under certain circumstances, mask positive results. It is extremely unlikely that you will experience a problem with your testing, however.

Internal testing has demonstrated a tendency for the antigen coated onto the microwells to denature and lose reactivity when washed with a solution prepared with reagent-grade water that has a pH of 4.5 or less. Most laboratory water systems deliver reagent grade water with a pH well in excess of 4.5.

We recommend that you determine the pH of the water used in your laboratory. In the event that your system demonstrates a lowered pH, you may add 0.1M sodium hydroxide to your reagent-grade water in a quantity sufficient to raise the pH to a minimum level of 5.0 prior to diluting the wash solution.

If your laboratory has been using this lot of reagent in conjunction with low pH water, please be assured that this is not a safety issue. Denaturization of the antigen would be flagged by low reactivity of the Positive Control and therefore yield unreportable results.

We apologize for any inconvenience and concern caused by this situation. If you have additional questions or concerns, please do not hesitate to contact your Technical Services Representative.

Thank you for your attention to this matter. We look forward to our continued partnership.

Most sincerely,

NOUVEAUX DIAGNOSTICS

Madeline LeBrun

Courtesy of Marilyn Brown

Modified Block

Modified block is exactly like semiblock, except that the paragraphs do not indent.

Laboratoire d' Esprit, S. A.
115 Rue St. Catherine, New Orleans, Louisiana, 12345
phone: (809)123-4567 fax: (809) 123-4568

December 1, 2005

Mr. Frank Smith
Laboratory Manager
Laboratory of the Americas
1234 SW 76th Street
New York, New York 10009

Dear Mr. Smith:

We have completed our assessment of the discordant samples submitted by your laboratory. These samples were confirmed to test positive using our reagents and negative using reagents supplied by Phoenix Diagnostics. To provide resolution to this discordance, we further characterized the antibody present in each sample.

The following data summarize the results of this investigation.

Sample ID	% of Flagellar Antibody	% Capsular Antibody
123	75	25
456	67	33

Sample ID	% of IgG Antibody	% of IgM Antibody
123	65	35
456	70	30

As the above data demonstrate, each sample contains a high percentage of antibody directed against the flagellum of the infectious agent and a significant percentage of IgM antibody.

Our Laboratoire D'Esprit Reagents utilize a rich blend of flagellar and capsular antigen as well as conjugate blended to detect both IgG and IgM antibodies. The Phoenix Reagent package insert indicates the wells are coated only with capsular antigen and the conjugate is not blended to detect IgM antibodies.

We believe that the enhanced detection provided by Laboratoire D'Esprit was able to discriminate and appropriately classify these samples as true positives.

continued on next page

Laboratoire d' Esprit, S. A.

We hope that these data have provided the answers to your questions and alleviated your concerns. If you should have additional inquiries, please do not hesitate to contact your Technical Services Representative. We are here to support your efforts in maintaining the highest standards in laboratory medicine.

We look forward to our continued partnership.

Most Sincerely,

LABORATORIE D'ESPRIT, S.A.

Marie DeCampagne
Regulatory Associate

Courtesy of Marilyn Brown

Full Block

In full block, everything positions left. This format is gaining popularity worldwide, probably because it eliminates the need to set tabs. The following letter, in full block format, is an initial request for information from a contract laboratory. Note that, since it does not address a specific person, it includes a reference line, rather than a salutation.

> Some writers are logical and to the point, which makes for an excellent letter. I like to take a good message and format it so that it lies on the page cleanly, balancing text and white space. The way a letter looks will always either enhance or detract from its message.
>
> **Constance McGoff, Secretary**

Ronway Laboratories

Ronway Laboratories 2323 Ronway Drive Jackson, Wyoming 83001

October 5, 2005

200 Northwoods Biosciences
2 Paragon Industrial Way
Ames, Iowa 30007

RE: Pharmaceutical Testing Services

I recently attended the PharmTech Worldwide conference in Dallas and was impressed by your exhibit and your range of services. Our firm, Ronway Laboratories, has recently entered into a joint venture agreement and is in the process of determining contract services. While we will be able to develop and manufacture our product on site, we are specifically seeking a laboratory to conduct acute toxicity studies for an oral dosage product. Briefly, we require information about the following:

- Determination of the maximum tolerated dose (MTD)
- Determination of no observable effect level (NOEL)
- Toxicity and potential target organs; reversibility of toxicity; and clinical monitoring parameters.

Would you please provide us with details about your services for acute toxicity studies, including a general statement as to costs? We understand that, without further information about our product, you will be unable to be exact in your estimation. However, once we review your capabilities, we will be in a better position to determine if your laboratory can provide the services we need. I look forward to hearing from you.

Sincerely,

RONWAY LABORATORIES

Gary Williams
Vice President, Scientific Affairs

Forms as Correspondence

In regulated industries forms frequently replace or serve as an adjunct to other correspondence. For instance, a Form 1571 is a necessary part of a New Drug Application (NDA). The 1571 allows clear presentation of information that, presented as text, would require considerably more space. As such, it is an efficient vehicle for delivering details. Note, however, that companies usually include a cover letter with an NDA submission. While it's not a requirement, it adds a human touch.

DEPARTMENT OF HEALTH AND HUMAN SERVICES
PUBLIC HEALTH SERVICE
FOOD AND DRUG ADMINISTRATION

INVESTIGATIONAL NEW DRUG APPLICATION (IND)
(TITLE 21, CODE OF FEDERAL REGULATIONS (CFR) PART 312)

Form Approved: OMB No. 0910-0014.
Expiration Date: January 31, 2006
See OMB Statement on Reverse.

NOTE: No drug may be shipped or clinical investigation begun until an IND for that investigation is in effect (21 CFR 312.40).

1. NAME OF SPONSOR

2. DATE OF SUBMISSION

3. ADDRESS *(Number, Street, City, State and Zip Code)*

4. TELEPHONE NUMBER *(Include Area Code)*

5. NAME(S) OF DRUG *(Include all available names: Trade, Generic, Chemical, Code)*

6. IND NUMBER *(If previously assigned)*

7. INDICATION(S) *(Covered by this submission)*

8. PHASE(S) OF CLINICAL INVESTIGATION TO BE CONDUCTED:

☐ PHASE 1 ☐ PHASE 2 ☐ PHASE 3 ☐ OTHER ______________ *(Specify)*

9. LIST NUMBERS OF ALL INVESTIGATIONAL NEW DRUG APPLICATIONS *(21 CFR Part 312)*, NEW DRUG OR ANTIBIOTIC APPLICATIONS *(21 CFR Part 314)*, DRUG MASTER FILES *(21 CFR Part 314.420)*, AND PRODUCT LICENSE APPLICATIONS *(21 CFR Part 601)* REFERRED TO IN THIS APPLICATION.

10. ***IND submission should be consecutively numbered. The initial IND should be numbered "Serial number: 0000." The next submission (e.g., amendment, report, or correspondence) should be numbered "Serial Number: 0001." Subsequent submissions should be numbered consecutively in the order in which they are submitted.***

SERIAL NUMBER
__ __ __ __

11. THIS SUBMISSION CONTAINS THE FOLLOWING: ***(Check all that apply)***

☐ INITIAL INVESTIGATIONAL NEW DRUG APPLICATION (IND) ☐ RESPONSE TO CLINICAL HOLD

PROTOCOL AMENDMENT(S):	INFORMATION AMENDMENT(S):	IND SAFETY REPORT(S):
☐ NEW PROTOCOL	☐ CHEMISTRY/MICROBIOLOGY	☐ INITIAL WRITTEN REPORT
☐ CHANGE IN PROTOCOL	☐ PHARMACOLOGY/TOXICOLOGY	☐ FOLLOW-UP TO A WRITTEN REPORT
☐ NEW INVESTIGATOR	☐ CLINICAL	

☐ RESPONSE TO FDA REQUEST FOR INFORMATION ☐ ANNUAL REPORT ☐ GENERAL CORRESPONDENCE

☐ REQUEST FOR REINSTATEMENT OF IND THAT IS WITHDRAWN, INACTIVATED, TERMINATED OR DISCONTINUED ☐ OTHER ______________ *(Specify)*

CHECK ONLY IF APPLICABLE

JUSTIFICATION STATEMENT MUST BE SUBMITTED WITH APPLICATION FOR ANY CHECKED BELOW. REFER TO THE CITED CFR SECTION FOR FURTHER INFORMATION.

☐ **TREATMENT IND 21 CFR 312.35(b)** ☐ **TREATMENT PROTOCOL 21 CFR 312.35(a)** ☐ **CHARGE REQUEST/NOTIFICATION 21 CFR312.7(d)**

FOR FDA USE ONLY

CDR/DBIND/DGD RECEIPT STAMP	DDR RECEIPT STAMP	DIVISION ASSIGNMENT:
		IND NUMBER ASSIGNED:

FORM FDA 1571 (1/03) PREVIOUS EDITION IS OBSOLETE. **PAGE 1 OF 2**

(continued on next page)

12. **CONTENTS OF APPLICATION**

This application contains the following items: *(Check all that apply)*

- ☐ 1. Form FDA 1571 *[21 CFR 312.23(a)(1)]*
- ☐ 2. Table of Contents *[21 CFR 312.23(a)(2)]*
- ☐ 3. Introductory statement *[21 CFR 312.23(a)(3)]*
- ☐ 4. General Investigational plan *[21 CFR 312.23(a)(3)]*
- ☐ 5. Investigator's brochure *[21 CFR 312.23(a)(5)]*
- ☐ 6. Protocol(s) *[21 CFR 312.23(a)(6)]*
  - ☐ a. Study protocol(s) *[21 CFR 312.23(a)(6)]*
  - ☐ b. Investigator data *[21 CFR 312.23(a)(6)(iii)(b)]* or completed Form(s) FDA 1572
  - ☐ c. Facilities data *[21 CFR 312.23(a)(6)(iii)(b)]* or completed Form(s) FDA 1572
  - ☐ d. Institutional Review Board data *[21 CFR 312.23(a)(6)(iii)(b)]* or completed Form(s) FDA 1572
- ☐ 7. Chemistry, manufacturing, and control data *[21 CFR 312.23(a)(7)]*
  - ☐ Environmental assessment or claim for exclusion *[21 CFR 312.23(a)(7)(iv)(e)]*
- ☐ 8. Pharmacology and toxicology data *[21 CFR 312.23(a)(8)]*
- ☐ 9. Previous human experience *[21 CFR 312.23(a)(9)]*
- ☐ 10. Additional information *[21 CFR 312.23(a)(10)]*

13. IS ANY PART OF THE CLINICAL STUDY TO BE CONDUCTED BY A CONTRACT RESEARCH ORGANIZATION? ☐ YES ☐ NO

IF YES, WILL ANY SPONSOR OBLIGATIONS BE TRANSFERRED TO THE CONTRACT RESEARCH ORGANIZATION? ☐ YES ☐ NO

IF YES, ATTACH A STATEMENT CONTAINING THE NAME AND ADDRESS OF THE CONTRACT RESEARCH ORGANIZATION, IDENTIFICATION OF THE CLINICAL STUDY, AND A LISTING OF THE OBLIGATIONS TRANSFERRED.

14. NAME AND TITLE OF THE PERSON RESPONSIBLE FOR MONITORING THE CONDUCT AND PROGRESS OF THE CLINICAL INVESTIGATIONS

15. NAME(S) AND TITLE(S) OF THE PERSON(S) RESPONSIBLE FOR REVIEW AND EVALUATION OF INFORMATION RELEVANT TO THE SAFETY OF THE DRUG

I agree not to begin clinical investigations until 30 days after FDA's receipt of the IND unless I receive earlier notification by FDA that the studies may begin. I also agree not to begin or continue clinical investigations covered by the IND if those studies are placed on clinical hold. I agree that an Institutional Review Board (IRB) that complies with the requirements set fourth in 21 CFR Part 56 will be responsible for initial and continuing review and approval of each of the studies in the proposed clinical investigation. I agree to conduct the investigation in accordance with all other applicable regulatory requirements.

16. NAME OF SPONSOR OR SPONSOR'S AUTHORIZED REPRESENTATIVE	17. SIGNATURE OF SPONSOR OR SPONSOR'S AUTHORIZED REPRESENTATIVE	
18. ADDRESS *(Number, Street, City, State and Zip Code)*	19. TELEPHONE NUMBER *(Include Area Code)*	20. DATE

(WARNING: A willfully false statement is a criminal offense. U.S.C. Title 18, Sec. 1001.)

Public reporting burden for this collection of information is estimated to average 100 hours per response, including the time for reviewing instructions, searching existing data sources, gathering and maintaining the data needed, and completing reviewing the collection of information. Send comments regarding this burden estimate or any other aspect of this collection of information, including suggestions for reducing this burden to:

Food and Drug Administration
CBER (HFM-99)
1401 Rockville Pike
Rockville, MD 20852-1448

Food and Drug Administration
CDER (HFD-94)
12229 Wilkins Avenue
Rockville, MD 20852

"An agency may not conduct or sponsor, and a person is not required to respond to, a collection of information unless it displays a currently valid OMB control number."

Please **DO NOT RETURN** this application to this address.

FORM FDA 1571 (1/03) PREVIOUS EDITION IS OBSOLETE. **PAGE 2 OF 2**

Envelopes

Envelopes should bear exactly the same information as the inside address. Word processors will actually transfer the inside address of a document to an envelope — it's a convenient feature that ensures consistency. Further, allowing the computer system to generate an address ensures that bar coding, if affixed, will not interfere with the address information.

Memo Elements

Companies generally make electronic memo templates available for writers. Memos may include logos and slogans or other conventions. Most memos, however, include the following elements, although not necessarily in the same order.

DATE:
TO:
FROM:
RE:
CC:

Note that if you are writing within a templated document, the date may appear automatically.

To:

To complete the addressee section, insert the addressee's name. When you are sending a memo to a few people, do the same, listing names one above the other either alphabetically or by department. To avoid pushing the memo text too far down on the page, you can choose to limit the direct address to two or three people and use a distribution list when sending a memo to a number of recipients. Some companies have predesigned distribution lists that you can print out and attach to your memos.

From:

Complete the *FROM:* component with your name. When the memo is complete, you can sign or initial it next to your name. Sometimes writers add a closing to their memos, as they would with a letter, but it's optional, since it serves to repeat information presented at the beginning of the memo.

RE:

Indicate briefly the purpose of the memo here.

cc:

It's a good idea to limit the cc list to a few people. A lengthy list can push text too far back in a memo. A lengthy cc list can be included at the bottom of the memo, or write "See page 2" to indicate that a cc list is included in a multiple-page memo.

Additional Pages

Some electronic formats automatically include pagination for memos running more than one page. If, however the memo format used by your company doesn't, you can type a short header with the reference line abbreviated, the date, and the subsequent page number.

Memo Formats

Treat memo text as you do letter text. That the document most likely won't leave the site doesn't mean it's not important. Many a critical decision has hung on the information relayed in a memo. Memos have been included as data in submissions and have been subject to FDA scrutiny as well. It's often a memo that fills in the gaps that takes up the slack. Give yours the attention they deserve.

Make your memos clean and to the point and format them appropriately. Some companies have style standards in place, and if that's the case where you work, your job is a little easier. If you are left to your own devices, however, it helps to understand the components of memos and what makes them effective.

The most common format calls for paragraphs that align to the left, as in a full or modified block format. Occasionally, writers will indent their paragraphs, as they would in a semi-block format letter. What you don't want to do, however, is mix apples and oranges — inconsistencies serve to confuse recipients.

The way your memos look says a lot about you as a writer. These pieces of correspondence become part of the history of your work achievements, the information you have shared, and your problems and resolutions over time. Make these documents do the best job they can.

> When I do an audit of a facility, I always look at the records it keeps. The flow of information in its internal systems tells me a lot about a company and the way it operates.
>
> **Monica Grimaldi,**
> **Certified Quality Engineer**

Contamination Alert Memo

In the following memo, a microbiology director alerts a quality assurance director about mold and yeast in the plant.

KIM LABORATORIES, SA Memorandum Page 1 of 1

DATE: August 21, 2005
TO: Joseph James
FROM: Michael Nolan

RE: Envirnomental Monitoring of the Production Facility

The results from the environmental monitoring conducted in August indicate that there is a high level of mold and yeast present in areas of the production facility. We have resampled several areas, and the results demonstrate that this situation continues to exist.

We have, therefore, determined several courses of action. Local media reports have shown that the unusually rainy and cool weather this spring and summer has caused ideal growth conditions for molds and yeasts. Therefore, although the results obtained from the Environmental Monitoring Program are above the proposed alert or specification levels in many cases, they are not unexpected. The high levels of molds and yeasts in the production facility have not affected our product. This has been demonstrated by the lack of contamination in all products tested for release.

I have discussed the situation with the Director of Operations, and we have targeted two rooms that have demonstrated particularly high counts for thorough sanitization. These are batching room 551 and the Pharmacy. Other actions we will take include establishing a better overall awareness and conducting more thorough sanitization of the entire facility. The Director of Operations is also looking into methods of improving the airflow in several areas and has taken action to reduce the humidity in the lotion production areas.

Courtesy of Michael Nolan

Staff Request Memo

The following memo, to a Director of Human Resources from a Quality Assurance Director, requests an addition to the quality staff. Working as a proposal, this is a memo of some length, and as such requires more than one page to adequately deliver the information.

Ronway Laboratories *MEMORANDUM* Page 1 of 2

DATE: March 25, 2005

TO: Maryann Sorensen

FROM: David Nettleton

RE: SOP Controller Staff Addition

As we discussed yesterday, an addition to the quality assurance staff is necessary. The rapid growth of our company has been accompanied by an even more rapid growth in documentation. I cannot stress strongly enough how critical it is for us to have a person bearing overall responsibility for managing our SOP database. So, in response to your request, here's a list of duties the new hire would assume. I'm sure once you see the scope of the duties, you'll agree that filling this position is wise, especially since we are negotiating two significant manufacturing contracts with Swiss-based Soloman Pharmaceuticals.

SOP Controller duties

1. The SOP Controller maintains the SOP database and tracks all SOPs, from inception, to preparation, review, revision, issuance, and withdrawal. The SOP Controller plays an active role in documentation control, and reports directly to the Documentation Manager.

2. The SOP Controller issues numbers and folders for new and revised documents, and routing records if needed, and enters the status of the new or revised SOP into the system. (Thus SOP MQ52.A may be active, but MQ 52.B may have been issued for development and review.)

3. The SOP Controller prepares SOPs that have gone through a final, unedited review or have been subject to a document concurrence meeting. This entails making minor mechanical, but not content, changes such as spacing, numbering, and so forth, and printing the final version on Original Document paper. The SOP Controller then secures final signatures. The SOP Controller solicits a copy of the initial training on the new or revised SOP, stamps the effective date (determined in conjunction with the preparer and training requirements), makes blue copies in appropriate numbers for specific manuals, and issues them. The SOP Controller retrieves previous versions of newly issued SOPs and destroys them. The SOP Controller files the original SOP and stamps the original of the previous version "retired." The SOP Controller includes the history of review, and document concurrence sheet if applicable. The SOP Controller prepares a file sheet generated by the issue and retrieval sheet and files the new or revised version.

(continued on next page)

Ronway Laboratories Page 2 of 2

4. The SOP Controller also issues numbers and folders for SOPs to be designated obsolete. When a final review shows reviewers to be in agreement, the SOP Controller retrieves existing copies of the SOP and destroys them. The SOP Controller stamps the original "obsolete" and files it with the existing original documents.

5. The SOP Controller periodically checks the manuals to make sure their integrity is intact. The SOP Controller issues new manuals as needed, and issues blue copies of relative SOPs to them, keeping records consistent with the system.

6. The SOP Controller makes the electronic template, bearing the SOP number and version, available to writers electronically.

One final note: We are in the process of rewriting the SOP on the SOP system to incorporate the duties of the SOP controller, so it's imperative that we move ahead as quickly as possible.

Electronic Correspondence

E-mail and facsimile (fax) equipment make transmission of information instantaneous. Use the guidelines for preparing letters and memos to help you prepare documents for electronic transmittal. Exercise some caution as well; the urge to "hit the button" and transmit can be strong, and once you've sent the communication, it is complete. Always take time to scrutinize what you've written; take a look at it in hard copy. Let it rest a bit if you can, and reread and edit before you send.

E-Mail

Some companies use e-mail exclusively for internal communication; it certainly cuts down on the amount of actual paper changing hands and going to file. E-mail can also be transmitted from the company effectively. It's far less formal than a letter, but it usually does the job it's intended to do very nicely.

> When tempted to push the "send" button, ask yourself, "Is this piece of writing something I'm proud of?" If it's not, resist and revise.

The following is an e-mail about batch records that traveled internally in a company. In it, the GMP supervisor details a problem and calls for a resolution.

From: ECandoken_@ronway.los.com

Sent: Wednesday, 16 Sept 2005 11:07 AM

To: Bhaverstraw_@ronway.los.com

Cc:

Subject: Batch Record Checks

Bob,

We have been making strident efforts to produce satisfactory batch records; our new SOP calling for more frequent supervisory checks on the records as they are produced attests to this fact. However, as we have heightened our efforts, it has become apparent where problems lie. Now we need resolution.

The records that our third-shift supervisor, Bill Bailey, reconciles consistently show calculation errors. One out of nine has to be corrected. I have spoken with Bill repeatedly over the months, and retrained him in the procedure. However, he either doesn't seem to understand the importance of precision or is unable to calculate accurately.

In addition, Jose, his manager, concurs that this constitutes a serious problem. Not only is the danger of inaccuracies in batch records ever present, the time we expend in making sure they are correct is costly. Please let me know what our next step should be. Thanks.

Facsimiles

Fax formats vary from company to company. Many use letterhead or variations thereof, which include the company street address, telephone number(s), and facsimile number. Most also include the following components and may include other elements.

Recipient's name and contact information

Sender's name and contact information

Fax number

Page numbers

Disclaimer

Text may accompany a fax as a separate letter on letterhead, or as part of the fax distribution information. The following fax is a simple request for information from an equipment manufacturer, and it does the job nicely.

Ronway Laboratories Facsimile

2323 Ronway Drive, Jackson, Wyoming 83001

TO:	JASON TANG
FROM:	JUNG WOOK
FAX NUMBER:	DATE:
908-131-0080	OCTOBER 28, 199
COMPANY:	FAX NUMBER:
CONTRACT ASSOCIATES	973-252-1234
PHONE NUMBER:	PAGES (INC. COVER SHEET)
908-131-0800	1

This message is intended only for the use of the addressee and may contain information that is privileged and confidential. Any dissemination of this communication by anyone other than the intended recipient is strictly prohibited. If you have received this communication in error, please notify us immediately by telephone or fax. Thank you.

Ronway Laboratories is expanding its stability function to better accommodate our new product line. We are now investigating environmental chambers and would like to know more about your microprocessor-based chambers for drug stability studies.
Specifically, we need a unit that has the following features:

- Temperature range from –10 to +70° C with a tolerance of 0.1° C
- A relative humidity range that can be adjusted from 50% to 95% with a tolerance of ±5%
- Digital LED display
- Audible-visual alarm and remote contact

Please provide us with information about your product. We look forward to hearing from you.

Forms for Fax Messages

Many companies use fax transmissions as integral records. Consider the following fax form. It is a controlled document — and it bears a form number. It is a notification of compound shipment from one contractor to another. It includes the name of the sponsor (at the top). Providing this fax helps ensure that the sponsor receives notice that the compound is to be shipped to a study site. The fax also goes to the study site so personnel there know the compound is coming. This is part of an audit trail for compound shipping. The receiving site must also confirm receipt when it occurs, must monitor compound use during the study, and must reconcile inventory at the end of the study. Thus the fax form and additional tracking records become part of the study file.

To:	The Wayward Company 200 University Avenue Livingston, New Jersey 07930 Telephone: (973) 992-1007
Fax number:	973 992-1008
To:	{study site}
Fax number:	
From: **Address:** **Telephone:** **Fax number:**	
Date:	
Number pages (incl. cover)	

The following clinical inventory has been shipped to the study site.

Shipping date: **Shipping method:**

Upon receipt, ***please complete the Confirmation of Clinical Inventory Receipt form and send it to The Wayward Company and the supplier, if other than Wayward.***

Product: **Protocol:**

Shipping Conditions	**Receiving Instructions**

NOTE: Study inventory must remain sealed within its original box until opened and inventoried at the study site. Dispensing of inventory must be according to approved procedures.

☐ **MSDS included**

Signature of sender

Date

FORM 12130-1

5

Policies, Plans, Manuals, Procedures, Methods, and Instructions

"There shall be standard procedures and such procedures shall be followed." This mandate permeates the regulations that dictate how companies may develop, test, and market therapeutic products. While the regulations tell what systems and procedures must be in place, they do not give the specifics for how these systems and procedures must occur. It's up to companies to determine "how it happens here" and put backbone documents in place.

A whole group of documents tell how things occur on a regular basis. It's safe to say, however, that each company may have a different setup for its master "how to" documents. Many companies have a flat document structure, with every "how to" document a standard operating procedure (SOP). Many other companies tier their documents with a quality policy at the top that tells of the firm's adherence to quality standards overall. Then, in the next tier, they put SOPs in place to delineate systems and processes that drive the business. Such systems and processes are "big picture" documents that define overall activities such as metrology of instruments and equipment. Next in line are instructions for carrying out specific procedures — activities that one or two people can perform, such as calibrating a scale or verifying temperature in a freezer or refrigerator. On a par with instructions are laboratory methods — documents that tell how certain regularly conducted experimental activities occur. At the base of the tier structure are support documents that result from the defined activities. Forms working in tandem with SOPs and instructions, for instance, capture data as systems and processes occur. Data collection in turn feeds into process reports and ultimately summary reports. This tier structure has been adopted by many companies complying with the regulations for medical devices, 21 CFR Part 820, and the International Organisation for Standardization (ISO). The reasoning is that the Food and Drug Administration (FDA) inspectors conduct inspections by a "top-down" method structured as the Quality System Inspection Technique (QSIT), so the top documents should present the "big picture." Companies then need only produce those documents that are required for the inspection; they don't need to make every document available unless asked. Companies that are ISO compliant also adhere to a top-down document organization, since ISO requires a quality policy to be in place.

This chapter explains the backbone process documents that must be in place for companies to operate compliantly. Chapter Six discusses data collection and routine reporting, while Chapter Seven discusses process reports, and Chapter Eight discusses summary writing.

Process Documents and Compliance

Problems with procedures are primary causes for FDA issuing forms 483 and warning letters. A lack of procedures, outdated procedures, incomplete procedures, contradictory procedures, overlapping procedures, and procedures that aren't followed don't escape notice. Without properly documented processes, a company has no proof that it is doing things consistently with the same results every time. For instance, even if a cleaning procedure is performed according to current Good Manufacturing Practice (cGMP) standards, lack of a document explaining how it occurs means the company can't substantiate its practices. Similarly, if a company can't produce documentation that tells how and when a scale is calibrated, it can offer little proof that the process occurs repeatedly the same way.

An engineer writes a plant procedure detailing how evacuation of the facilities takes place, step-by-step, in the event of an emergency. A microbiologist writes procedures for sampling and testing water systems. A Quality Assurance (QA) department produces documents that show systematic and regular sampling activities. A Quality Control department spells out the procedures for testing samples at regular intervals. And a Regulatory Affairs department puts documents in place that tell how to prepare regulatory submissions, manage complaint files, and prepare and release disclosure writing, such as journal articles, posters, abstracts, and slide sets. In total, process documents such as these tell how the company functions.

Good process documents are about more than compliance, however. These are the documents that tell how business activities occur. They ensure that the processes are both safe and effective and integral to the ongoing development and production of the firm's products. Consider a clinical trial. Once a product is approved for testing in human subjects, as the result of an Investigational Device Exemption (IDE) or an Investigational New Drug (IND), it behooves the sponsor to make sure procedures are in place to drive the trial successfully. The cost is enormous to the company if it has to repeat or augment a trial; it's better to do it right the first time. The sooner the trial meets its endpoints, the less the company has to spend.

The same logic applies throughout the various stages of development, from initial discovery through product in the market. It's costly to repeat studies because they failed to have an adequate number of replicates, or because data were not properly recorded or assessed. It's costly to discard an adulterated product during manufacturing because personnel followed an obsolete procedure that doesn't yield consistent results. And it's costly to address citations resulting from a pre-approval inspection or general GMP inspection that has found the company's processes to be inadequate or faulty. The bottom line is this: Procedures are integral to quality operations, and companies that know this have them in place, and review and revise them as necessary on a regular basis, typically every two years.

Common to most process writing is establishing the purpose of the activity, the scope of the process, terms integral to the process (definitions), the responsibilities for carrying it out, the procedure steps or activities themselves, references relative to the process, such as a regulatory dictate, and identification of documentation that must be compiled as part of the process.

How documents look varies from company to company. Most companies use templated formats that ensure consistent organization of information, so that people who use the documents know what information to find where. Templates allow ease in preparation and consistency in appearance — and appearance is important. Often these are the first documents an inspector or auditor sees.

In working templates, numbers adjust automatically and headers and footers capture meta data, such as the document number, version, effective date, pagination, and confidentiality statements. The company's backbone documents must also have histories of changes, from the original concept for the document through its life, with a record of every version and the changes each was subject to, ending with the reason a document goes into retirement.

With electronic record keeping, meta data is captured in an audit trail, so the system can tell when the concept for the document occurred, who approved it, who wrote the document, who approved the document, and every change that happened to it throughout its life. Typically, electronically controlled documents have a watermark or other indicator that says "valid only on print date," with the original document residing in the electronic system or as controlled copies, often on special paper, in manuals, such as laboratory method manuals that must be available in laboratories. Note that electronic systems must comply with 21 CFR Part 11, *Electronic Records; Electronic Signatures* and undergo validation. When a company determines to use electronic signatures, it must notify FDA, typically within 30 days of going electronic, of the intent and practice. Note also that Part 11 does not supercede the documentation requirements of the other regulations. The predicate rules still apply.

Written Procedures; Deviations

There shall be written procedures for product and process control designed to assure that the drug products have the identity, strength,

> quality, and purity they purport or are represented to possess. . . . These written procedures, including any changes, shall be drafted, reviewed, and approved by the appropriate organizational units and reviewed and approved by the quality control unit.
>
> Written procedures and process control procedures shall be followed in the execution of the various production and process control functions and shall be documented at the time of the performance. Any deviation from the written procedures shall be recorded and justified.
>
> 21 CFR 211.100

Backbone process documents benefit from detailed numbering that identifies segments specifically and individually. A handy way to control these reports is to use a numbering system to show sequence and relationship. Whereas there are many ways to number elements in documents, one that works well and is widely used is an Arabic system with decimal subparts. With this method, main headings of individual sections are numbered 1, 2, 3, and so forth. Information within each section is numbered if that information is broken into subparts, such as 2.1, 2.2, and 2.3. Each subpart in turn may have its own breakdown. If this is the case, the system adds another decimal and a third digit follows: 2.1.1, 2.1.2, and 2.1.3. This system makes specific reference easy. Actually, regardless of the type of document, this system works — from simple procedures to quality policies to regulatory submission documents. Here is an SOP template that shows the numbering sequence as well as other standard components of SOPs.

	STANDARD OPERATING PROCEDURE **SOP 00000 1**
	Title

Effective date: 00 Mon 00

1.0 OBJECTIVE AND SCOPE (STYLE HEADING 1)

1.1 (Style Text #d Level 1)

(Style Text Level 1)

- (Style Bullet Level 1)

1.1.1 (Style Text #'d Level 2)

(Style Text Level 2)

- (Style Bullet Level 2)

1.1.1.1 (Style Text #d Level 3)

2.0 DEFINITIONS

2.1

3.0 MATERIALS AND EQUIPMENT

3.1

4.0 RESPONSIBILITIES

4.1

5.0 PROCEDURE

Subhead

5.1

6.0 REFERENCES

6.1

7.0 DOCUMENTATION

7.1

CONFIDENTIAL Page 1 of 1

Courtesy of Synaptic Pharmaceutical Corporation, a Lundbeck Company

Preparing to Write a Document

The best process documents result when people who actual perform or participate in the processes do the writing. If you are writing a process document, there are some preliminary steps that you need to take before you actually put words on paper. Ask yourself these questions:

Why do we need this document?

Is there an existing document that can be modified to include the information?

Is anyone else working on a similar document?

Will this document impact other documents? If so, will they require modification?

What is the system for document control?

What do I need to do to get started? (document template/number/version)

Do I need a coauthor? Who can provide input during the drafting process?

Creating Multiple Related Documents

Consider carefully the information to goes into the document. Top level process documents such as policies, plans, and SOPs that involve many people and require extensive review prior to initial issue or a subsequent version often do best if you identify information that is likely to change and create a separate document for it, which you can then cross-reference in the top level document. For instance, if you are writing an Emergency Plan, and the plan identifies a Safety Committee, consider creating a separate reference document to identify the committee, rather then including the specifics in the plan. That way, when committee members change, you won't have to revise the more inclusive plan, but simply the reference document.

The Training Component

All process documents require review and approval through the document control system. It's important that no document of this ilk becomes effective unless the people who carry it out have received training. This means a document may undergo review and receive approval, but before it becomes active, the people who will be responsible for carrying out the procedure must undergo training, and that training itself must be documented. Indeed, training activities themselves require established procedures to ensure that employees are qualified to do their jobs. Further, all necessary materials and equipment must be in place before a procedure is implemented.

Tense and Voice in Process Documents

Because process documents tell how things happen with regularity, either descriptively or sequentially, they work best in the present tense. The present tense is not just for identifying what is happening now or what is so at a present point in time. It is also the tense for habitual action. Consider a statement like this: "I work at the world's number two pharmaceutical company." In this sentence, "I work" indicates an ongoing activity, not necessarily an action at the moment. Similarly, "I eat breakfast at 8 a.m." signifies habitual activity. Thus, habitual activities that you delineate in process documents deserve to be in the present tense.

While it's true that most regulations use "shall," you need to understand that this future tense construction is a form of legalese and not the easiest to read and follow. Indeed, inspectors have often said clear, direct language that tells what happens and how to do something makes for the most comprehensive procedures.

You must also assess the audience. Is the document an overview that delineates an inclusive process or system that affects more than a few people? For instance, a quality manual that tells how quality is assured, who has responsibility, and what happens overall is a "big picture" document. Plans for emergencies, animal husbandry, chemical hygiene, and other processes are other "big picture" documents that tell how things happen. An SOP on Hazardous Medical Waste can describe a process to which laboratory and operations people must adhere, so it also presents the "big picture." So does an SOP on managing clinical inventory because it involves sponsor personnel and site personnel and a series of interrelated activities.

Other documents give instructions and tell how to do activities specifically. These are not "big picture" documents, but "how to" instructions for activities that one or two people can carry out and that have defined start and stop points. Examples are master formulas, laboratory tests, and work instructions for routine activities such as testing air purity, performing routine maintenance on a batching tank, or loading labeling into a bar coder.

Generally "big picture" documents work well in the third person. That is, they are most readable when they identify people by title and responsibility and tell how things happen. "The Chief Executive Officer (CEO) supports all quality activities." Direct "how to" documents, on the other hand, are generally most readable when they speak right to the people who carry out the process, using the imperative voice, which is "you" understood. "Fill the bath with a saline solution," is an example of the imperative voice.

Sometimes, documents may mix the third person and the imperative voice. For instance, a chemical hygiene plan may be written to the chemists in the company, but refer to upper management and committees in the third person.

> Third person:
>
> > The Chemical Safety Committee (CSC) evaluates the Emergency Plan periodically. The CSC also meets to assess all emergency activities and determine measures to prevent recurrence.

In the same document, instructions for chemists may work best directed right to them.

> Imperative voice:
>
> > Report all spills immediately.

General Writing Guidelines for Process Documents

1. Keep the Language Simple

Write the way you talk. Choose a short word over a multisyllabic one or a phrase. Instead of "utilization," say "use"; instead of "avoidance of," say "avoid."(See also Chapter Nine, Developing a Clear Style, and Chapter 10, Building Strong Sentences.)

2. Limit the Number of Ideas in a Step

Limit steps in a procedure to no more than three logically connected actions. If actions aren't related, they need to be in separate steps.

Don't write, "When the last package has cleared the line, shut the machine down, complete the packaging record and sign and date it, and get the signature of the QA inspector."

Instead, break the information into separate steps:

1. When the last package has cleared the line, shut the machine down.
2. Complete the packaging record and sign and date it.
3. Secure the signature of the QA inspector.

3. Don't Include Restrictions if They Don't Matter

Instead of writing, "Use a 4-mL beaker" if size or condition is unimportant, simply say, "Use a suitably sized beaker."

4. Be Consistent

If you write, "Turn the knob counterclockwise" in one part of the document, don't say, "Turn the dial" in another.

5. Put Information Down in the Order in Which the Activity Occurs

Rather than write, "Press the start button as soon as the ready light comes on," say, "When the ready light comes on, press the start button."

6. Be Precise

Avoid using terms that are open to interpretation. Don't say "a representative number" when "at least ten" is more definitive. Similarly don't say "respond quickly," since "quickly" is imprecise. Say rather, "respond within one minute."

7. Avoid Using the Conditional Unless a True Condition Is Part of the Process

Don't write, "You should record the temperature every 24 hours." Rather say, "Record the temperature every 24 hours."

Words such as "should," "could," and "would" can imply condition. If you tell someone, "You should do a calculation," you are, in fact, offering a choice. In directives, simply say, "Do a calculation." No margin for interpretation then exists.

However, if a process document, such as a master formula, truly calls for a condition, do elucidate, because it's appropriate. "The mixture should pass through a number-ten screen."

8. Avoid the Passive Voice Where You Can

The following sentence starts out with a passive construction in the first clause, followed by an active construction. "The knob should be turned counterclockwise slowly, until the gauge reads 'ready.'" A rewrite makes it all active: "Turn the knob counterclockwise until the gauge reads 'ready.'" (See also Chapter Nine, Developing a Clear Style, Chapter 10 Building Strong Sentences, and Chapter 11, Managing Verbs in English.)

9. Make Documents Consistent with Each Other

If processes call for a supervisory or quality assurance check, handle information so it appears the same way in all documents requiring similar activities. The following is an excerpt from a cleaning SOP. The company has many SOPs requiring cleaning rooms and floors and recording the information in a logbook. Each document for each procedure includes the cleaning of the room and floor in these words, with quantities and other variables adjusted for the job at hand.

5.1 To clean the room and the floor, do the following.
 5.1.1 Prepare a solution of 10 gallons of hot water and 150 cc of LpHse detergent.
 5.1.2 Mop the floor.
 5.1.3 Rinse the floor with hot water and squeegee dry. Let thoroughly air dry.
5.2 When cleaning activity is complete, do the following.
 5.2.1 Put a "Cleaned" sign on the door.
 5.2.2 Record the activity in the Cleaning Log Book.
 5.2.3 Notify the QA Inspector that cleaning is complete.

10. Distinguish Between "Warning" and "Caution"

Many people think "warning" and "caution" are interchangeable terms. But FDA recommends using "warning" and "caution" the way they are in the labeling regulations. "Warning" is the agency's preferred term for words and graphics that alert people carrying out a process to possible injury, death, or other serious adverse reaction. "Caution" is the term for the words and graphics that alert a user to a possibility of a problem associated with equipment or instrumentation. Here's a caveat: If your company has defined these terms, adhere to the distinctions established in the company.

11. Don't Get Creative

Don't write, "X the form." Say rather, "Check the box on the form."

12. Anticipate the Unexpected

Because people make mistakes, components may be faulty, and equipment may not always perform as it should, despite adequate training and controls, it's best to allow for unexpected eventualities in documents. If for instance, a laboratory method says "the mixture must pass through the screen in three minutes" to demonstrate acceptable consistency, say what should happen if it doesn't.

4.6 The mixture must pass through the screen in three minutes.

4.6.1 If the mixture doesn't pass through the screen in the specified time, discard it according to the approved procedure for handling chemical waste.

4.6.2 Repeat steps 4.1 to 4.5.

13. Remember the Purpose of the Document

Finally, remember that you are not writing a training document. Don't write, as a step, "Turn the computer on," or "Hit enter." The regulations require that all personnel be trained in procedures before carrying them out. If they can't turn the computer on without instruction, something's awry.

Writing Manuals and Policies

Many companies put top level documents in place to spell out the overall quality policy for the company. While these top level quality documents may take the same form as SOPs, for many companies quality manuals are preferable. The quality manual is standard in medical device companies and those driven by ISO and other regulations. For other companies, quality manuals simply make good business sense because quality manuals spell it all out from top management down. With the advent of discovery occurring more and more in small companies and larger pharma seeking mergers and acquisitions to build their pipelines, quality manuals — and quality agreements — are receiving more attention as tools for ensuring good business practices.

Quality Manual Components

Quality manuals are usually bound editions that are subject to review, approval, and version control, just like other process documents. Components reflect the company's activities, so they may vary from company to company. In sum, all the components of a quality manual give the total picture of "how things happen here."

Here's an outline of typical medical device quality manual components with select text excerpts. Notice that it refers to activities occurring "according to approved procedures," which, for the company are second tier documents.

Quality Manual

Preliminary Information

Cover page with Company logo, Document Title, Document number, version, and effective date.

Table of Contents – usually completed when the document is done so that the entries hyperlink to specific sections of the document

1.0 Policy Statement

This section gives an overview of the quality framework to which the company adheres in the development and marketing of its products. It states the focus of the company (usually end users/customers) and provides the governing regulations that drive operations. It then states the employee dedication to the ideals set forth within the quality manual.

2.0 Introduction

This section identifies the company by name, type, and product line, and gives the company address and trading information, if publicly owned. Here is the place to identify management commitment to quality as well, as the following excerpt does:

> The Netgo Quality Manual reflects management's position about quality and is the top level document in the Netgo quality system. This manual describes the quality management system in total and identifies its implementation for the development, manufacture, and distribution of Netgo products.

3.0 Management Responsibility

This section delineates the following:

3.1 Quality Policy– Top down responsibility from Chief Executive Office (CEO) and upper management and buy in throughout the organization by all employees.

3.2 Organization – Responsibility and authority for maintaining the quality system. Note that responsibilities are identified by title, not specific name.

3.3 Resources – Who does what? Specifically, who gives technical assistance, performs monitoring and auditing? Who verifies activities and records data? Is personnel sufficient and are they adequately trained?

3.4 Management Representation -- Who is the liaison with top management and people engaged in quality activities? And who is the liaison with external parties engaged in contract activities? Here's an excerpt:

> The Vice President of Quality Assurance (VP/QA) is responsible for and has the authority to establish, implement, and maintain the quality system through involvement with individual departments within each company facility to ensure compliance with regulatory and established internal standards. The VP/QA reports on the performance of the quality system to the CEO.
>
> The VP/QA does the following to ensure effective representation to the CEO:

1. Determines audit schedules and manages internal audits to audit the facility in compliance with FDA 21 CFR Part 211 Good Manufacturing Practices, BS EN 46001, ISO 9001, ISO 13485, and the EU Medical Device Directive;
2. Ensures that an effective corrective action system is in place and optimally operative;
3. And reviews audit and corrective action results with the CEO.

3.5 Management Review

How often do the CEO and direct reports review the quality system to ensure continuing effectiveness?

4.0 Quality System

4.1 Documentation -- General Overview

Level One – Policies, such as the quality manual
Level Two – Standard Operating Procedures – confirmation that they meet the requirements of the binding regulations
Level Three – Work instructions
Device Master Records (DMRs)
Drawing/specifications
Operation methods (OMs)
Product Requirement Specifications (PRS)
Methods for manufacturing and inspections
Quality Control (QC) procedures
Batch Records
Test Methods
Technical file data
Visual standards
Level Four – Quality records

How do documents in various levels develop and what do they cover?

4.2 Contract Review – General Overview

Review
Amendments
Records

4.3 Design Control – General Overview

Design and Development Planning
Organization and Technical Interfaces
Design Input
Design Output
Design Review
Design Verification
Design Validation
Design Changes

4.4 Document and Data Control – General Overview

Document and Data Approval and Issue
Document and Data Changes

4.5 Purchasing – General Overview

Evaluation of contractors
Purchasing Data
Verification of purchases product(s)
Control of purchased product(s)

4.6 Product Identification and Traceability

4.7 Process Control

4.8 Inspection and Testing – General Overview

Receiving Inspection and Testing
In-Process Inspection and Testing
Final Inspection and Testing
Inspection and Test Records

4.9 Control of Inspection, Measuring, and Test Equipment – General Overview

Control procedure
Inspection and Test Status
Control of Non-Conforming Product

4.10 Corrective and Preventive Action – General Overview

Corrective Action
Preventive Action

4.11 Handling, Storage, Packaging, and Delivery – General Overview

Handling
Storage
Packaging
Preservation
Delivery

4.12 Quality Records

This section presents an overview of how records develop and how the company maintains them. Here's an excerpt.

> Procedures are in place for identifying, collecting, indexing, accessing, filing, storing, maintaining, and disposing of all items the quality records identify. Persons engaged in all product-related activities generate and maintain records to support quality-related activities. Records provide evidence that the product is in compliance with regulatory requirements. Records are legible and identifiable to the product(s).
>
> Netgo retains quality records for a period of time at least equivalent to the lifetime of the product, but in no case, less than two years from the date of shipment. The company maintains device history records for each batch of product; these provide traceability required by the Product Identification and Traceability section of this document and identify the quantity manufactured and the quantity released for distribution. All device history records are authorized and verified according to approved procedures.

4.13 Internal Quality Audits

This section discusses how audits are scheduled, conducted, and documented, according to approved procedures

4.14 Training

This section presents an overview of the training processes approved an in place within the company. (Here's an excerpt.)

Training plays an important role in producing high quality Netgo products; therefore, documented procedures are in place for identifying training needs and providing training for personnel whose work affects quality.

All personnel are qualified by means of education, experience, and specialized training. In-house personnel or approved third party persons or organizations, conduct specialized training. This includes any training necessary for special processes or working under controlled environmental conditions. Satisfactory completion of training is demonstrated by methods such as examination, certification or documentation of attendance. All training is documented and Human Resources maintains records in the HR files of the DocAware Electronic Record Keeping System. Personnel who are required to work under special environmental conditions or who perform special processes are appropriately trained or supervised by a trained individual.

Identifying and Providing Training - Each supervisor identifies training needs by comparing the needs of the business with existing employee skills. The supervisor arranges for the training of personnel who have a direct effect on quality (through either formal training courses or on the job training). Training and training needs are assessed and documented by the supervisor through the employee annual review process.

Assessing and Developing Training - New training needs may be developed when the following activities occur:

1. When hired, the employee's skills are assessed - job description, education, experience, resume, and so forth.

2. When new process or procedure requires special training.

3. When quality management systems and new standards are developed and implemented.

4. When a need exists, by encouraging employees to train for higher level jobs or additional operations.

5. When initiated by the employee, as part of continual employee development. This training includes internal and external courses.

6. When refresher or re-certification courses are provided.

Training Records - When an employee is hired, applicable training is performed and documented. On the job training is also documented.

4.15 Servicing

4.16 Statistical Techniques

Needs Identification
Procedures in Place
Sampling Inspection

4.17 European Directive Requirements

4.18 FDA Requirements

4.19 Clinical Evaluation

Plans

Plans are not far afield from quality manuals. In essence they delineate what happens repeatedly to satisfy the dictates of the regulations and to make sure that the company is following safe and effective processes in the day-to-day activities. Usually plans are all-inclusive for a specific area. An example is an animal husbandry plan, a document that companies certified by the American Association for Accreditation of Laboratory Animal Care (AAALAC) typically put in place. Other plans may be for Emergency Action, Respiratory Protection, Preventive Maintenance, or any number of other topics.

A common plan in companies that practice Good Laboratory Practices and satisfy The Occupational Safety and Health Administration regulations is the one for Chemical Hygiene. The Occupational Safety and Health Standard in 21 CFR Part 1910 explicitly calls for such a plan: "'Chemical Hygiene Plan' means a written program developed and implemented by the employer which sets forth procedures, equipment, personal protective equipment and work practices that (i) are capable of protecting employees from the health hazards presented by hazardous chemicals used in that particular workplace and (ii) meets the requirements of paragraph (e) of this section."

The following is an excerpt from the general provision section of a biotech's chemical hygiene plan.

Plan Components

Like other process documents of length, plans cover a lot of territory. They are tedious to write and require controls for review and revision. Planning the sequence is a matter of determining "what we must include in our plan." Here's a general outline that will work for most plans. Remember, too, that the document metadata — such as pagination, document number, version, effective date, and so forth — need to be captured.

- Cover page
- Statement of Intent
- Table of Contents/table of figures if any
- Overview
- Introduction
- Responsibilities
- General Requirements
- Specific Requirements
- Glossary
- Appendices

The following is an excerpt of the general provision section of a brotech's chemical hygiene plan.

2 GENERAL SAFETY REQUIREMENTS

Synaptic Pharmaceutical Corporation, a Lundbeck Company, maintains procedures relevant to safety and health considerations when laboratory work involves the use of hazardous chemicals. Where this Chemical Hygiene Plan (CHP) does not adequately address the scope of hazards, supervisory personnel develop procedures or work instructions for work area specific operations. All procedures and instructions receive periodic review.

2.2 GENERAL SAFETY PRINCIPLES

To minimize hazards and maintain basic safety in the laboratory, adhere to the following guidelines:

- Always wear eye protection and a lab coat in the laboratory.
- Examine the known hazards associated with the materials to be used. Never assume all hazards have been identified. Carefully read the label before using an unfamiliar chemical. Review the Material Safety Data Sheet (MSDS) for special handling information. Determine the potential hazards and use appropriate safety precautions before beginning any new operation.
- Be familiar with the location of emergency equipment, such as fire alarms, fire extinguishers, emergency eyewash and shower stations and know the appropriate emergency response procedures.
- Avoid distracting or startling other workers when they are handling hazardous chemicals.
- Use equipment and hazardous chemicals only for their intended purposes.
- Be alert to unsafe conditions and actions and call attention to them so that corrective action can be taken as quickly as possible.
- Inspect equipment for leaks, tears and other damage before handling a hazardous chemical. This includes fume hoods, gloves, goggles etc.
- Avoid tasting or smelling hazardous chemicals.

2.3 Protective Clothing and Laboratory Safety Equipment

2.3.1 General Considerations

Selecting protective clothing, equipment and methods of operation requires forethought. In addition, always consider reducing potential exposures to hazardous chemicals, when possible, by the following means:

- Substituting a less hazardous substance
- Scaling down size of the experiment
- Substituting less hazardous equipment or process (e.g., safety cans for glass bottles)
- Isolating the laboratory person or the process/experiment
- Improving local and general ventilation (e.g., use of lab hoods)

The MSDS lists the personal protective equipment recommended for use with the chemical. The Laboratory Supervisor or the CHO can assist in determining which personal protective devices are required for each task. Appropriate personal protective equipment will be provided to employees

2.3.2 Laboratory Ventilation

Always use laboratory chemical hoods when working with all hazardous chemicals that can become airborne. As a rule of thumb, use a hood when working with volatile substances with a Permissible Exposure Limit (PEL) or Threshold Limit Value (TLV) of 50 ppm or less

Every laboratory ventilation hood used for the control of air contaminants is tested at least once annually to ensure that adequate air flow provides continued protection against employee over-exposure. The Safety Officer (SO) coordinates this testing. Laboratory hood airflow is demonstrably adequate when the average face velocity equals 85 to 115 feet/minute with the hood sash at a working height (12 inches). Other local exhaust ventilation, such as instrument vents, also undergo testing. The CHO determines any additional criteria for minimal acceptable flow. The PSO records and maintains results of laboratory ventilation tests. Results must be posted on the hood along with the date of testing.

Note: It is extremely important that you do not use fume hoods that are out of service.

2.3.3 Protection of Skin and Body

Skin and body protection means wearing protective clothing over all parts of the body which could become contaminated with hazardous chemicals. Select personal protective equipment (PPE) on a task basis, and check it to ensure good condition prior to use (e.g., no pinholes in gloves.)

2.3.4 Clothing Requirements in the Laboratory

Where there is no immediate danger to the skin from contact with a hazardous chemical, it is still prudent to select clothing to minimize the risk of exposure. Wear clothing with long legs to cover areas that lab coats do not. Avoiding short trousers or skirts minimizes exposure risk. In addition, wear socks and closed shoes in the laboratory at all times. Sandals and perforated shoes are not permitted. In addition, confine long hair for your safety and to prevent compromising the integrity of the lab work. In addition, do not wear jewelry that can catch in laboratory equipment. Most important, always wear a laboratory coat over street clothes. Laboratory coats prevent contact with dirt, chemical dusts and minor chemical splashes or spills. Always wear clean laboratory coats and dispatch soiled ones to the hamper bins in the change room.

Note: If you suspect your lab coat has become contaminated, remove it immediately, bag it, and wash all exposed skin surfaces thoroughly. Refer to the MSDS sheet for any additional measures. Call the CHO to collect the contaminated lab coat.

2.3.5 Additional Personal Protective Equipment

Additional protective clothing may be required for some types of procedures or with specific substances, such as when carcinogens or large quantities of corrosives, oxidizing agents or organic solvents are handled. This clothing may include impermeable aprons and gloves as well as plastic coated coveralls, shoe covers and arm sleeves. Protective sleeves are prudent when wearing an apron. These are either washable or disposable. The choice depends on the degree of protection required and the areas of the body which may become contaminated. Rubberized aprons, plastic coated coveralls and arm sleeves offer much greater resistance to permeation by chemicals than laboratory coats and, therefore, provide additional time to react.

For all work where contamination is possible, seal all openings in the clothing. Tape works well for this purpose. Wear head covering to protect hair and scalp from contamination.

Wear chemical-resistant gloves whenever the potential for contact with corrosive or toxic substance and substance of unknown toxicity exists (e.g., nitrile or rubber gloves). Select gloves on the basis of the materials being handled, the particular hazard involved and their suitability for the work to be done. Refer to the MSDS for specific gloves required. See Appendix B for examples. An important consideration in glove selection is "grip." Glassware is especially slippery when wet, and broken glass is a major source of injury. Latex lab gloves do not provide the needed adhesion and should not be used. Before each use, check gloves for integrity.

Replace non-disposable gloves periodically, depending on frequency of use and their resistance to the substance handled. Dishwashing gloves are available in the glassware kitchen. Check the stock room if there are none available.

Note: If you suspect any of your protective gear has become contaminated, remove it immediately, bag it, and wash all exposed skin surfaces thoroughly. Refer to the MSDS sheet for any additional measures. Call the CHO to collect the contaminated gear.

2.3.6 Eye Protection

Eye protection is required for personnel and visitors in all laboratories. Wear safety glasses with side shields, goggles or face shields in the laboratory based upon the physical state, the operation or the level of toxicity of the chemical used. Wear goggles in situations where bulk quantities of chemicals are handled and chemical splashes to the face are possible. When handling highly reactive substances or large quantities of hazardous chemicals, corrosive, poisons and hot chemicals, wear goggles with a face shield.

Caution -- Use Correct PPE for *In Vivo* Laboratory Work
INS 00034 gives more specific requirements for PPE relative to *in vivo* work in specific laboratories.

2.3.7 Eyewashes and Safety Showers

All laboratories where bulk quantities of hazardous chemicals are handled and could contact the eyes or skin resulting in injury have access to eye wash stations and safety showers. The SO does the following to make sure these are accessible at all times:

- Keeps all passageways to the eyewash and shower clear of any obstacle (even a temporarily parked chemical cart)
- Checks eyewashes weekly to be certain that water flows freely
- Checks showers monthly to ensure that access is not restricted and that the start chain is within reach
- Tests the flow through the safety showers annually to ensure sufficient flow (approximately 60 gallons per minute)

2.3.8 Fire Safety Equipment

Fire safety equipment easily accessible to the laboratory includes a fire extinguisher. In addition, the chemistry laboratory has a Class D fire extinguisher. The BSO makes sure that these additional extinguishers are in place and operative.

2.4 Food and Drink in the Laboratory

The Company does not permit food or drink in the laboratories. You may not store or dispose of food or drink items in the laboratories. Smoking or applying cosmetics in laboratories is not allowed.

2.5 Housekeeping

Safety results from good housekeeping practices. Use the following guidelines to maintain an orderly laboratory:

- Keep work areas clean and uncluttered with chemicals and equipment.
- Clean up work areas upon completion of an operation or at the end of each work day, including floors.
- Dispose of wastes according to SOP 00004 Hazardous Chemical Waste, SOP 00005 Regulated Medical Waste, and SOP 00006 Radioactive Materials.
- Discard non-contaminated glass in the designated waste receptacle.
- Clean spills immediately and thoroughly, as per the Emergency Plan.
- Do not block exits, emergency equipment and controls, and do not use hallways and stairways as storage areas.
- Assure hazardous chemicals are properly segregated into compatible categories. Appendix C lists Chemistry Division Codes of Practice.

2.6 Chemical Handling and Storage

Use chemicals with caution according to the MSDS for the substance. Assume that all substances of unknown toxicity are toxic and handle accordingly. Always use the minimum amount of chemical required for your work, and avoid retaining chemicals at your workstation longer than necessary to complete your tasks.

2.7 Transferring of Chemicals

When transporting chemicals to and from the laboratory, avoid dropping or spilling chemicals. Carry glass containers in specially designed bottle carriers or a leak resistant, unbreakable secondary container of sufficient size to contain the entire volume in the event of a spill. Transport chemicals on a cart; use a cart that is suitable for the load, has high edges to contain leaks or spills, and large wheels capable of going over grooves such as at the elevator entrance.

2.8 Signs

Laboratories that use hazardous materials have visibly posted signs by the external doorway to the laboratory. These contain contact numbers for knowledgeable personnel in case of an emergency. These numbers are always current; the PSO updates them when contact numbers change.

2.9 Unattended Hazardous Operations

If it is necessary to leave work unattended, complete an Unattended Hazardous Operation Form, which is available from the laboratory supervisor or CHO. Post the form in the vicinity of the unattended work, such as on the bench or hood. Make sure it does not interfere with the normal operation of the equipment. Complete the following information.

- Your name
- Your home telephone number or a number where you can be reached
- The date(s) your equipment will be operating unattended
- Identification of the instrument or experiment
- Reagents used
- Solvents used
- Check any services used
- Check any hazard which may apply
- Indicate action(s) to be taken in the event of an emergency (i.e., turn off power and water source)

Select a link for the form and instructions below.

General Unattended Hazardous Operations Form
Chemistry Unattended Hazardous Operations Form Instructions

You may also obtain copies of the form from the Chemical Hygiene Officer.

Standard Operating Procedures, Instructions, and Methods

SOPs, instructions, and methods are difficult to write because they must ensure consistent results every time an activity occurs, but at the same time they must not be so restrictive as to cause problems in performance and outcome. Some companies have a wide range of SOPs and do not distinguish laboratory methods and work instructions from SOPs; others make the distinction. Whatever system is in place where you work is, of course, the one to adhere to.

From company to company, you will most likely find fair commonality in what goes into standard procedures, whether they are SOPs, methods, or instructions. For instance, there is usually an SOP that explains the documentation system. This document typically includes initiating, writing, routing, and revising documents, as well as the criteria that must be met for approval and issue of the new document and retrieval of any previous versions, and final archiving. In addition, the system includes a history of each SOP, from the first version to the current one, whether the system is manual or electronic.

Components of SOP Instructions and Methods

A solid SOP, instruction, or method usually includes many of the following elements or a combination of them, but, depending on the company, may exclude some and include others. These documents reflect the needs of the organization itself. Templates should be flexible enough to accommodate variables. For instance, if a document does not require a section such as "Definitions," the template can adapt to omit that component.

- Title
- Purpose/Objective
- Scope
- Objective
- Responsibilities
- Equipment
- Frequency
- Caution
- Warning
- Procedure steps
- Calculations
- References
- Documentation
- Appendices

Changing Environments

There are many variables in procedures. While some companies list all related documents, others don't because these can be associated electronically. Consider for instance a cleaning SOP; a related document is the one that tells how to prepare the United States Department of Agriculture (USDA)–approved cleaning agent. Does the cleaning SOP need to reference the related SOP? Is it sufficient to say the solution is prepared according to an approved procedure? Or is it clear by a document log that these two are associated?

While this is not a book on electronic record keeping, some issues may be worth mentioning here, since procedures are often signed manually. If the documents are controlled in a closed system and associated documents carry the approval signatures and historical information, the procedure itself doesn't have to carry this information. Progressive companies have determined that to move toward electronic record keeping they do best to eliminate from the documents themselves conditions that can make the transition troublesome. They understand that documents with handwritten signatures will require scanning and Quality Control accuracy verification to move into an electronic system. A signature-free electronic document can go into a system electronically, provided there are proper controls in place.

Suffice it to say that the conventions for documents are changing because technology is. Companies seeking to move to electronic record keeping need to first determine the requirements for validation, audit trails, security, and accountability as delineated in 21 CFR Part 11 *Electronic Records; Electronic Signatures*. For healthcare providers, the binding regulations are part of the *Health Insurance Portability and Accountability Act*, 45 CFR Parts 160, 162, and 164, which address electronic records. They can then move toward transitioning their legacy documents into the system.

Range of Policies, SOPs, Instructions, and Methods

The Code of Federal Regulations (CFR) both states explicitly and implies what procedures a company should have in place. In addition, companies must stay abreast of shifts in requirements and industry standards. If a company manufacturing a product similar to another company's learns that the competitor has been cited for a procedural violation, that should be the signal for the company to look at its practices and documentation and adjust them as necessary.

A good procedure system has documents that interrelate with each other and cover all facets of a company's operations. Thus a procedure for receipt and quarantine of incoming components will work in tandem with an SOP on Quality Assurance (QA) sampling for testing, and subsequently with a document detailing laboratory testing of the components.

Procedures cover a webwork of activities, from personnel responsibilities and training, to equipment and facility maintenance. They detail processes involved in the receipt, sampling, laboratory testing, and acceptance or rejection of components, drug product containers, and closures. They explain how to manufacture and package products. They tell how to handle out-of-specification (OOS) laboratory results or manufacturing deviations. They detail how to control changes to processes, facilities, or equipment. They define labeling practices and materials reconciliation. Procedures outline holding and distribution activities and cover such activities as maintaining complaint files and handling field alerts and product recalls, as well as returns and salvaged products. Procedures also explain how to operate and clean equipment. In sum, SOPs provide the pieces that make up the big picture of how a company runs. No list will adequately delineate every procedure a company must have because the operations of each company will vary.

Many companies actually put procedures in place that are not specifically required for their type of operation. Consider, for example, a non–Good Laboratory Practices (GLP) discovery company that puts an SOP for Notebook Management and a corresponding instruction for preparing laboratory notebooks in place. For the company it makes good sense because these documents help protect intellectual properties. The same argument holds for putting a disclosure document in place. Who can publish and what's the process? Why risk revealing patent information, or worse, making an unsubstantiated claim about a product in development. The goal is to have procedures in place that cover all facets of the operation and all eventualities to ensure good business activities.

FDA Has Its Say

Want to avoid a warning letter? Say what you do and do what you say. The following excerpts from warning letters all point to deficiencies in documentation.

> "Written sanitizing procedures do not address the action to be taken."

> "[There is a] failure to establish and follow appropriate written procedures."

> "The analysis procedure does not contain a specification for an acceptable retest on out-of-specification results for stability retesting."

"The quality control unit was found to be deficient in assuring that written procedures are being followed."

"Some written procedures are out-of-date or obsolete."

"Some laboratory operations had no documented, approved procedures in place."

"[There is a] failure to establish and/or follow adequate written procedures for production and process control."

["There is a] failure to establish and maintain procedures for implementing corrective and preventive maintenance."

www.fda.gov/foi/warning_letters

Top Level SOPs

Top level SOPs, which some companies call policies, tell how a major system operates or a major process occurs. Top level SOPs generally impact many people. An SOP on the management of the company's documentation is such a document. It tells, in essence, what happens when the company determines to create a document, how that creation happens, who reviews, and who approves. It also tells how certain documents are distributed, tracked, and retrieved. Such a top level document may be all inclusive, or it may have companion instruction documents, such as those for document numbering, maintaining central records, and document archiving. Second tier documents, which many people call work instructions or simply instructions, usually provide information or describe procedures that one or two people can carry out. How the company sets up the documents is an individual choice. The important factor is that all significant activities are covered.

The following SOP explains a documentation control system. Related documents include a reference document of minimum required signatures that tells who signs what documents, the forms necessary to issue and retrieve documents from manuals and sites, and an instruction explaining the company's use of electronic signatures.

Standard Operating Procedure	SOP 00023 1
E-Document MaxSystem	

Effective date: 00 Mon 00

1.0 OBJECTIVE

1.1 This document describes the procedure for generating, reviewing, revising, approving, issuing, and retrieving documents in the E-Document MaxSystem.

2.0 SCOPE

2.1 This procedure applies to all areas of the company engaged in the development, manufacture, and marketing of the Company's products.

3.0 DEFINITIONS

3.1 Active document. A document that has received final signatory approval and satisfied training requirements, if any.

3.2 Approvers. Persons who give final approval to a document. Approvers are the designated signatories by document category for controlled documents. REF 00001 Minimum Required Signatures identifies approvers for types of documents.

3.3 Authors. Persons who prepare actual documents. Generally there is one primary author, but some documents may have multiple authors.

3.4 Disabled document. A document that is no longer active.

3.5 Document Reviewers. Persons who review a document for content and correctness. Document reviewers always include document approvers, but may include any number of other persons who are impacted by the document.

3.6 Effective date. A date issued to documents such as standard operating procedures, methods, instructions, and forms. The effective date signifies that all requirements have been met to make the document active.

3.7 Meta data. The historical record of each document, including the concept approval, the review and approval, posting to the hierarchy, superseding of versions, and disabling.

3.8 Not Issued. A designation for a proposed document that received a number in the E-MaxSystem, but was subsequently not completed and the concept voided.

3.9 Procedure manuals. Binders to which approved SOPs, instructions, and methods are issued. These are assigned to specific areas such as laboratories and are by controlled by numbers and a log in the E-MaxSystem.

Standard Operating Procedure	SOP 00023 1
E-Document MaxSystem	

Effective date: 00 Mon 00

3.10 Superseded document. A former version of a currently active document.

3.11 System Administrators. The persons responsible for overall management of the E-Maxsystem. Systems Administrators import documents into the system, post them to the hierarchy, and disable them as necessary.

3.12 Watermarks. The watermark on printed documents that reads "Simons Laboratories document: valid only on print date" or "Simons Laboratories document: controlled copy."

4.0 MATERIALS AND EQUIPMENT

4.1 SOP Manuals

4.2 Yellow paper for controlled printed document copies that reside in controlled manuals.

5.0 RESPONSIBILITIES

5.1 System Administrators issue numbers; bring documents into the system, and post them to the hierarchy; import and export documents to and from the system; issue documents to and retrieve them from manuals, sites and contract organizations; maintain binders.

5.2 System Administrators train new users and disable user access rights as appropriate.

5.3 Human Resources notifies System Administrators of new hire start dates and termination.

5.4 Authors generate and revise documents and arrange any relevant training.

5.5 Reviewers evaluate documents from their areas of expertise and provide content and mechanical critiques.

5.6 Approvers provide final e-signatures.

6.0 PROCEDURE

System Users

6.1 System Administrators train new users in the system according to *SOP 00014 Training.*

Standard Operating Procedure	SOP 00023 1
E-Document MaxSystem	

Effective date: 00 Mon 00

6.2 Employees must undergo training in system use, security, and accountability before they may use the system. They must demonstrate their ability to use the system before they are enabled to do so.

Note: REF 00022 is a flowchart of the E-Document MaxSystem use process.

6.3 System Administrators train users and verify that they are able to use the system. New users sign User Accountability statements.

6.4 All users have two passwords known only to them; one allows access to the system, the other serves as an electronic signature.

6.5 System Administrators enable users to access the system and disable users who are transferred or leave the employ of the company.

6.6 In the event a user forgets a password, the System Administrator may provide a temporary log on password. Once logged on, the user changes the password.

6.7 All users receive notification to change passwords at three month intervals.

Document Generation

6.8 When a new document is needed, the primary author reviews the document hierarchy and documents in progress to see if a comparable document is in the system.

6.8.1 If a comparable document exists, the author examines a copy of that document to determine if it can be modified to accommodate the new information.

6.9 When a new document or a revision to an existing document is needed, the primary author advises a System Administrator.

6.10 The System Administrator enters the information into the e-system which either assigns a new number or a subsequent version to an existing number.

6.10.1 For revisions, the current version remains active while the new version undergoes revision and review.

6.10.2 System Administrators export documents for revision to the authors.

6.10.3 The author downloads a document template for the new document.

Standard Operating Procedure	SOP 00023 1
E-Document MaxSystem	

Effective date: 00 Mon 00

6.11 The author drafts the new document, using the appropriate format/template, or revises the existing one, and subjects it to preliminary peer review for content and correctness.

6.12 The author forwards new or revised documents to a System Administrator and includes a list of reviewers.

6.13 The System Administrator imports the document, and it goes into the review cycle.

Review, Revision, and Approval

6.14 All new documents and revised documents go into review.

6.15 Periodic review is mandatory for certain documents, such as SOPs, instructions, methods, and associated forms. The system flags such documents that have reached a two-year active cycle, and a System Administrator puts them into review.

6.15.1 System Administrators may confer with original authors to determine proper reviewers.

6.16 SOPs, instructions, methods, forms, and select other documents that are no longer required for operations may be put into review for the purpose of document retirement.

6.17 Off-site signatories may access the Company's Intranet E-Compliance MaxSystem site to review documents.

6.18 Reviewers examine the document, make comments as appropriate on the document, and approve or reject the document.

6.18.1 Upon occasion, reviewers may determine the document is unnecessary, and document coordinator logs records the document as "not issued" in the document number log.

6.19 The owner or author evaluates comments on approved documents and incorporates them as appropriate.

6.19.1 If a reviewer rejects a document, the author may opt to discuss proposed changes with the reviewer(s).

Standard Operating Procedure	SOP 00023 1
E-Document MaxSystem	

Effective date: 00 Mon 00

6.19.2 Revisions require a new draft to go into review. The cycle continues until reviewers agree that the document is sound.

6.20 When reviewers sign off on a document, it goes into the approval cycle.

6.21 Approvers or their alternates sign off on the document.

6.21.1 If the document delineates a process that requires training, a System Administrator prints training copies of the documents, and the document author arranges training.

6.21.2 Training occurs according to SOP 00014 Training.

6.22 When all the requirements for the document have been met, a System Administrator posts the document to the hierarchy where it receives an effective date.

Note: All coordinate documents referenced in the posted document are electronically "associated" to it in the hierarchy.

6.23 Previously approved manually signed documents undergo Scanning according to INS 00016 Document Scanning, and receive a QC check. These are then brought into the MaxSystem with two System Administrator Signatures as legacy documents and posted to the hierarchy.

6.23.1 The hardcopy goes to the Company's document archive.

6.24 Electronic documents prepared by offsite system users are brought into the E-Document Maxsystem on the seat on the Company's website.

Hard Copy Issue, Retrieval, and Retention

6.25 System users may print out copies of documents. These print out with a watermark bearing the print date, time, and this phrase: "Valid only on print date."

6.26 The document coordinator posts new and revised forms electronically to the E-Document MaxSystem hierarchy when they are approved and removes them when they are retired or superseded. Posted forms and reference documents are "read only." Authors requiring forms can "save as" with a new document name, which allows data entry.

Standard Operating Procedure	SOP 00023 1
E-Document MaxSystem	

Effective date: 00 Mon 00

6.27 The document coordinator prepares an Issue and Retrieval form for document copies disseminated to procedure manuals, clinical sites, or contractors. These may include, but are not limited to SOPs, methods, and instructions, Material Safety Data Sheets (MSDS), and select reports.

6.28 A System Administrator updates the Document Distribution feature and enters data on documents distributed to manuals, clinical sites, or contractors.

6.29 A System Administrator prints out yellow copies of controlled hard copies for manuals and forwards PDF electronic copies to clinical sites and contractors.

6.30 A System Administrator prints out a distribution report for Document Manuals and confirms placement of the new or superseded document in the manual and destruction of any previous versions.

6.31 Sites and contractors retain the issued documents according to approved procedures and confirm destruction of previous versions, if any.

6.32 System Administrators make meta data available for all documents, as required, such as for a system audit or an inspection.

7.0 REFERENCES

7.1 21 CFR Part 11 *Electronic Records; Electronic Signatures*

Electronic Record Keeping

Electronic document management systems require an infrastructure of documents. Documentation begins when the need for a system is first identified. A company needs to have documentation for the following activities in place to make the transition. If they don't exist, it's time to create them. They must undergo review and approval in the existing system.

Facilities Security. Building security is the first line of defense for electronic records. An SOP or policy must address facility access for employees, access denial for separated employees, visitor and contractor access, and how to restore lost access ability.

Network Security. This is the second line of defense for electronic records. This document needs to address network user access to accounts, password requirements, mandatory password change intervals, screen saver password use, remote access, internet access, and virus protection.

Computer System Back Up. Electronic records on networked servers, individual workstations, and parts of systems must have back up. Back up should allow data recovery with no more than 24 hours of data lost. This document needs to tell what is and what is not backed up, how back ups are managed, and how media labeling, reuse, and destruction occurs.

Data Archiving. Archiving means copying original data to another medium. Archiving makes more storage space available. These data must be retrievable for inspections in years to come. This document tells how the company archives its data.

Computer System Event Recording. This document tells how modifications to the system are tracked. Working with this document should be logs that create an audit trail for the system. Typically each server and each system user has an event logbook as well to record the date and description of the event, an explanation of the testing, tester, and date of completion.

Computer System Change Control. This document tells how changes to the system are managed. Each application in the system must have a log to record modifications. Each application may itself have a system-specific change control procedure. The change control procedure covers request initiation and approval, modification testing and approval, and request closure.

Computer System Disaster Recovery. This document tells how systems become operable again after the loss of hardware, software, or data. Disaster recovery may require equipment or assistance from vendors.

Electronic Signatures. This document satisfies the requirements of 21 CFR Part 11, *Electronic Records; Electronic Signatures; Final Rule,* which calls for a written policy that holds users accountable for their electronic signatures and that states that electronic signatures are the equivalent of handwritten signatures.

Training. Training is required for going electronic. It is also required in the predicate rules, and most companies have a training policy or SOP in place. Nevertheless, it needs review, and possibly adjustment, to make sure it is compatible with the needs for training system users. Forms work in tandem with the training SOP to create training records.

Electronic Record Retention. Electronic records require the same set of retention times as do paper records, and this document needs to spell them out.

Computer System Validation. This document tells how commercial off the shelf (COTS) software and all the system components undergo validation, a requirement of 21 CFR Part 11.

Control of Electronic Mail. This document needs to spell out its policy for e-mail back up, archiving, and retention.

Computer Software Procurement. This document tells how the company evaluates potential vendors to select the optimum product to meet the end users' requirements.

Vendor Auditing. This is a quality assurance document that identifies how the company examines the software development life cycle (SDLC) practices, regulatory experience, vendor history, and product history.

Systems Inventory Management. Any new COTS system must be added to the company's inventory. This document tells how that occurs.

In addition to these documents, the actual electronic document management system needs a complement of documents to support it.

Document Management. An overall document that explains what the system does and who manages it needs to be in place.

Additional documents. Additional documents may include the "how to" for any number of peripheral activities. If the system interacts with another system, such as regulatory submission software, there must be a document that tells how this happens. And depending on how the company sets up its system, it may require additional documents that tell who signs what record document issue and retrieval.

Accountability Statements. The company may secure hand signed statements from users acknowledging their understanding that electronic signatures are legally binding.

HR Handbook and HR Policies. A new system may impact such diverse documentation as the company's handbook if most of the company's employees use the system. HR may need to alert system administrators of new hires needing training and advise about separated employees so that their ability to access the system can be terminated.

Letter to FDA. Usually within 30 days of a system going live, the company notifies FDA in writing that it is employing electronic signatures.

Validation Documents. When a company brings in an electronic record keeping system, it needs to create a validation packet of documents, from user requirements through system release. And whenever the system is upgraded and there are additional validation activities, there must be documentation to attest to those activities. (See Chapter Seven for a list of validation deliverables.)

An Operations Procedure and Training

The following SOP delineates a key system for a company engaged in discovery. Employees of the company using radioactive materials (RAMs) must adhere to this procedure. Notice that the documentation section (8) lists the documentation that must occur as this procedure is carried out, from inventory logbook entry to notebook entries and more.

Immediately following this procedure is a PowerPoint slide training presentation that parallels the SOP. Note that training for the RAM SOP occurred before the document received its effective date. Once the training is confirmed and the SOP is effective, new hires and transferees receive training before they can carry out the procedure. The slide set is also used for update training. Of course, if the SOP undergoes revision, so does the slide set.

STANDARD OPERATING PROCEDURE

SOP 00006 1

Radioactive Materials

Effective date: 18 Aug 03

1.0 OBJECTIVE AND SCOPE

1.1 This procedure delineates the process for receiving, classifying, handling, using, packaging and disposing of low-level radioactive materials (LLRM) in a manner that is safe for personnel and in compliance with local, state and federal regulations.

2.0 DEFINITIONS

2.1 ChemTrak. A validated electronic database for tracking radioactive materials

2.2 Waste Broker. Any representative of the disposal company

3.0 MATERIALS AND EQUIPMENT

3.1 Wheeled carts

3.2 Alcohol swabs

3.3 Glass scintillation vials

3.4 Scintillation cocktail

3.5 Scintillation counter

3.6 USDA-approved decontamination spray

3.7 Calibrated survey

3.8 Waste containers; cardboard biohazard boxes; biohazard bags

3.9 Drums: steel, plastic, fiberboard

3.10 Plastic drum liners

3.11 Absorbent material, such as vermiculite

3.12 Socket wrench; crescent wrench

STANDARD OPERATING PROCEDURE

SOP 00006 1

Radioactive Materials

Effective date: 18 Aug 03

4.0 RESPONSIBILITIES

4.1 The Radiation Safety Officer (RSO) has overall responsibility for receiving, handling and preparing for disposal of radioactive materials.

4.2 Group leaders approve the requisition of radioligand compounds.

4.3 Researchers determine need for radioactive materials, order them for use at Synaptic and sign for incoming shipments. They dispose of waste or notify operations personnel to do so.

4.4 Shipping and receiving personnel or the RSO notifies researchers when radioactive materials arrive on site.

4.5 Purchasing personnel order radioligand compounds.

4.6 Operations personnel control waste packing inventory, maintain the low level radioactive waste (LLRW)/chemical waste storage room and assist in disposing of waste as required.

4.7 The waste broker provides paperwork and picks up drums for disposal.

5.0 SAFETY PRECAUTIONS

5.1 All radioactive materials require cautious handling. Persons handling such materials must wear protective gear, such as safety glasses, lab coats and disposable gloves.

5.2 Any physical contact with radioactive materials requires immediate attention. The Material Safety Data Sheet (MSDS) for the material delineates the appropriate action.

6.0 PROCEDURE

Radioactive Material Receipt

6.1 The Group Leader determines the need for a radioligand compound.

Synaptic | STANDARD OPERATING PROCEDURE
SOP 00006 1

Radioactive Materials

Effective date: 18 Aug 03

6.2 Researchers, at the direction of the Group Leader, complete the requisition form, secure the approval of the Group Leader and forward it to purchasing, and purchasing personnel order the radioligand compound.

6.3 When a shipment of radioactive material arrives at Synaptic, the RSO notifies the researcher who ordered the material to receive it.

6.4 Upon receipt, the RSO or the receiving party examines the package for integrity. It must be unopened and undamaged. The RSO verifies that the reference date is acceptable.

6.4.1 If a package is in questionable condition, the RSO uses a portable survey meter and scans the package. If the conditions are unacceptable, the RSO directs the return of the shipment to the supplier.

6.4.2 If there is significant contamination over 6600 disintegrations per minute (DPM), the RSO contacts the Nuclear Regulatory Commission (NRC) both by phone and in writing.

6.4.3 The RSO then notifies the Synaptic purchasing group and the supplier that the shipment has been rejected.

6.4.4 The RSO records the rejection in the Inventory logbook.

6.5 The RSO signs for an acceptable shipment and transfers it to the receiving area.

6.6 The material is transferred to the work area, where it is kept in isolation while the receiving researcher performs an initial wipe test by passing an alcohol swab in a "S" pattern over an area of approximately 100 cm^2 in three places: the package exterior, the outside inner container, and the vial itself. The wipe test must take place within a three-hour period.

6.6.1 In the rare event that the wipe test does not occur within the three-hour period, the radioligand compound is kept isolated until such time that the wipe test can be performed.

6.6.2 The reason for the delay is documented and retained with the file for the compound.

6.7 The researcher places the swab in a glass scintillation vial with approximately 7 mL of scintillation cocktail and puts the vial on the scintillation counter. The scintillation counter reads the amount of radiation present.

6.7.1 If there is significant contamination (over 6600 DPM), the researcher notifies the RSO who contacts the NRC both by phone and in writing.

6.7.2 Any contamination is eliminated with a USDA-approved decontamination spray.

6.8 The researcher logs the product, product quantity, number of containers and amount of radioactivity into the ChemTrak software. ChemTrak assigns a unique four-digit conference code number to the shipment and creates a printout.

6.9 The researcher verifies that the data on the printout is accurate, signs and dates the printout, and files it with the ChemTrak records for the shipment.

Radioactive Material Use

Note: The RSO ensures that the MSDS, if available, and the certificate of analysis (COA) for the compound are on site and accessible before releasing the compound for use.

6.10 The researcher places the material in the refrigerator or freezer as appropriate for the isotope.

Note: The refrigerator or freezer is maintained in a centralized, secure location.

6.11 As the material is used, the researcher logs the quantity used into ChemTrak, produces a printout and verifies, with a signature and the date,

STANDARD OPERATING PROCEDURE

SOP 00006 1

Radioactive Materials

Effective date: 18 Aug 03

that the data are accurate. All printouts are filed with the ChemTrak records for the shipment.

6.12 Researchers record the amount of material used experimentally in their laboratory notebooks.

6.13 After use of any radioactive material, researchers must conduct a wipe test survey for the entire laboratory. The RSO designates the area to be tested. Contamination in any one location over 200 DPM must be decontaminated and rewiped until DPM levels are acceptable.

Radioactive Waste Disposal

6.14 The RSO ensures that the waste packing inventory is adequate at all times. Inventory levels must not fall below a one-week supply.

6.15 Operations personnel maintain the radioactive waste area and prepare drums to accommodate waste.

6.15.1 Drums for liquid are plastic and rest on six inches of absorbent material within a steel drum.

6.15.2 Drums for dry waste are steel, and have two clear plastic liners.

6.15.3 Drums for Hydrogen 3 (tritium) are fiberboard, and have two clear plastic liners.

6.15.4 Drums for scintillation plate assays (SPA) and liquid scintillation vials (LSVs) are steel, with two clear liners and six inches of absorbent material in the inner bag.

6.16 The RSO attaches color-coded labels to the top of the drums, and indicates the opening date, closing date, isotope type and barrel number. The following are the color codes for isotopes in use at Synaptic.

- Beige – Phosphorus 33
- Light orange – Phosphorus 32
- Bright orange – Iodine 125

- Light Green – Sulfur 35
- Bright Green – Hydrogen 3

6.17 When it is time to dispose of low level radiation waste (LLRW), the researcher logs the waste information into ChemTrak, produces a printout, verifies the data, and signs and dates the printout. The printout goes into the files for the compound.

6.18 The researcher (or RSO, if requested) puts the waste into a container, places the container on a wheeled cart, and brings the LLRW to the LLRW/chemical waste storage room.

NOTE: The LLRW/chemical waste storage room has limited, controlled access.

6.19 Upon entry and exit of the LLRW/chemical waste storage room, log in on the Synaptic Waste Room log with the following information:

6.19.1 Date

6.19.2 Name

6.19.3 Time In

6.19.4 Time Out

6.19.5 Comments, if necessary.

Note: If there is any deviation from standard procedure, notify the CHO immediately.

6.19.6 If the condition of the room is unsatisfactory, personnel immediately contact the RSO for correction.

6.20 Personnel add the waste to a working drum labeled for the type of radioactive waste (liquid, solid, scintillation plate assay [SPA] or liquid scintillation vials [LSVs]).

6.20.1 If there is inadequate capacity, the RSO starts a new drum.

6.21 When a drum is full, the RSO seals it with a wrench.

Synaptic	STANDARD OPERATING PROCEDURE **SOP 00006 1**
	Radioactive Materials

Effective date: 18 Aug 03

6.22 Once the drum is sealed, the RSO performs a wipe test.

6.22.1 If contamination levels are unacceptable, the drum is sprayed with a USDA-approved decontaminant and the test redone until levels are acceptable.

6.23 The RSO removes the drum to a secure location to await shipment from the premises or to be placed in decay storage for handling according to the approved biohazardous waste procedure.

Note: Radioactive isotopes, with the exception of Tritium, can decay in storage, depending on the half-life, which is generally not more than 2 ½ years. When radioactive isotopes demonstrate decay, they can be disposed of as biohazardous waste and incinerated.

6.24 The RSO prints out the drum contents from ChemTrak, verifies the data, and signs and dates the printout. The RSO indicates the results of the wipe test and the drum contents and retains the paperwork in the file for the compound.

6.25 The RSO completes the LLRW Decay to Background Disposal Worksheet and retains it with the paperwork for the compound.

Waste Removal from the Premises

6.26 The RSO calls the waste broker for pick up as needed. Upon affirmation, the waste broker sends a waste profile form.

6.27 The RSO completes the form, identifying the isotope, the drum type and activity and sends it, via fax or mail, to the waste broker. The RSO retains a copy.

6.28 The supplier signs the NRC form and the waste broker brings it to the site, where the RSO signs it to release barrels.

6.29 Once on site, the waste broker labels the drums with the following information:

- Radioactive Waste
- Company name

- Company address
- NJDEP solid waste registration number
- Date of shipment
- Identification of the contents
- Weight

6.30 The RSO archives copies of the pick up paperwork.

7.0 REFERENCES

7.1 10 CFR 20 Standards for the Protection against Radiation (NRC)

7.2 NJAC 7:26 Hazardous Waste Regulation Program

7.3 40 CFR Part 260 Hazardous Waste Management System

7.4 MSDS sheets for radioactive isotopes

8.0 DOCUMENTATION

8.1 Inventory logbook

8.2 ChemTrak printouts

8.3 Laboratory notebooks

8.4 LLRW room log sheet

8.5 Waste Disposal logbook

8.6 NRC forms

8.7 Supplier correspondence

8.8 Drum labels

Courtesy of Halim Hasan

Managing
Radioactive Materials

Synaptic

SLIDE 1

Radioactive Materials

- Who
- What
- Why
- When
- How

Synaptic

SLIDE 2

Radioactive Materials: Who

- The Radiation Safety Officer (RSO) has overall responsibility
- Responsibility for adhering to this procedure rests with Synaptic personnel who handle and use Low Level Radioactive Materials (LLRM)

Synaptic

SLIDE 3

Radioactive Materials: What

- **Hazardous Chemicals**: A chemical for which there is statistically significant evidence based on at least one study conducted in accordance with established scientific principles that it may induce acute or chronic health effects in exposed employees

Synaptic

SLIDE 4

Radioactive Materials: What

- **Health Hazards:** Chemicals that are carcinogenic, toxic or highly toxic agents, reproductive toxins, irritants, corrosives, sensitizers, hepatotoxins, nephrotoxins, neurotoxins, agents acting on hematopoietic systems and agents that can damage lungs, skin, eyes or mucous membranes

Synaptic

SLIDE 5

Radioactive Materials: What

- Radioactive Materials
  - Phosphorous 33
  - Phosphorous 32
  - Iodine 125
  - Sulfur 35
  - Hydrogen 3
  - Carbon 14

Synaptic

SLIDE 6

Radioactive Materials: Why
• NJAC 7:26-3A New Jersey Regulated Medical Waste Program
• US Nuclear Regulatory Commission 10 CFR 19, 20 and 30
• 29 CFR OSHA Regulations
• US Department of Environmental Protection (US EPA)
• NJ Department of Environmental Protection and Energy (NJ DEPE)
• NJ Department of Health (NJ DOH)
Synaptic

SLIDE 7

• When it is ordered
• When it is received
• When it is used
• When it is disposed

SLIDE 8

SLIDE 9

SLIDE 10

Radioactive Materials: How
• Receiving LLRM
• Shipping and Receiving or the RSO notifies you to receive material
• Examine package
• Verify reference date is acceptable (RSO)
Synaptic

SLIDE 11

Radioactive Materials: How
• Receiving LLRM
• Scan package with portable survey meter (RSO)
▪ Unacceptable packages are returned to supplier (contamination over 6600 disintegrations per minute [DPM]; rejections logged into inventory logbook)
• Shipping and Receiving or the RSO signs for acceptable shipments
Synaptic

• Procuring LLRM

• Complete requisition form
• Secure approval of Group Leader
• Forward to purchasing

SLIDE 12

Radioactive Materials: How
• LLRM in the Work Area
• Acceptable shipments transferred
• Perform a wipe test within 3 hours*
▪ S-pattern over 100 cm2 area: exterior, outside inner container, actual vial
*If delayed, document and retain with file for the compound
Synaptic

SLIDE 13

Radioactive Materials: How
• LLRM in the Work Area
• Place swab in glass scintillation vial with 7 mL (approx.) scintillation cocktail and place on counter (to read amount of radiation)
▪ For significant contamination (over 6600 DPM) notify RSO
▪ Eliminate any contamination with USDA-approved decontamination spray
Synaptic

SLIDE 14

SLIDE 15

Radioactive Materials: How
• LLRM in the Work Area
• ChemTrak assigns unique four-digit conference code number
• Print out and verify data
• Submit records for shipment to RSO
Synaptic

SLIDE 16

Radioactive Materials: How
• LLRM Use
• MSDS, if available, and the Certificate of Analysis (COA) for the compound
• Place material in refrigerator or freezer
• Log quantity used into ChemTrak; produce printout and verify data
• Record amount of material used in experiments in laboratory notebooks
• Affix stickers to corresponding page in notebook
Synaptic

SLIDE 17

Radioactive Materials: How
• After use of any radioactive material, conduct a wipe test for the entire laboratory
• The RSO designates the area for testing
• Decontaminate any contamination in any one location over 200 DPM
Synaptic

SLIDE 18

Radioactive Materials: How

- Disposal in Drums
  - Liquids: plastic with six inches absorbent material in steel drum
  - Dry waste: steel with two clear plastic liners
  - Hydrogen 3: fiberboard with two clear plastic liners
  - Scintillation plate assays (SPA) and liquid scintillation vials (LSVs): steel with two clear liners and six inches of absorbent material in inner bag

Synaptic

SLIDE 19

Radioactive Materials: How

- Color Coding on Drums
  - Beige – Phosphorus 33
  - Light orange – Phosphorus 32
  - Bright orange – Iodine 125
  - Light green – Sulfur 35
  - Bright green – Hydrogen 3
  - White – Carbon 14

Synaptic

SLIDE 20

Radioactive Materials: How

- Disposal
  - Log waste information into ChemTrak and laboratory notebook
  - Log into chemical waste storage room; log in on Synaptic Waste Room log
    - Date
    - Name
    - Time In
    - Time Out
    - Comments

Synaptic

SLIDE 21

Radioactive Materials: How

- Disposal
  - Add waste to working drum (appropriately labeled)
    - Notify RSO if there is inadequate capacity
  - RSO completes the disposal and arranges removal from premises

Synaptic

SLIDE 22

Courtesy of Synaptic Pharmaceutical Corporation, a Lundbeck Company

Instructions

Instructions often work in tandem with SOPs that delineate larger systems. Instructions may provide the "how to" for a process that one or two people carry out, or they may simply give information. The following instruction presents the company's position on electronic signatures. It is a straightforward document and the company has it in place to satisfy the a requirement of Title 21 of the Code of Federal Regulations (CFR), Part 11. Note that the reference section of this simple document cites the regulation specifically.

INSTRUCTION	INS 0115	1
Electronic Signatures	21 Jul 05	

1.0 OBJECTIVE AND SCOPE

1.0 This document defines the acceptable use of electronic signatures. It applies to all departments within the company using the electronic data systems, such as LIMS and E-Records, which employ electronic signatures. Employees must be authorized to use the systems.

2.0 EQUIPMENT AND MATERIALS

4.1 Validated electronic systems

5 RESPONSIBILITIES

3.1 Company authors and reviewers using electronic data systems sign documents electronically.

4.0 PROCEDURE

4.1 An electronic signature consists of a user name and password that uniquely and securely identifies approvers.

4.2 An electronic signature is equivalent to a handwritten signature. All system users sign statements of accountability attesting to their understanding that their electronic signatures are legally binding.

4.2 Documents approved with electronic signatures are available electronically. These documents are available only on date of print.

5.0 REFERENCES

5.1 21 CFR Part 11 *Electronic Records; Electronic Signatures; Final Rule* Federal Register Vol. 62, No. 54, Thursday, March 20, 1997.

Simons Corporation CONFIDENTIAL Page 1 of 1

A QA Process

The following instruction explains the process whereby QA Inspectors clear a packaging line prior to a run. This activity always takes place after production personnel have prepared the line. The next step is for the line leader and then the QA Inspector to affirm that the line is ready to run. The result of this procedure is the completion of the QA Packaging Line Start-Up Clearance form, which becomes part of the batch record.

<table>
<tr><td>*Netgo Labs*</td><td colspan="2">STANDARD OPERATING PROCEDURE
INS 1234 2</td></tr>
<tr><td colspan="2">**Pre-Run Clearance of
Packaging Lines 1, 2, and 6**</td><td>**Effective: 15 Mar 05**</td></tr>
</table>

1.0 PURPOSE

1.1 To define the procedure for affirming correct staging of materials and cleanliness of packaging equipment and packaging lines prior to packaging line start up.

2.0 SCOPE

2.1 This procedure applies to automated packaging lines 1, 2, and 6 in the Jurgen Street Location.

3.0 RESPONSIBILITIES

3.1 The Director of Quality Assurance or designee holds overall responsibility for this procedure.

3.2 Production personnel notify Line Leaders and Quality Assurance Inspectors when a packaging line is clean and ready for use.

3.3 Line Leaders check the line prior to start up.

3.4 Quality Assurance Inspectors, as assigned, verify activities.

4.0 PROCEDURE

4.1 The line leader enters product description, product code, and lot number into the Line Log Book, and dates and signs it.

4.2 Upon notification by production personnel that the materials are staged and the line is ready for use, the line leader checks the materials against the Packaging Specification Sheet and completes the Packaging Line Start-Up Clearance form (Attachment A), signs and dates it, and gives it to the QA Inspector.

4.3 The QA Inspector checks the materials against the Packaging Specification Sheet and completes the QA section of the Packaging Line Start-Up Clearance form and signs and dates it.

4.4 If the line is not clear of foreign material or if the stated labeling is incomplete or incorrect, the QA Inspector notifies production to reclear the line or restage the labeling.

CONFIDENTIAL Page 1 of 2

Netgo Labs	STANDARD OPERATING PROCEDURE **INS 1234 2**
Pre-Run Clearance of Packaging Lines 1, 2, and 6	**Effective: 15 Mar 05**

4.5 Upon reclearing or restaging, the Line Leader and QA Inspector reinspect the line and complete and sign the Packaging Line Start-Up Clearance form.

4.6 The QA Inspector enters the start-up time into the Line Logbook and initials and dates it.

4.7 When the line receives QA clearance approval, the line leader initiates the run.

4.8 The Packaging Line Start-Up Clearance form remains with the Packaging Record.

5.0 DOCUMENTATION

5.1 Packaging Specification Sheet

5.2 Packaging Line Start-Up Clearance form FRM 01245

5.3 Line Log Book

Laboratory Methods

Laboratory methods are the result of a company's efforts to develop new products or test existing ones. Companies either rely on United States Pharmacopoeia (USP) methods, which they qualify for the development of their products, or they develop in-house methods.

A suitable in-house method may reflect many attempts at development. Often, indicators may be that a method will be acceptable, but efforts to validate it reveal failure to meet some of the criteria. Then it's back to the drawing board, and the test development begins again. Once an adjusted or completely revised method has been developed, the validation process begins once more. This process continues until validation is successful.

Often companies choose to develop in-house methods is because a suitable one may not be available in the USP. If a company does, in fact, develop an acceptable in-house method, the next edition of the USP may include a suitable method. When that happens, the company must evaluate the USP method against its own established method.

In a Nutshell: Laboratory Controls

General requirements

Specifications, standards, procedures, sampling plans, or other lab control mechanisms, including specification changes, must be established; they are subject to review by Quality Control. Laboratory controls include determination of conformance and calibration of instruments, apparatus, gauges, and recording devices at appropriate intervals.

Testing and release for distribution

There must be a procedure for sampling each batch for final specifications. The accuracy, sensitivity, specificity, and reproducibility must be established and documented. Reprocessing of rejects may be performed; reprocessed products must meet relevant criteria.

Stability testing

A written testing program must include sampling size, test intervals, reliable, meaningful tests, and storage conditions. Data maintenance must provide adequate number of batches for each drug product tested for expiration date determination.

Special testing

Written procedures for sterile and/or Pyrogen free products are required.

Reserve samples

Each lot of each shipment of active ingredient must be kept for at least 1 year after expiration date. OTC products must be kept for at least 3 years after distribution of last lot containing active ingredient. Double the quantity necessary for all tests must be kept on hand. Results must be recorded and maintained.

A General Test Method

The following method applies to many materials and products and represents a standard laboratory test procedure for determining Organoleptic properties. The Raw Material Analysis Report in Chapter 6 refers to this method. This interconnection shows how documents work together and form a paper trail of a product's history.

F & MG Laboratory Test Method	Number: 0001-A
Determining Organoleptic Properties	Supercedes: New
	Effective date: 02 Mar 05

1.0 SCOPE

1.0 This procedure delineates the method to determine and report the organoleptic properties of raw materials and finished dosage forms. The organoleptic tests determine color and appearance.

2.0 DESCRIPTION

2.1 Visual test

3.0 CHEMICALS, APPARATUS, REAGENTS

3.1 Glassine weighing paper

3.2 Microspatula

4.0 TEST

4.1 Color

4.1.1 Using the microspatula, place approximately 2 cm of sample or 1 dosage unit on a piece of glassine weighing paper.

4.1.2 Determine the color of the sample in specific terms. Refer to the laboratory color chart.

4.2 Appearance

4.2.1 Using the sample prepared in 4.1.1, visually examine the appearance. Compare color to the Laboratory Color Chart.

4.2.2. For raw material, determine the appearance of the sample in specific terms, such as granular, crystalline, or fine powder.

4.2.3 For finished dosage forms, determine the appearance of the sample in specific terms, such as round, oval, scored, bisected, or imprinted with FMG logo.

5.0 RESULTS

5.1 Consult with the supervisor when unsure about the color or appearance of the sample.

5.1.1 Record the results on the analysis report.

F & MG Laboratory Test Method	Number: 0001-A
Determining Organoleptic Properties	Supercedes: New
	Effective date: 02 Mar 05

5.0 DOCUMENTATION

5.1 Analysis Report

6.0 REFERENCES

7.1 Current USP

7.2 Laboratory Color Chart

Page 2 of 2

Courtesy of Monica Grimaldi, Certified Quality Engineer

A Product-Specific Test Method

The following laboratory method details how to test acetaminophen. Unlike the organoleptic method, which applies to many products, this document is product specific. Notice that in the organoleptic method, the calculations section has an N/A entry. In this method, however, the calculation is an integral part of the method.

F & MG Laboratory Test Method	Number: 0002-A
Acetaminophen Tablets, All Strengths	Supercedes: New
	Effective date: 10 Mar 05

1.0 SCOPE

1.0 This method describes the procedure for determining the purity of acetaminophen tablets, all strengths.

2.0 DESCRIPTION

2.0 Ultraviolet assay method

3.0 CHEMICALS, APPARATUS, REAGENTS

3.1 Methanol (UV Grade)
3.2 2 × 500 mL volumetric flask
3.3 100 mL volumetric flask
3.4 50 mL volumetric flask
3.5 UV spectrophotometer at 244 nm
3.6 1 cm cell
3.7 Analytical balance

4.0 TEST

4.1 Standard Preparation

4.1.1 Accurately weigh approximately 50 mg in-house reference standard and record the weight.

4.1.2 Place in a clean, dry 500 mL volumetric flask with 10 mL methanol and dissolve.

4.1.3 Dilute to volume with purified water and mix well.

4.1.4 Pipet 6.0 mL of this solution into a clean, dry 50 mL volumetric flask.

4.1.5 Dilute to volume with purified water and mix well.

4.2 Sample Preparation

4.2.1 Weigh approximately 120 mg of raw material and record the exact weight.

4.2.2 Place the raw material in a clean, dry 500 mL volumetric flask with 10 mL methanol and dissolve.

F & MG Laboratory Test Method	Number: 0002-A
Acetaminophen Tablets, All Strengths	Supercedes: New
	Effective date: 10 Mar 05

4.2.3 Dilute to volume with purified water and mix well.

4.2.4 Accurately pipet 5.0 mL of this solution into a clean, dry 100 mL volumetric flask.

4.2.5 Dilute to volume with purified water and mix well.

4.3 Procedure

4.3.1 Concomitantly determine the absorbances of the final solutions in 4.2.3 above with the final standard solution prepared in 4.1.4 in a 1 cm cell at a wavelength of 244 nm using purified water as a blank.

5. 0 CALCULATIONS

5.1 Calculate the purity of the sample according to the following equation:

$$\% = \frac{\text{Absorbance SAM}}{\text{Absorbance STD}} \times \frac{\text{Wt. STD}}{500} \times \frac{6}{50} \times \text{purity of STD} \times \frac{500}{\text{wt of SAM}} \times \frac{100}{5} \times 100 =$$

6.0 RESULTS

6.1 Record the results on the Raw Material Analysis Report.

6.2 Notify the supervisor of any results that do not meet specifications.

7.0 REFERENCES

7.1 Current USP

Courtesy of Monica Grimaldi, Certified Quality Engineer

Process Flow Charts

Presenting information visually helps people remember a process. While documents such as SOPs and methods may be clear in themselves, and people may understand them because they have undergone training before they carry out the process, a flow chart provides a handy confirmation of the actions. Such a flow chart may be a stand alone document or be included in the appendix of the document. The following flow chart is an adjunct to an SOP that delineates shipping from a biotech company to contract research organizations (CROs) conducting studies for the company. The purpose of the SOP is to maintain control of the company's compounds, and the flow chart provides an overview of what happens.

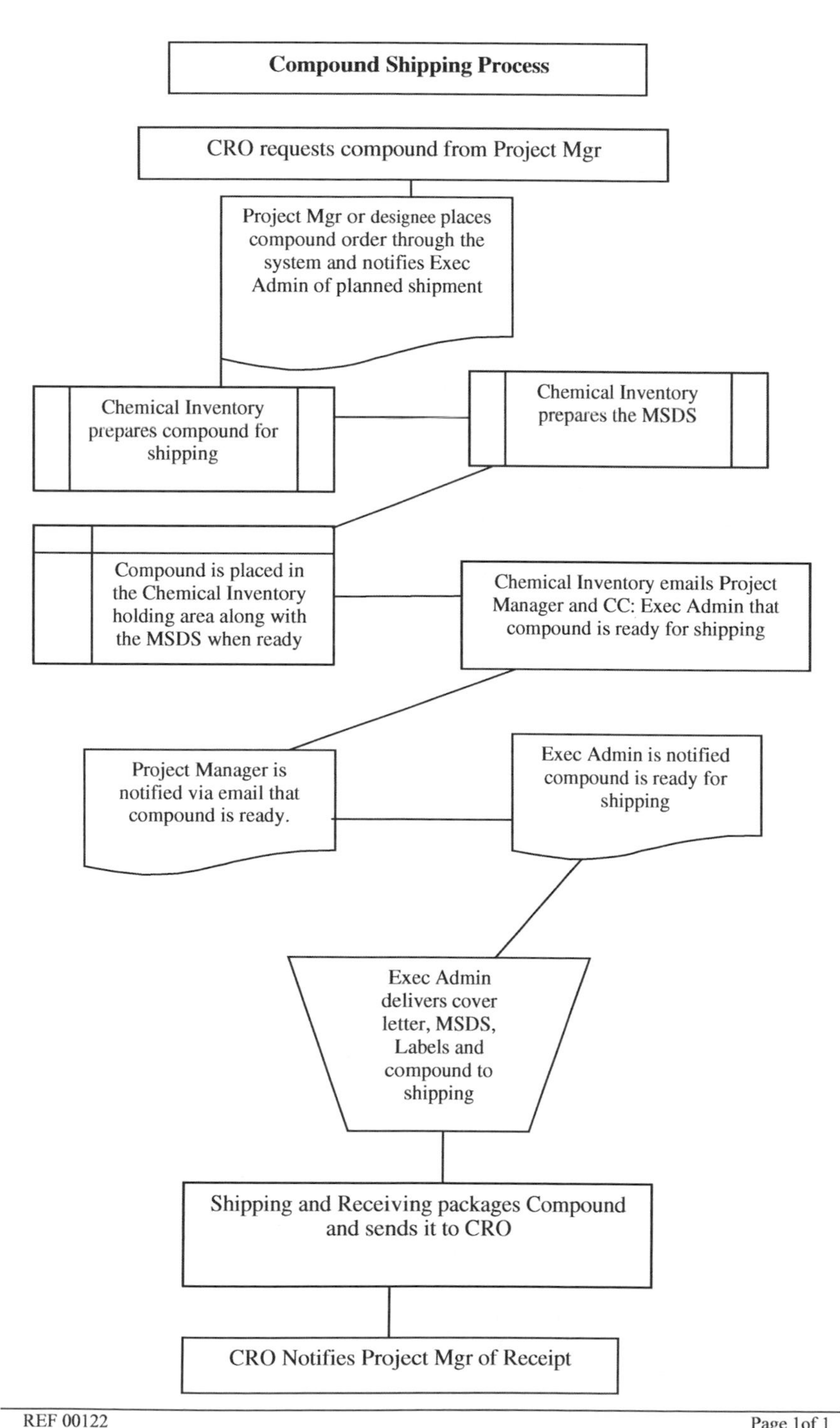
Compound Shipping Process
CRO requests compound from Project Mgr
Project Mgr or designee places compound order through the system and notifies Exec Admin of planned shipment
Chemical Inventory prepares compound for shipping
Chemical Inventory prepares the MSDS
Compound is placed in the Chemical Inventory holding area along with the MSDS when ready
Chemical Inventory emails Project Manager and CC: Exec Admin that compound is ready for shipping
Project Manager is notified via email that compound is ready.
Exec Admin is notified compound is ready for shipping
Exec Admin delivers cover letter, MSDS, Labels and compound to shipping
Shipping and Receiving packages Compound and sends it to CRO
CRO Notifies Project Mgr of Receipt
REF 00122
Page 1of 1

Courtesy of Synaptic Pharmaceutical Corporation, a Lundbeck Company

6

Routine Reporting

The purposes of reports vary, but historically mankind has relied on reports to make decisions, to progress, to accomplish goals. If you think about it, people look at the weather report to see what kind of day tomorrow will be. They read news reports to see what's going on in the world. They follow financial reports to determine the best places to put their money. Reports serve the same function in business. Reports tell readers what's going on so they can make knowledgeable decisions.

Much of the reporting that occurs in the pharmaceutical, device, and biologics industries is routine. The regulations require reporting, and even when they don't, it makes good business sense because it presents a history of activities and supports projects as they develop. Many routine reports ultimately feed into additional documentation as source data or as components of summary documents, often in appendices. A software vendor evaluation, for instance, may confirm that the product satisfies user requirements, or it may not. When it doesn't, it becomes part of the background for the quest for a software product that will prove satisfactory. When a vendor evaluation does confirm that a product is suitable for the company's needs, the document and the user requirements provide the foundation for purchasing a product, validating it, and releasing it to production. This is but one example of routine reporting that may feed into a larger process.

As another example, consider a laboratory notebook, a controlled vehicle for routinely capturing laboratory observations. Significant findings first recorded in a laboratory notebook may be the catalyst for discovery that leads, years later, to a product in the marketplace. In such a case, the initial notebook observation must be confirmed. Confirmation of the initial observation leads to further controlled experimentation and the development of a potential product. This in turn may lead to toxicology studies, which may occur at the company developing the product or by a contractor. When a contractor conducts studies, all the activities must occur according to approved procedures and be documented.

Successful studies may lead to an application for conducting trials in humans or animals, such as an Investigational Device Exemption (IDE) or an Investigational New Drug (IND). Upon approval, the investigational product enters clinical trials, and the documentation grows. Successful and well-documented clinical data then can lead to approval to market a product. Once approved, production can begin. For each drug batch or device, there's a record, so the product that goes into the market has a complete development history. Then the product is monitored in the market, and studies continue to ascertain efficacy throughout its expiry date. All the documentation that occurs — and it is massive — can be traced back to the initial observation in the laboratory notebook. The routine reporting that goes on through discovery, development, marketing, and expiry of the product is critical to the product's success.

> Each manufacturer shall establish procedures for identifying training needs and ensure that all personnel are trained to adequately perform their assigned responsibilities. Training shall be documented.
>
> **21 CFR Part 820.25(b)**

The compilation of all data concerning systems and activities relative to the company's intellectual properties are essential to the health of the company itself. The documentation overall must present the big picture of what's going on. The routine reporting is the nuts and bolts. Don't overlook either such records as training verification; these records are reports of activities that have occurred and serve to confirm that the people who carry out the day-to-day activities in an organization understand the significance of what they're doing and the need for accuracy and consistency.

Using Established Formats

It is increasingly popular to use electronic data capture. Inputting data electronically can lead people to think that the process is not really writing, but it is. Comments and observations that go into all reports, electronic or otherwise, are critical and can often spell the difference between seamless and effective activities and bumpy ones with fits and starts stemming from poorly recorded or missing data.

Well-thought-out formats for routine reports help make writing easier and consistent, whether the document is electronic or paper. Most companies have their own predesigned formats for a vast number of documents such as clinical site data collection, Material Safety Data Sheets (MSDS), certificates of analysis (COAs), laboratory methods, batch records, master formulas, deviations, procedures, audit reports, and more.

Standard formats require that the people writing the reports receive formal training in systems that create them as well as the satisfactory completion of the reports themselves. For example, if a company creates a new or changes an existing out-of-specification (OOS) report format, the people who will be writing them must receive retraining in completing the report within the system.

Creating Your Own Format

It may happen that there is no existing format for a report you set out to write. In such a case, you will need to make the data manageable yourself. Organize your report so that you have a clear and logical point or points at the beginning. Then present additional information in support of your point. (Chapter 3 discusses organizing and developing information.)

Routine Reports

Some routine reporting occurs at regular intervals. Most auditing, for instance, follows a schedule, but additional audits may occur when there are findings that require addressing. Some routine reporting occurs only when a specific activity happens, such as validation or revalidation or engaging a new vendor or contractor. In such events, part of the routine reporting may be testing or identification of user needs. Routine reporting is typically dictated by standard operating procedures (SOPs) or instructions, but may also be part of the company's best practices, as in writing minutes, agendas, and trip reports.

The goal in standardized reporting is accuracy and consistency. This is important since much routine reporting, as we've seen, feeds into future activities. Thus, when you are preparing a standardized report, deliver repeating information consistently. For instance, if you are preparing a stability report and the terminology in other stability reports reads, "There are no observable trends," don't write, "No observable trends are reported." The logic is this: Consistency makes reading related reports easier. Readers know what to expect and get what they expect.

Similarly, it's best to create a record of the information immediately because information committed to memory has a way of changing with time, even short lapses. Collecting patient data, for example, occurs best when the data go into a report on the spot, not a few hours later when the clinical research associate (CRA) is off site. Even if you're at a conference that you know you must report about upon your return, it's best to take notes as the event progresses. That way, you'll have the information at the ready when you do write it down.

> Complete every section of a standardized report. Leaving a section blank can be interpreted to mean you overlooked an element. Indicate that you have considered all elements, even though your report did not require one or two.
>
> **Monica Grimaldi, Certified Quality Engineer**

Following here are examples of routine reports typical of those in the pharmaceutical, device, and biologics industries.

Raw Materials Report

It's important that companies verify the quality of the raw materials they bring on site for manufacturing. If they don't, they risk serious consequences in product development or manufacturing if the raw materials are imperfect. Typically manufacturing sites accept raw materials and place them in quarantine. Then quality assurance (QA) brings a sample to quality control (QC) for testing. A QC analyst performs the testing, identifying the tests and attributes of the materials. If the materials meet established criteria, they are approved for use. Rejection means they cannot be used. QA then releases approved raw materials into production or notifies the supplier of a rejection.

The format for this report parallels formats of other documents within the company and gives the visual message that the company concerns itself with quality and consistency across the board.

F&MG RAW MATERIAL ANALYSIS REPORT

Product: Acetaminophen	**Product Code:** 456		**RM Number:** 12050
Manufacturer: Excelsior	**Mfg Number:** 42		**PO Number:** 11982
Amount Received: 2 kg	**Unit:** 16	**Testing Initiated:** July 14, 200-	
		Testing Complete: July 16, 200-	
Specification Number: 250-4	**Supercedes:** 250-3		**Effective:**
Prepared by: **Date:**	**Checked by:** **Date:**		**Authorized by:** **Date:**

Test	Method	Specification	Results	Analyst & Date	Checked & Date
Identification	A. USP <197K>	Sample exhibits maxima and minima at thc same wavelengths as the std.	Sample exhibits maxima and minima at the same wavelengths as the std.		
	B. USP <197U>	Sample exhibits maxima and minima at the same wavelengths as the std.	Sample exhibits maxima and minima at the same wavelengths as the std.		
	C. Thin Layer Chroma-tographic test USP <201>	Spots have the same R_f value and intensity as the std.	Spots have the same R_f value and intensity as the std.		
Appearance	Organoleptic TM 001-A	White powder or crystaline material	White powder		
Melting Range	USP <741>	Between 168! and 172!	Between 168! and 172!		
Water	USP <921> Method I	Not more than 0.5%	Not more than 0.5%		
Residue on Ignition	USP <281>	Not more than 0.1%	Not more than 0.1%		
Chloride	USP <221>	NMT 0.014%	NMT 0.014%		
Sulfate	USP <221>	NMT 0.02%	NMT 0.02%		
Readily Carbonizable Substances	USP <271>	Solution has no more color than Matching Fluid A	Solution has no more color than Matching Fluid A		

Approved: ☐ Rejected: ☐

Signature:____________________________ **Date:**____________________________

Form 07023 Page 1 of 1

Courtesy of Monica Grimaldi, Certified Quality Engineer

Certificate of Analysis

The following certificate of analysis shows repetition of the format of the previous raw materials report. Such consistency helps people who prepare these records to complete them efficiently and helps readers know where to look for what. Moreover, when a firm undergoes an audit or inspection, the consistency helps move the activities through to completion and sends the message that the company has its systems and related documentation under control.

F&MG CERTIFICATE OF ANALYSIS

Product: Acetaminophen	Product Code: 456	Unit: 16
Specification Number: 250-4	Supercedes: 250-3	Date:
Prepared by: Date:	Checked by: Date:	Authorized by: Date:

Test	Method	Specification	Results	
Identification	A. USP	Retention time of the major peak in the chromatogram of the Assay preparation corresponds to that of the standard preparation	Conforms	
	B. Thin Layer Chromatographic Test USP <201>	Spots have same R_f value and intensity as the standard	Spots have the same R_f value and intensity as the standard.	
Dissolution	USP <711>	NLT 80% (Q)	90.0 97.0 95.0 93.6 97.8 94.5 **AVE. 94.7**	
Uniformity of Dosage Units	USP <905>	85.0% - 115.0% RSD ! 6.0%	100.0 101.0 102.0 103.0 102.6 **AVE: 101.1** **%RSD : 1.44%**	100.9 101.4 102.7 100.3 97.8
Assay	Method 2361	90.0% - 110.0%	100.0%	

Approved: ☐ Rejected: ☐

Signature: ______________________ Date: ______________________

Quality Control Manager ______________________ Date: ______________________

Form 07016 Page 1 of 1

Courtesy of Monica Grimaldi, Certified Quality Engineer

Investigations

FDA is clear in what investigations must cover. Investigation reports always involve the input of more than one writer, and the writing is factual, not subjective. A document may stand alone, but just as often may trigger a domino effect when an initial investigation leads to an expanded investigation, which in turn leads to a discarded product batch, or worse, a field alert or recall. The cause of an OOS or a processing deviation may be as simple as a miscalculation in the laboratory (analyst error), and as complicated as a series of errors in production that have gone undetected.

> The palest ink is better than the strongest memory.
>
> **Chinese proverb**

Investigations, whether into an adverse event at a clinical site or a product that fails to meet stability, share common ground. They identify what happened, who did the investigation, how it proceeded, and the outcome.

Laboratory Investigation

The following investigation report template provides room to grow as the investigation demands. Here, individual components expand as data are added to each. The summary box is the last element to receive completion, since what it ultimately says depends on how far the investigation goes and what it yields. When the investigation is complete, the summary tells at a glance what the investigation covers and what it has found. The analyst completing the report begins by completing sections I and II. The OOS report calls for the concurrent completion of a form by a supervisor. (The form follows the investigation template.) This report may conclude at step V if the analyst and the supervisor can assign cause, such as analyst error, and retesting confirms the findings.

If, however, the laboratory cannot assign a cause to the OOS, the investigation receives a quality alert action number, and the investigation expands to include sampling and processing. When this occurs, the quality unit begins its investigation. Upon QA verification of the investigation findings, the laboratory may complete sections VI and VII.

When the laboratory component of the investigation is complete, after either step V or VII, the analyst or supervisor writes the summary. If the investigation has extended into processing, the laboratory investigation ultimately becomes an attachment to a quality unit report.

Pcatz Labs | **LABORATORY INVESTIGATION** | **NUMBER** 00000

SUMMARY

I. INVESTIGATION DATA

Product (incl. Strength):

Batch Number:

Code Number:

Type of Test:

Laboratory Test Site:

Laboratory Analyst:

Laboratory Supervisor:

Manufacturing Test Site:

II. INITIAL EVENT REQUIRING INVESTIGATION

Test:

Date:

Test Results:

Analyst:

IV. SUPERVISORY ASSESSMENT

Notebook Reference:

Instrument:

Checklist complete (Form 3667)

V. RETESTING

Reinjection of original solution:

Injection reserve:

Redilution from stock:

VI. CONCLUSION

FORM 3666c Page 1 of 2

Pcatz Labs

LABORATORY INVESTIGATION

NUMBER
00000

a. **Assignable cause/identify**

b. **No assignable cause**

QUALITY ALERT ACTION NUMBER:

V. RETEST

Test:
Date:
Test Results:
Analyst:

VI. TESTING REVIEW

Notebook Reference:
Instrument type and number:
Findings:

Analyst:____________________________________ **Date:**__________

Supervisor:____________________________________ **Date:**__________

Director of Analytic Development:________________________ **Date:**__________

Director of Quality Assurance:________________________ **Date:**__________

Process Investigations

A deviation means that a procedure critical to a process was unsatisfactory. It's never enough to suspect where a deviation has occurred; it's imperative to identify what happened and determine the impact on the product. And that is not sufficient in itself. The investigation must produce documents that identify the product, lot number, and batch; the course of the investigation; the findings; and the decisions reached as a result of those findings.

> Scientists find themselves at their desks more often than at the bench these days. In fact, laboratories have changed dramatically to accommodate the writing that gets done in the course of a workday. Automation has assumed much of the day-to-day bench work, but the amount of writing a scientist must do has increased dramatically. Says one research and development scientist, "When I determined to become a scientist, I didn't realize how much writing was involved. Now it seems that the real product of my laboratory is documentation. No matter how good the science is, if it's not committed to paper, it might as well not exist."

The following investigation report works in tandem with the previous laboratory investigation, if it is the result of a quality alert from an OOS result in the laboratory. The investigation can begin outside the laboratory, however, if the source of the deviation is known to be in sampling or processing, but it will most likely trigger laboratory testing to ensure the integrity of the batch. Like the laboratory investigation, this format permits the sections to expand as data are added.

The results of the investigation may be the rejection of a batch or its release. Determining the impact on related batches is a critical part of the investigation as well. If testing is on a product already in the marketplace, a field alert or recall may be in order. Whenever a batch has been identified as adulterated and, thus, rejected, testing needs to continue to determine why the failure occurred and what corrective action is necessary.

Pcatz Labs

PROCESS INVESTIGATION

NUMBER
0000

SUMMARY

I. INVESTIGATION DATA

Product (incl. strength):
Batch Number:
Code Number:

II. INITIAL EVENT REQUIRING INVESTIGATION

Laboratory OOS:

QUALITY ALERT ACTION NUMBER:

Other:

III. SAMPLING INVESTIGATION

IV. PROCESSING INVESTIGATION

V. CONCLUSION

Prepared by:______________________________ **Date:**__________

Vice President of Quality Assurance:______________________________ **Date:**__________

Director of Quality Assurance:______________________________ **Date:**__________

FORM 3668c

Page 1 of 1

Assessments and Evaluations

A good deal of the routine reporting that occurs is the result of assessment and evaluation. By self-assessment, companies determine what actions they need to take to maintain forward mobility. By auditing potential and existing vendors, suppliers, partners, clinical sites, and other entities, companies make sure that the services they contract out or purchases they make have integrity and suitably meet the company's specific requirements.

Identifying Requirements

When a company seeks to implement change, it must first assess what needs to happen. If a company wants to implement a new process that requires new equipment, for instance, it usually identifies the company's need for it according to the SOP for change control and then evaluates available equipment. Similarly, if a company seeks to purchase and install a Laboratory Information Management System (LIMS) or an electronic document management system, the process starts with a needs assessment and an identification of user requirements. User requirements typically identify what the system must provide to satisfy the needs of the users. It is the precursor to selecting a vendor. The following is an example of a user requirements document.

Requirements for Electronic Document Control	QA 1401 1

1.0 PURPOSE

1.1 This document delineates the requirements for an electronic document control system to meet the needs of intended users.

2.0 BACKGROUND

2.1 Netgo corporation is evaluating developers/vendors of commercial off-the-shelf (COTS) software to provide electronic document control of Netgo documents including those provided by contract research organizations (CROs) and clinical sites.

3.0 INTENDED USERS

3.1 Users are to be select full time Netgo employees, part time employees, consultants, and contract personnel involved in product development and product approval.

4.0 SYSTEM REQUIREMENTS

4.1 The system will ideally have undergone developer validation for compliance with 21 CFR Part 11 Electronic signature; electronic records and ISO 9000.

4.2 The system should be web-based so that remote, controlled access is possible at all times.

4.3 The system must allow for minimal redefinition of the Company's manual document control system.

4.4 The system must require minimum internal Information Technology (IT) management. The vendor must provide a readily-accessible help desk available during normal Netgo hours as well as system service when necessary.

4.5 The system must show evidence of the ability to grow with the Company and adapt to changing needs. That is, it should be compatible with software such as that for electronic document submission, scanning, and corrective and preventive action (CAPA).

4.6 The developer must provide a client list to demonstrate its ability to provide document control in companies similar to Netgo.

4.7 The system must provide proof of user accountability.

Requirements for Electronic Document Control	QA 1401 1

4.8 The system must be password-driven, with mandatory password changes at regular, pre-determined intervals.

4.9 The system must be secure and have a viable back up system.

4.10 The system must provide an audit trail of all documents, from inception to retirement.

4.11 The system must provide automatic document numbering compatible with the current Synaptic manual document system.

4.12 The system must allow for export of documents and control of exported documents. Ideally printed documents will say "Valid only on print date."

4.13 The system must allow incorporation of all legacy documents.

4.14 The system must be "user-friendly"; that is, it must be such that new users can readily learn the program with minimal training.

4.15 The system must allow for multiple roles, as necessary, such as reviewers, authors, system administrators.

4.16 The system must be cost effective; that it, it must demonstrate its ability to satisfy the user requirements without excessive initial cost, or update or maintenance fees over its life cycle.

The user requirement list is a tool for selecting a vendor and is the initial part of a validation. If several vendors are possible candidates, the list helps narrow the selection. Identification of a preferred vendor may trigger a reworking of the user requirements, particularly if the product has options that the company didn't know were available but would find useful. Contractual arrangements, purchase, planning, installation, setting functional specs, doing hazard analysis, testing, training, and release follow. But the impetus was the initial user requirements document. Once an electronic system goes into production, another piece of reporting must occur if the system employs electronic signatures: notification to FDA within 30 days of going live. Companies often use the exact wording that 21 CRF Part 11 includes.

Clinical Site Monitoring

When companies sponsor clinical trials, they must make sure that the clinical sites function in accordance with good manufacturing practices (GMPs). This they do by vigilance. A plan for assessing compliant activities calls for regular monitoring and accompanying documentation. The following template provides a way for sponsors to evaluate clinical sites while a trial is in progress. When an evaluation finds a situation that warrants improvement, the company has an identifier. Such an evaluation also reaffirms adherence to a protocol and serves as a marker for the forward motion of the study. The evaluations in total help present a picture of things as they really are. A company using a template like this would have companion documents for other activities such as initial site assessment and site closeout.

JSG Laboratories

VISIT DATA

Sponsor: Site monitor:	Protocol number: Visit date(s):

Investigator:	Site:

Investigator address:

Persons in attendance:	Title:

ENROLLMENT STATUS TO DATE

Number of subjects screened:	Number of subjects completed:
Number of subjects enrolled:	Number of subjects terminated early*:
Number of subjects active:	
	* Identify early termination(s) and reason(s) in the Narrative section.

Respond to questions in the following sections regarding site activity since the last site visit. Note that responses in fields marked with an asterisk (*) require explanation in the Narrative section.

GENERAL REVIEW

Yes	No*	N/A	Not chkd*	
☐	☐*	☐	☐*	Overall compliance with protocol/regulatory requirements
☐	☐*	☐	☐*	Staff and/or study facilities unchanged
☐	☐*	☐	☐*	Satisfactory enrollment rate

CRA signature/date

Sponsor signature/date

FRM 00001-1

Page 1 of 3

REGULATORY REVIEW

Yes	No*	N/A	Not chkd*	
☐	☐*	☐	☐*	Monitor Log signed
☐	☐*	☐	☐*	Progress reports submitted to IRB/IEC appropriately
☐	☐*	☐	☐*	Protocol changes submitted to IRB/IEC at appropriate times
☐	☐*	☐	☐*	All SAEs and/or safety letters on file
☐	☐*	☐	☐*	All SAEs and/or safety letters submitted to IRB/IEC
☐	☐*	☐	☐*	Lab certifications and normal ranges current
☐	☐*	☐	☐*	Study file currently reflects the status of the study
IRB/IEC re-approval due date:				

CRF AND CHART REVIEW

Yes	No*	N/A	Not chkd*	
☐	☐*	☐	☐*	New SAEs
☐	☐*	☐	☐*	Subjects' original source records available for review
☐	☐*	☐	☐*	CRFs consistent with source documents
☐	☐*	☐	☐*	Consent form signatures, dates, and documentation acceptable

LABORATORY/CLINICAL SUPPLIES REVIEW

Yes	No*	N/A	Not chkd*	
☐	☐*	☐	☐*	Lab specimens processed and shipped appropriately
☐	☐*	☐	☐*	Lab specimen storage conditions acceptable
☐	☐*	☐	☐*	Lab supplies adequate
☐	☐*	☐	☐*	Other clinical supplies and forms organized and adequate

NARRATIVE

GENERAL COMMENTS ☐ Not applicable

ENROLLMENT ☐ Not applicable

REGULATORY ☐ Not applicable
(Comment on changes with the study files and on any pending or outstanding items.)

CRF AND CHART REVIEW ☐ Not applicable
(At a minimum, identify the subjects reviewed at this visit and comment on the following: visits monitored, consent versions and dates signed using ddmmmyy format, SAEs, protocol deviations/violations, and early terminations.)

SAEs ☐ None noted

STUDY DRUG ACCOUNTABILITY ☐ Not applicable

CRA initials/date

Form 252

STUDY EQUIPMENT ACCOUNTABILITY	☐ Not applicable
LABORATORY SPECIMENS	☐ Not applicable
CLINICAL SUPPLIES	☐ Not applicable
STATUS OF PRIOR ACTION ITEMS	☐ Not applicable
CURRENT ACTION ITEMS	☐ None
ATTACHMENTS *(Do not attach CRFs or site documents.)*	☐ None
☐ Confirmation letter	☐ Follow-up letter

CRA initials/date

Audit Reports

Regular auditing is mandatory, and the reports that ensue are routine. The purpose of auditing is to make sure the area under audit is operating as it should, according to approved procedures. Audits may be internal or external. The company may look at its own operations or, externally, the operations of contractors or suppliers.

Laboratory Audit

Laboratory Audit

Date:____________ Auditor:________________________

Laboratory/Area:____________________ Audit time:______________________

Item	√	**Laboratory Hazard Communication** ***OSHA 1910.1450***	Comments
1		Laboratory Hazard Communication extends to all laboratory personnel, including chemists, technicians, engineers, and custodians	
2		The Chemical Hygiene Officer holds periodic meetings; these meetings are documented	
3		The Chemical Hygiene Plan is controlled and up to date. The Safety Committee reviews the plan periodically.	
4		The Chemical Hygiene Plan identifies the following: • Laboratory training protocol • Safety Committee and Chemical Hygiene Officer • Procedures for working with hazardous chemicals • Criteria for implementation of control procedures • Specific measures for performance of fume hoods and PPE • Prior approval requirements for special lab operations • Provisions for medical consultations/exams • Provisions for work with "particularly hazardous substances" • Provisions for work with "select carcinogens"	
5		Training records are current and complete.	
6		Special required training conducted for any lab employee working with a chemical covered by a specific OSHA Standard.	
7		Chemical inventory complete with MSDS available for each chemical	
8		Procedures/methods/instructions accessible to laboratory personnel	
9		Employees use equipment properly.	
10		Employees are knowledgeable in emergency procedures.	

11		Equipment calibrated/serviced according to approved procedures.	
12		Emergency Plan/exit instructions accessible	
13		There is documented review to identify presence of OSHA Select Carcinogens, highly toxic/hazardous chemicals, and so forth.	
14		Employees knowledgeable in medical contact	
15		Medical supplies	
16		Exposure assessments, monitoring, and/or medical assessments of potential laboratory exposure documented and available to employees and their representatives. Monitoring results reported within 15 days	
17		Chemicals are properly segregated/stored:___ oxidizers segregated, ___proper flammable storage, ___refrigeration,___ gas cylinders chained/capped when not in use,___ bonding & grounding during flammables dispensing, ___no large containers stored overhead,__ ventilated storage for high toxicity/carcinogens, ___special circumstances noted (perchoric acid, picric acid, peroxides and peroxide as formers, hydrofluoric acid, etc.-flame arrestosrs on safety cans.	
18		Laboratory waste disposal practices documented.	
19		_____ of_____ lab/chemical/sample containers properly labeled.	
20		Documented assessment of laboratory hood function is conducted per Chemical Hygiene Plan requirements by both lab and IH personnel.	
21		Routine housekeeping/safety checks made in laboratory • Eye wash/shower operation • Fire extinguisher location and suitability PPE availability/condition(face shields, aprons, shields) • Spill clean up materials • Old/unlabeled materials/samples disposed of routinely • Area chemical monitors/alarms • Emergency respirators • Clutter/tight working spaces	

Additional Items*			

*You may also include additional items or comments as signed and dated attachments.

Action items

____
____
____
____
____
____
____
____

Auditor Signature:__________________________ Date:__________________

Vendor Audit

The following is a report of a vendor's software development life cycle application. Before a company purchases software it must evaluate the suitability of the vendor. The following evaluation is a preliminary assessment, one that is objective in tone, states the facts, and makes a recommendation based on those facts.

Netgo Vendor Qualification Audit Report

Auditee	1-2-3 Imaging Corporation 123 Anywhere Street Anycity, CA 98020
Product/Service	Imaging Database Application
Netgo Audit Team	Astrid Reynolds, Lead Auditor Mark Wright, Quality Assurance Carol Simonsen, Information Systems
Date of Audit	02 Sep 2004
Date of Report Issue	12 Sep 2004

Table of Contents

Summary

Netgo performed a site audit at 1-2-3 Imaging Corporation in Anycity California on Jan 2, 2004. This was a qualification audit to assess the vendor's software development life cycle (SDLC) practices and 21 CFR Part 11 compliance in the development of the Image Database application. This software application tool allows for the recall of images and their related measurements for review by FDA. Audit results indicate that the vendor is capable of providing an acceptable off-the-shelf application.

This report summarizes the general aspects of vendor's capabilities, and the audit observations as a result of the facility tour and documents reviewed. The detailed audit findings are organized in the sequence of the areas audited.

Audit Team

The following comprised the Netgo audit team:

Carol Simonsen, Information Systems (IT)
Mark Wright, Quality Assurance (QA)
Astrid Reynolds, Lead Auditor.

The following vendor staff facilitated the audit tour and documentation review:

Kristine Ogozalek, Audit Host
Timothy Osterhaus, Quality Control
Martin Chow, Quality Assurance

Overview

The vendor has over 200 employees and offices located in the USA Europe, and Asia.

The vendor's Quality Assurance Unit (QAU) consists of two areas: Quality Assurance and Software Quality Control. The Quality Assurance group is responsible for auditing, document management, and project audits. The Software Quality Control group is responsible for product testing.

The vendor receives audits by clients and FDA. In June 2004 FDA audited the vendor and issued three 483 citations. A review of these citations revealed them to be minor, and the vendor has adequately addressed them.

We took facility tour of two key areas: the Technology Laboratory and the Testing Laboratory. The Technology lab is where the software is developed. The Testing lab is where the software is tested and released. The tour ended with a demonstration of the Imaging Database application.

We reviewed the Imaging Database application Version 3.2 validation documentation. The vendor engaged a consulting firm to assist in validation. Imaging Database application Version 3.3 validation is now in progress.

Netgo Corporation *Page 3 of 4*

We reviewed training records for quality, core developers, application developers, and testing staff. Training records were unavailable for staff not located at the Anycity facility. For most staff, the training record summaries did not include evidence that training on all required SOPs was performed. However, upon request, the vendor provided the training records that indicated training on SOPs not included on the training record summaries. The vendor acknowledged that this is an evolving system and needs some additional process controls.

Many of the vendor's newly issued SOPs were found to be very comprehensive; however, there was not much evidence available to be able to evaluate effectiveness. New leadership in the vendor's QAU in recent years has resulted in improvements to their quality system.

Recommendation

The audit team recommends that any contracts with this vendor include a provision for the review of the completed Imaging Database application Version 3.3 validation documentation.

Audit Observations

1. The process for managing software defects and enhancements in production for both core and application-level software is not clearly documented in the change control procedure.

2. The SDLC SOPs are deficient in the following areas: source code control procedure, product version scheme, design review, build notes, release notes, hazard analysis (risk assessment), and file accountability. The vendor is currently reviewing and revising these SOPs.

3. Training record summary reports do not reflect all SOP training that has taken place, making it difficult to ascertain that individuals have met established training requirements.

4. There is no procedure to instruct the use of the validation templates; furthermore, the validation template we reviewed was not change controlled.

5. The specification test case traceability matrix was not a deliverable in the Imaging Database application Version 3.2 validation project report – dated November 2002 that was conducted by a consultant; in addition, the Version 3.3 validation has progressed to testing without a traceability matrix.

6. Test cases observed from the Imaging Database application Version 3.2 validation documentation do not have documentation to indicate they were approved prior to execution.

7. The SOPs governing the application validation are not thorough and do not meet industry standards. These SOPs do not address user requirements, specifications, and test cases to include positive and negative scenarios.

Courtesy of David Nettleton, Software Validation Specialist

Meetings

> Institute a vigorous program of education and training. Eliminate fear and encourage open communication. Keep people abreast of new developments. Involve them in decision-making, and make them part of the team. These are the keys to successful operation, the ones that unleash potential and get results.
>
> **Gary Gough, Vice President, Human Resources**

Some routine, but important, reporting occurs as the result of meetings. This type of writing is akin to journal keeping. It tells, via agendas, what the plan is for a meeting, and then captures, via minutes, what happened at the meeting. These can be critical documents because they capture the plan for topics of discussion and the decisions reached as a result. They are often the first records of thoughts and ideas that launch significant projects.

Agendas

The best results generally come from preplanning; rarely are they serendipitous. Good plans usually involve the cooperation of a number of parties, and good communication, both oral and written, is the means to a desirable end.

> There are many ways of going forward, but only one way of standing still.
>
> **Franklin D. Roosevelt**

Good agendas keep people focused and on track. They also require sequencing. What is the purpose? What is to be addressed first? What comes next? When does it conclude? And agendas that work address time restraints. The following agenda is for a meeting to address a 483 received by the firm. A predetermined plan will allow the group to address each issue. The group leader has summarized the citations and broken them down into units: documentation, cleaning validation, and training. This agenda includes some questions the attendees will address. By their nature agendas cannot anticipate every issue that will arise. Often the agenda serves as a catalyst for broaching new ideas. At the actual meeting, notations on white board, flip chart, or computer will capture solutions as the group arrives at them.

JSG Laboratories

Page 1

AGENDA

November 30, 2004
9:00 a.m. to 11:00 a.m.
Chemistry Conference Room

Executive Conference Room Group Leader: Yale Jen

A. Key points of the 483

1. Failure to note two manufacturing deviations (no impact on related batches, but not properly documented).

2. Written procedure for operating the filling line not accessible in the area.

3. Lack of cleaning validations for batching rooms 14 and 16.

4. SOP for cleaning rooms 14 and 16 inadequate; they do not call for QA supervisory signatures.

5. Packaging personnel improperly dressed, in violation of the SOP. Two operators noted wearing loose jewelry.

6. Training records do not show periodic GMP refresher training.

B. Corrective Action(s)

1. Documentation

Immediate action:
Make SOPs available to operators.
Revise documents in question.

Long-term action:
Assessment of SOPs to see if the citation applies to other documents.

Discussion:
Who will do it?
When will drafted documents be ready for review?

JSG Laboratories ***Page 2***

---

2. **Cleaning Validation**

Immediate action:
Assess validation requirements.
Complete the validation process.

Long-term action:
Assessment of all cleaning processes and validation records
Establish permanent validation team.

Discussion:
Who will be on the validation team?
Who will be the validation lead?
Who will write the protocol?
When can validation take place?
What are the deliverables?
How long will it take?
What will be the impact on production?

3. **Training**

Immediate action:
Retraining for dress code.

Long-term action:
Review of the training program and recommendations for revision.
What will be needed to revamp the training program? (Staffing, support services, etc.

4. **Additional Issues:** Discussion

Minutes

Many people find themselves taking minutes in the course of their careers. It's unwise to think that administrative staff will always take minutes. It's not uncommon, for instance, for executive committees to share the note taking efforts, rotating who compiles the minutes each time the group meets.

If you're asked to serve as a transcriber at a staff meeting, you'll be responsible for recording an official record — the minutes — of what transpires. Minutes are distributed to committee members and other related parties to refresh their memories about what occurred at the meeting and to provide a history of verbal encounter. This means the opinions of the minutes-taker do not figure into the text, unless they represent opinions voiced at the meeting itself. The following minutes record the events at a quality assurance departmental meeting, and the primary issue is the status of validation efforts for an electronic document control system. QA meeting minutes will record progress of the validation as it progresses and help keep the project on target.

Spitzbergen Products Inc. ***Memorandum*** *Page 1 of 2*

Date: April 15, 2005

To: QA Staff

From: Li Lonxing

Subject: Minutes of the April 13, 2005 QA Staff Meeting

Attenders: Susan Clark, Vincenzo DiPaoli, Jose Figueroa, Valeri Gorbechek, Li Longxing, Lata Miganta, Su Li Chan, Dina San Juan

Topic: Electronic Documentation System Validation Update

1. Jose Figueroa called the meeting to order on Tuesday, 13 April, at 9:30 a.m. in the Quality Assurance Conference Room.

2. The minutes of the 30 March meeting were unanimously approved.

3. Su Li Chan, the validation Lead, reported that system testing for validation of the electronic document control system is complete.

4. All SOPs for the system and facilities have been approved in the manual system. These will be brought into the new system as legacy documents as soon as the system goes live.

5. Valeri Gorbechek asked about the schedule for bringing all legacy documents into the system. Chan said there is a comprehensive plan, and system administrators will bring the documents into the system as scanned documents. These documents require two signatures. A QC check will occur on each document by a third party prior to document importation into the system.

6. What remains now is general user training and user accountability verification. Chan said she and Vincenzo DiPaoli have undergone training on the system, as has the two system administrators. Once training is complete the system will be released into production (go live), scheduled for May 15.

7. Validation itself will most likely be complete thirty days after the go live date when the system is assessed and the final report prepared.

8. Valeri Gorbechek will schedule initial training in the system, beginning with QA staff; training sessions will take place in the QA Conference Room, the Manufacturing Conference Room, and the Human Resources Training Room.

9. Su Li Chan and Vincenzo DiPaoli have put together a comprehensive system of overheads for the training sessions and will serve as in-house trainers.

Spitzbergen Products Inc. ***Memorandum*** *Page 2 of 2*

10. Hands-on training will then occur at individual workstations; this is to confirm that system users can log in to and use the system.

Other Business

11. Susan Clark reported that the investigation of the out-of-spec results for batch 6579 was not attributable to analyst error. An expanded investigation showed that it was caused by an unreported spill in the weigh room prior to batching. The operator on duty is undergoing retraining and the batch has been discarded.

12. Dina San Juan reported a violation of dress code in the filling area. Despite the dress code, she has observed several operators wearing jewelry and has spoken with them.

13. Jose Figueroa reported that the new QA auditor will begin work in another week. He will undergo GMP training for the first three days, then work under Lata Miganta's direction for the next month.

14. The meeting adjorned at 11:10 a.m.

Trip Reports

Whenever you attend a seminar or trade show, you may be required to complete a trip report. Sometimes trip reports are simply prepared like questionnaires — the writer fills in the data as directed by a report form. Just as often, however, a trip report, like the following one, will take a memo format, because it offers the writer flexibility in delivery. Notice how the headings break the report into readable segments.

KIM LABORATORIES, *S. A.* **Memorandum** Page 1 of 2

TO: Gao Kui
FROM: Andrea Cox
SUBJECT: Trip Report – Pharmaceutical Exposition and Conference
DATE: January 27, 2005
CC: Roger Chung

Overview

At the Pharmaceutical Exposition and Conference on September 22, 23, and 24, I served as cohost with Shirley Liu and Gene Yang at our exhibit, and I attended several seminars as well. Our exhibit, dedicated to the marketing of the contract manufacturing services we provide, was well-visited, and we made some excellent contacts, which Shirley will pursue.

Below are brief descriptions of several of the more informative seminars I attended. I have handouts from each if anyone is interested. In addition, my expense report to obtain reimbursement for my travel expenses is attached.

Therapy for Products and Processes: Coping with Failure

This session was a discussion of out-of-specification (OOS) investigations. Topics covered included the importance of conducting OOS investigations as quickly as possible and including all of the proper elements in the investigation. The presenters stated that investigations initiate within 24 hours of a testing failure and QA should be informed of investigations within 10 days. Much of the discussion was related to the Barr Decision.

Sterile Processing Operations: Current Formulation, Scale-Up, and Compliance Issues

The most interesting segment of this session was a presentation given by an FDA representative, Austina Tremont. Her presentation concentrated on sterility assurance, the number one concern for the FDA with products produced through aseptic processing. She discussed the importance of properly designed media fills, appropriate interventions, adequate process controls, and a rigorous environmental monitoring program. The elements of an appropriate Sterility Assurance Program are deemed more important than negative sterility results from end product testing. Toward the end of her presentation, she mentioned the impending issuance of an updated draft to the Guideline on Sterile Drug Products Produced by Aseptic Processing.

KIM LABORATORIES, *S. A.* **Memorandum** Page 2 of 2

Planning for and Managing FDA Inspections

Former FDA District Director Darren Chestnut presented this session. He provided excellent insight into the mindset of inspectors responsible for conducting FDA investigations. The session covered all aspects of an FDA inspection, including preparing for one and responding to any adverse findings. The depth of the material covered was impressive. It was enlightening to hear the material from the FDA's viewpoint.

Auditing for Drug Products, APIs, and Excipients for cGMP Compliance

As a QA Auditor, I found this seminar particularly informative. The session covered several types of audits, including those of contract testing laboratories, manufacturers of APIs, manufacturers of excipients, and related businesses. Several very useful hints were presented that can be applied to audits of any business. The scope of the session was extensive, and the material will prove to be useful in the performance of my job. Again, I have detailed handouts from these and other sessions and will make them available to whoever wishes to see them. Certainly, I will discuss these seminars at our next QA meeting.

Courtesy of Michael Nolan

7

Process Reports

> The secret of getting ahead is getting started. The secret of getting started is breaking your complex overwhelming tasks into small manageable tasks, and then starting on the first one.
>
> **Mark Twain, Author**

Process reports cover a wide scope. In fact, the argument can be made that all writing about therapeutic products is a form of process reporting. Process reports tell how something happens, has happened, will happen, or can happen. Most process writing explains the need and criteria for certain action and moves logically through a chronological sequence of steps. A protocol for a study of a product in development describes a process that has not yet happened. Amendments to the protocol show changes in the study process. And the final report tells what has occurred and what the outcome is.

For a clinical trial, a protocol tells what is going to happen. Like nonclinical study reports, trial protocols can be amended to reflect changes in the process. Among myriad documents required for compliant trials is informed consent for patients and donors, if they are part of the process. Informed consent tells patients and donors what will happen during the trial and what the possible outcomes are. When trials are complete, assessment reports describe what has happened and what the outcome actually was.

Other process documents, such as feasibility and probability reports, assess proposed actions or the outcome of future actions. When the activity is complete, an assessment report evaluates what has happened and what the end result was. The results of reports like these are tools companies use to make informed decisions.

Many process documents tell how things happen repeatedly. Chemical hygiene plans, animal husbandry plans, and emergency plans are but some process documents that tell how things happen repeatedly in a controlled manner so that the results are always the same. Product manuals tell how to use a product or operate a device. Quality manuals tell how things occur overall, from top management's responsibilities to the regulations that drive the company's operations. Standard operating procedures (SOPs), methods, and instructions are "how to" documents as well. These documents are the most common ones companies produce, since they are all subject to periodic review and revision. Chapter Five discusses these types of documents in detail.

Other process reports call upon information gleaned during routine reporting. A report on a toxicology study for instance will tell why and how the study was conducted, and the findings will be based on the data collected during the study itself. As another example, the initial qualification of the installation and operation of a filling line will require a set number of data collection forms to be completed during testing. The data in turn are the criteria that determine that the filling line meets qualification standards. If all or part of that equipment later undergoes modification, requalification is necessary, and also perhaps the generation of additional testing forms.

Many "how to" process documents call for the generation and/or completion of additional documents such as forms for recording data or verifying testing. And some process reports are components in a series that delineate a larger process. Each step in a validation process, for instance, requires documentation that is subject to review and approval. In total, all the documentation becomes the validation packet for the system.

Guidelines for Process Writing

Process writing has long been recognized as one of the most difficult forms of delivering information. Writers have difficulty on several counts. (1) They know a process so well that they omit the obvious. (2) They don't know a process well enough and omit the details they don't fully comprehend. (3) They include so much detail that the document is incomprehensible and they lose sight of the purpose. Or (4) they simply can't control the language to deliver information clearly.

Why are you writing? What is the reason for the document? Is this a one-time summation of completed activities? Is it a proposal or protocol for future action? Is it one document in a series? You must first decide the purpose of the report.

Next, determine what format you're going to use. Will a letter or memo suffice? Is a more detailed report required? Does the information require a standard format? Once you've determined these preliminaries and gathered your facts, you're ready to produce the document.

Most reports are a series of related components that fit together to support a single purpose. If you keep the overall purpose in mind, you can address each segment individually. You don't have to "begin at the beginning and go to the end" like Alice in Wonderland. You can begin writing anywhere.

If you know you're going to have to produce a report at the end of a project, for instance, you may be able to write some sections as the project takes place. Then, when it's time to produce the finished document, you can drop certain information in place, fine tune it, and make it work as part of the overall document.

In process writing, you must control the tenses you use. If you are writing about something that has occurred, such as a study, the past tense is appropriate. However, when a report has a finding, that finding does best in the present tense. "The results of this study showed that the product effectively inactivated six strains of bacteria" is the past tense. Look what happens when the tense changes to the present. "The results of this study show that the product has effectively inactivated six strains of bacteria." The reason for the present tense here is that the report is signed and dated, and this information is true at this point. That is not to say future studies will find otherwise.

For detailing activities that will occur in future time, the future tense is appropriate. In such documents you are writing about what will happen.

And, for activities that occur repeatedly, the present tense is best. Here's a caveat: Even though the regulations are written in the future ("There shall be records...."), it's not the best tense to use. If the process you are writing about requires data collection or activity verification, the present tense works best. And if the process includes procedure steps, these do well in the present tense. (See Chapter 11 for more information about controlling verb tenses.)

Supporting Details

Many process reports are lengthy. Some validations, for instance, can run more than 100 pages. Such documents require close attention, because it's easy to get bogged down in detail and lose sight of the purpose. Data that do not serve to support a purpose do not belong in the report.

Reports may incorporate information about what has happened, the current status, and even give recommendations for future action. As you organize information for your document, you'll most likely see a pattern: Some facts and figures may serve to support larger points, and so forth. Once that happens, it's useful to define the key point or points and explore elements that enhance that point. Some writers find it useful to keep folders of information that supports each major point. That way, data won't overlap or repeat unnecessarily, and the sequence of the report will be clear. The order of delivery, of course, will depend on the type of report that's in the works, but process reports always introduce, develop, and conclude in some manner. And they rely on data compiled over time.

Proposals and Protocols

Companies must plan activities before they occur. Business plans tell about activities for the next fiscal year. Project plans tell how specific activities are to occur. Proposals suggest a plan. And protocols delineate intended action. What these documents have in common is that they are defining activities that have not yet occurred. Once the activities are underway, the plans may be modified to address issues unforeseen at the time of the plan's approval and implementation.

Proposals

A proposal lays out a suggestion or a plan for future action. The purpose of a proposal is to persuade the reader(s) to act, and proposals are either solicited or unsolicited. A solicited proposal is one that is written in response to a request. When a problem comes to light, for example, a director or manager may call for several proposals that can satisfy a need. Each solicited proposal will offer a means.

Consider a company preparing to engage a contractor to carry out a study. The preliminaries for entering into such an arrangement may include soliciting a proposal from one or more contractors. A proposal will spell out what the company needs to know in terms of site, compliance with the regulations, method of conducting the study, cost, and time frame, among other things. A proposal becomes a yardstick for measuring suitability.

An unsolicited proposal springs from a perceived need and the desire to draw attention to it and effect a resolution. For instance, a department head may determine she needs to increase her staff, but realizes that her budget is insufficient. So, she may then write a proposal, suggesting a new hire to be shared between her department and another. In a proposal such as this, the writer delineates the plan and the advantages to both departments.

Short proposals can be in letter or memo form; longer ones work well in standard report formats. All proposals, however, identify a need and related issues, give concrete and specific information, and have a limited scope. A proposal that works answers questions like these:

- Why do we need this action?
- Why should we accept this proposal?
- Why is this the best plan?
- How will it work?
- How will it affect other systems?

A Training Proposal

> If you fail to plan, you plan to fail.
>
> **Anonymous**

The following proposal logically argues for improving training activities and record keeping. The audience is the Human Resource Director of a small, newly formed company, so the writer has included an excerpt from the CFR to reaffirm the message that training and record keeping are critical in this industry. A report like this could easily have been prepared as a memo, but the writer here elected to include a cover sheet; doing so gives the information a more formal presentation.

CONDUCTING SOP TRAINING AND TRACKING AND MAINTAINING RECORDS

A Proposal for Kim Laboratories, SA
Submitted to Erin Horton
By Christian Mitchell

July 17, 2004

CONDUCTING SOP TRAINING AND TRACKING AND MAINTAINING RECORDS

Introduction

In the highly regulated environment that is the pharmaceutical industry, training is imperative. The 21 CFR is implicit in its mandate:

> Each person engaged in the manufacture, processing, packing, or holding of a drug product shall have education, training, and experience, or any combination thereof, to enable that person to perform the assigned functions. Training shall be in the particular operations that the employee performs and in current good manufacturing practice (including the current good manufacturing practice regulations in this chapter and written procedures required by these regulations [SOPs] as they relate to the employee's functions). Training in current good manufacturing practice shall be conducted by qualified individuals on a continuing basis and with sufficient frequency to assure that employees remain familiar with cGMP requirements applicable to them. 211.25 (a).

For us at Kim Labs, it means we must take training seriously. The issue of training verification has been raised, particularly in relation to SOPs. In short, how do we confirm that training has taken place before a new or revised SOP becomes effective? Two SOPs right now have been approved and await effective dates, yet training is complete. This gap reflects lack of communication between activities. To better coordinate the conjoint SOP issuance and training efforts, I believe it would be in our best interests to establish a formal training program. The program needn't be complicated. Training does go on regularly, but record keeping is not centralized, and often knowledge of training does not reach beyond the area in which it occurs. What follows is a simple plan for ensuring that all required training takes place and that SOP training is verified before SOPs are issued. It also encompasses all cGMP training, and provides ready access to all employees' training histories.

A Centralized Database

A centralized database can house all completed training records. It can be set up by employee name, and information of training entered chronologically. Thus a warehouse employee may have safety training for lifting on Thursday, July 11, and forklift instruction on July 17. A history of each employee's training then becomes readily available. Training in specific SOPs, OSHA, GLP, cGMPs, and other required topics should be included, as should training that takes place off site, such as special seminars and conferences. The database does not need to be limited to those employees subject to regulated practices, but can serve as a record of the training histories of all personnel. Thus, a secretary learning new computer skills would build a training history in the system as well.

Database Location

The database can reside in Human Resources. The logic is threefold.

(1) Human Resources retains employee records, and training records are part of an employee's history; the records are useful in evaluating performance and effort, and in determining which employee should be considered for further training, such as seminars, and the like.

(2) Training records provide proof that employees are trained to do what their job descriptions require. The records will also prove useful in updating job descriptions when necessary.

(3) Personnel needing training information can request it from one centralized location.

Training and Documentation

When training takes place, it should be recorded on a standardized form. Kim Labs has several training logs, all different. The standardized form should include the Topic of Training, including an SOP number if the training is for that purpose, the date, the location, the trainer, the names of the trainees, and their signatures as verification that they comprehend the subject matter. A special "comments" section can be included also.

If the training is in accordance with an SOP, a form verifying training has been completed should be sent to Regulatory Affairs, so that an effective date can be issued. A copy should be made and retained in the area where the training took place. And the original should be forwarded to Human Resources, where it will be kept secure. The information will be logged into the database from the originals. It's necessary to keep the originals; FDA has been known to ask for training records.

Materials and Personnel to Maintain the System

A Human Resource administrative assistant can maintain the database. Once established, record keeping becomes a simple procedure. Training will continue on a regular basis, except that two forms will be developed and used:

- A new, standardized training form.
- An electronic training verification form for submission to Quality Assurance. This form will be issued to the trainer when an SOP has been signed off on (on original document paper), but has not received an effective date. This is Quality Assurance's signal to assign effective dates to, and issue, new or revised SOPs.

Time Frame and Interim Practice

Training will continue as it is currently being done. Copies of training logs that relate to SOPs will be forwarded to Regulatory Affairs as verification. Within the month, we can issue a new SOP (a simple amendment to a document now in circulation) that details the system. We can develop the two standardized forms, and train against the SOP as the database is established. We should be able to accomplish this endeavor by the end of July.

Conclusion

I believe it's imperative that documentation systems work in conjunction with each other, but that none be impeded by the strictures of another system. Establishing a centralized database to house training records thus makes sense. Further, minimal effort will be required to maintain it once it is established.

Protocols

A protocol is a plan. It tells how a study, either clinical or nonclinical, is going to occur. A protocol is a roadmap for the people conducting the study, and it is a controlled document that is part of the study file. The protocol has clear objectives and drives the study from start to finish. More important, the protocol helps ensure that the study is meeting regulatory requirements.

Before a clinical study can begin, the protocol must be written and approved by the sponsor and then reviewed and approved by an Institutional Review Board (IRB). When a nonclinical study takes place, it follows the protocol approved by the sponsor and study director.

Note: Sometimes companies use the term protocol to identify certain procedures and activities. So documents that function as SOPs may carry titles such as "Stability Protocol" or "Protocol for Receiving Inspections."(See Chapter 5 for guidelines for preparing procedures.)

A protocol must make sure that the regulatory requirements are met. For clinical protocols, that means adherence to Good Clinical Practices delineated in Title 21 of the Code of Federal Regulations (CFR) Parts 50, 54, and 56. Parts 58, 210, 211, 312, 600 to 680, 812, 814, and 820 apply depending on the product. 21 CFR Part 11, which addresses electronic record keeping, may also apply, as may 45 CFR Part 46 and Parts 160, 162, and 164, which deal with electronic record keeping and patient data. For nonclinical studies, it usually means compliance with Good Laboratory Practices in 21 CFR Part 58. Companies generally design templates that identify the components that the protocol must include as spelled out in the regulations. Then when it's time to write the protocol, the information can drop into sections.

The clinical protocol must include information of preclinical studies and previous trials, if any. It must present information about the safety and efficacy of the drug, device, or biologic that previous studies have confirmed. It must also state why the trial is being conducted, based on previous activities and findings, and what the expected outcome is. It needs to describe the study and how it is to be conducted and identify the healthy or patient volunteers, their treatment schedule, and the methods for assessing results.

A nonclinical study protocol tells the purpose of the study and identifies tests, controls, and substances by name and CAS or code number. The name and address of the sponsor and testing facility, and the start and end dates of the study are also requirements. There must be a statement of justification for the test system and, as applicable, animal data. The procedure for test system identification is also a requirement, as is a description of the experimental design. If diet of study animals is a factor, that needs inclusion, as does the route, method, and frequency of administration. The protocol must also delineate the type and frequency of test, analyses, and measurements and identify the records that the study will generate. Finally the protocol must identify the proposed statistical methods.

Components of a Clinical Protocol

Background and Rationale
Objectives
Trial Design
Assessment Schedule
Study Product (Drug, Device, Biologic)
Preparation, Packaging, Labeling
Treatment/Dosing Schedule
Storage, Dispensing, Reconciliation and Return
Data Collection
Statistical Analysis
Ethical Consideration
Informed Consent
Confidentiality
Monitoring
Publication of Results

When a study is underway and it becomes apparent that the protocol needs modification, the change is made through an amendment to the original protocol. The amendment is retained with the original protocol. Typically amendments identify what is to be changed, specify where the protocol is to be changed, and explain the reason for the change. All amendments must be signed and dated.

Periodic and Progress Reports

Periodic and progress reports are predictable: They come at regular intervals in a company's operation or in the sequential development of projects. Some reports, like assessments of a month's activities, are routine and can be quite short. Others, such as the summary of the development of a product or the decision to purchase a specific piece of equipment in the initial stages of a project, require more lengthy presentation.

These reports offer information about work activities on a daily, weekly, monthly, or other time-sequence basis. A department manager may require monthly reports from her staff, or a marketing director may demand quarterly sales reports, whereas a manager of a laboratory that has several working groups may mandate periodic reports to verify activities. Annual reports, monthly sales reports, stability reports, quarterly budget reports, status reports of product development stages — these are progress and periodic reports. They all answer such questions as these:

What has occurred in the time span since the last report?

What's the current status?

What's the current activity?

What's the purpose of the current activity?

What will result from the activity?

What are the plans for the next period, if any?

Elements of these reports are flexible. Many companies have their own templates that give them a distinct look. Common to most are headings that control content, even when the report comes in memo form. Longer reports may include cover sheets, control numbers, statements of confidentiality, and a standard sequence of elements determined by company preference.

> Facilities where documents are kept must minimize the risk of damage or destruction from the natural elements: environmental conditions, such as inordinate humidity or excessively high or low temperatures; mold or insect and rodent damage.

A Monthly Report

If you are required to present a monthly report, you can do so easily by grouping your activities. The following monthly report works well because it presents a brief overview that tells the reader, in this case a Quality Assurance Director, what activities have been completed during the time period and what their current status is as of the end of the month. Notice that the writer elects to put much of the information into a chart to facilitate reading. The writer also includes herself in the document, using "I" and "we," since the report is about her accomplishments for the month.

Facilities where documents are kept must minimize the risk of damage or destruction from the natural elements: environmental conditions, such as inordinate humidity or excessively high or low temperatures; mold or insect and rodent damage.

KIM LABORATORIES, SA **Memorandum** *Page* 1 of 2

Date: August 31, 2005

To: Leo Zeng

FROM: Erin Marias

RE: August Monthly Report

Overview

I worked on a diverse number of projects in August. Several customer projects required attention including Biometrics, Andromedia, and Progeny. I also reviewed batch records and put in significant hours on other projects. A summary of my activities follows.

Batch Record Review

Type of Record	Company	Product	Lot	Status
Purification	Andromedia	PDQ-06	981110	Review completed; record sent to customer.
Purification	M&FG	KIM456	K99715	Clarification of a Line Clearance Form needed. Amino acid sequencing testing results are incomplete.
Reprocessing	M&FG	KIM467	99718	Review completed; waiting for FDA approval.
Filling	Progeny	Media	PM991126	Review completed; record sent to customer.
Filling	Progeny	PY-LH	PM991128	Clarification needed on LAL testing.
Purification	Andromedia	PDQ-06	981111	Review in progress.

Additional Projects

- **Review of "Validation of a Endpoint-Turbidimetric Assay for the Determination of Endotoxin"**

  I completed a review of this report during the first week of August.

- **Biometrix Vaccine Project**

  I attended the organizational meeting on August 14th and provided QA input. I then prepared a short report highlighting the decisions we reached conjointly and distributed it by e-mail on August 15th.

KIM LABORATORIES, SA ***Memorandum*** *Page* 2 of 2

- **Kim Labs DI and WFI Water Systems**

  We have successfully revalidated the water system using the sampling plan submitted to FDA. The validation report has been submitted and document review has begun. I expect review to be complete by next week. In addition to the revalidation of the water system, we are modifying the routine monitoring program for the system to establish a more stringent program.

- **Validation Report Archives**

  We have initiated the archiving of validation reports in a central location in QA. QA has not yet obtained numerous procedure and process validations for archiving, so I am now drafting a memo requesting the validation reports. The memo will include a list of the validation reports dealing with procedure and process validations as well as any other pertinent reports.

- **Review of the "Qualification of the Purified Water System Modifications"**

  This report required significant changes since it was written several months ago. The final version has been received by QA and the audit of this report is complete.

Courtesy of Michael Nolan

An Equipment Evaluation Report

This report came early on in a project — the writers had the task of evaluating tablet presses for a product the company was preparing to manufacture. The report accomplishes its goal: It discusses the evaluation process and makes a recommendation for purchasing equipment. (This example does not include the appended equipment test results, because to do so would require the addition of another 14 pages.)

TABLET PRESS EVALUATION

FOR PELLETS-IN POWDER PRODUCTS

by Elton Davis and Mia Chu

October 25, 2004

SUMMARY

The tablet press evaluation is complete. Of the three presses we studied, the Esther 2000 press, manufactured by Esther/Ruth, best suits our needs. It offers a high level of control sophistication, ease of operation and maintenance, and special features, yet is competitively priced.

INTRODUCTION

Processing pellets-in-powder products requires equipment that, through sophisticated computer controls and special design features, minimizes weight variation and improves the content uniformity. The development of the modified release dosage form, comprised of pellets distributed in a powder matrix, presents formulation challenges; it also presents these processing challenges, especially during the compression of tablets:

- eliminating the weight variation, and
- achieving uniform distribution of pellets within the tablets.

All modern tablet presses rely on the principle of volumetric filling of die cavities as a means to control the tablet weight; during manufacturing, good powder flow is critical. This company needs a press that will enhance the flow of problematic powders from the hopper into the feed frame and, consequently, into the die cavity, without promoting the segregation of pellets from the fine particles of powder matrix due to vibration or paddle movement.

The Samuel 720LX tablet press we use for acetaminophen and diflunisol lacks the sophistication for handling pellets-in-powder products, has complicated software, and, for the last year, has experienced unsatisfactory performance.

RECOMMENDATION

After careful evaluation of the Naomi 642, Samuel 9002X, and Esther 2000 press, we have determined the Esther 2000 press to be the most suitable equipment for our manufacturing purposes and budget criteria and recommend its purchase.

PROJECT OVERVIEW

At the August 2nd project meeting, we agreed to initiate and complete a survey of three presses within a three-month time frame. Our scope was greater than the requirements for our current product, since we are continuing to develop pellets-in-powder products. However, we concentrated on the ability of each press to process our current product with minimal weight variation and uniform distribution of pellets within tablets.
Besides evaluating the ability of different tablet presses to handle our current product, we looked for the following features:

- ease of operation, cleaning and change-over
- available inventory of spare parts
- software sophistication and user-friendliness
- manufacturer's assistance in IQ/OQ/PQ process validation guidance

- manufacturer's training and routine servicing
- history of successful compression of pellets-in-powder products
- price

We evaluated and tested these three tablet presses:

Naomi 642 manufactured by Naomi Riga, Inc.

Samuel 9002X manufactured by Samuel Industries

Esther 2000 press manufactured by Esther/Ruth

We manufactured three batches, each approximately 25,000 tablets. For actual compression testing we ran the presses at relatively high speed (50–80 RPM). We were interested in the capability of the equipment to tightly control the weight variation of tablets.

We then evaluated the ability of the equipment to minimize the segregation of pellets from the powder. The method (attached) calls for disintegration of cores in water using a modified USP method for dissolution in baskets, and weighing the pellets that remain in the basket after all powder matrix has completely disintegrated and none remains in the basket.

All three tablet press manufacturers recommended using a "D" type tablet press configuration for our product. Complete test results for the individual machines are attached.

SUMMARY OF TABLET TESTING RESULTS

Tablet Press	**Naomi 642**	**Esther 2000**	**Samuel 9002X**
Batch Number	PB 9836	PB 9837	PB 9838
Tablets: target weight 850 mg			
Avg Weight (mg)	855.8	856.4	861.9
Wt Range (mg)	827-875	845-869	844-874
RSD %	1.4	0.82	1.02
Pellets			
Avg Wt/Tab (mg)	232.2	246.8	242.0
Range Wt/Tab (mg)	213-246	231-260	216-270
RSD%	3.9	3.3	6.4
% Pellets/Tablet	27.1	28.8	29.3
Theo. % pellets/Tab	26.1	29.0	29.6

As the table clearly shows, the tablets manufactured on the Esther 2000 press are superior to those manufactured on the other two presses. We believe the superiority of the Esther press can be attributed to the unique dosing system.

FEATURES OF THE ESTHER 2000 PRESS

All three presses have a sophisticated computerized weight control system via measurement and adjustment of compression forces. What sets them apart are the unique features—designed to improve the quality of compressed tablets, simplify operation control, and enhance the processing of problem powders. The special features of the Esther 2000, however, contributed to our decision to recommend its purchase.

- Automatic pressure control with three-way tablet tester, which feeds back data to PLC Control for the automatic adjustment of tablet weight and hardness or thickness. The controls and software are designed and manufactured in the USA.

- Automatic two-stage controlled feed system, which consists of a powder supply hopper with a level sensor and lower feed shoe.

- Material flows from the hopper to the lower shoe only upon receiving the signal from sensors on the PLC Control; therefore, the head pressure is eliminated.

- Completely sealed tableting zone; the feed shoe is designed to contain, not recirculate, excess powder. The powder does not travel around the turntable.

- The Esther 2000 provides the documentation and support necessary to conduct the Installation Qualification/Operational Qualification/Performance Qualification procedures as well as the process validation.

- In addition, Esther/Ruth provides extensive training to the operators at the site for one week at no cost. The Esther/Ruth company will also provide refresher training at a nominal fee.

PRICE

The price for the 32-station Esther 2000 tablet press, model E-7375-EMU with a Level II control system, single-tablet rejection system, and automatic teflon powder lubrication system is $925,000.

CONCLUSION

Of the three tablet presses, the Esther 2000 produced tablets that met our specifications more precisely than the other tablet presses did. Further, the Esther 2000 is less expensive and comes with one week of operator training, as well as qualification and validation assistance.

Validation and Qualification

Validation and qualification reports generate whenever the company determines to bring new equipment on board, to modify equipment, or to institute or change a process. These activities usually occur in accord with industry standards for installation qualification, operation qualification, and process qualification (IQ/OQ/PQ) and process validation. These activities are mandated by regulations. FDA has published its own guidelines and supports additional industry guidelines, such as Good Automated Manufacturing Practices (GAMP) to help companies achieve compliance.

When a company determines to introduce configurable off-the-shelf (COTS) software for record keeping, it does so according to industry standards and 21 CFR Part 11 Electronic Records; Electronic Signatures, Final Rule; or 45 CFR Parts 160, 162, and 164. Industry standards are in place for Part 11 and companies are complying. 45 CFR Part 160, 162, and 164 are new parts of the Health Insurance Portability and Accountability Act (HIPAA), which was first passed into law in 1996. Part 160, General Administrative Requirements; 162, Administrative Requirements; and 164 Security and Privacy were issued in 2003, with an effective date of April 2005. These regulations have parallel requirements to Part 11 and drive computer software validation (CSV) for electronic record keeping.

Qualification and validation of equipment and processes require an infrastructure of standard procedures that tell how to do it. So does CSV. In addition, both require forms for data collection and process verification, as well as training and training records. Companies must also prepare procedures for using equipment or software or conducting a process once the systems have been qualified or validated.

Computer Software Validation

It has taken much dialog between FDA and industry to establish what needs to happen to validate software. And what needs to happen requires diligent documentation. Software validation begins with identifying user requirements and ends with a validation packet of all the documents generated in the process.

Additional documentation includes vendor information, contracts, invoices, receipts, training confirmation, and testing verification. All components act in total to affirm validation. Revalidation may require some or all of the documents again, depending upon the extent of the revalidation.

Computer Validation Project Plan

The following is an example of a validation plan. Note that the template for the document is the same as for all the other documents that comprise the validation, and this one is the second of ten deliverables for the project.

Simons Laboratories, Inc.	**VALIDATION**
E-DocComply Project Plan	**VL 00014-2**

1.0 PURPOSE

1.1 This plan outlines the resources required for the validation of E-DocComply software. VL 00014-1 E-DocComply User Requirements identifies the requirements for intended use, and validation will affirm that the system satisfies the end users' needs.

2.0 VALIDATION GUIDANCE

2.1 Validation will occur according to SOP 00389 COTS Software Validation.

3.0 PLANNED VALIDATION DELIVERABLES

- User Requirements
- Project Plan
- Installation Protocol
- Installation Report
- Functional Specification
- Hazard Analysis
- User Testing Protocol
- User Testing Report
- System Release Report
- System Review Report

4.0 HISTORY

4.1 This is the initial validation of the system.

5.0 SCOPE

5.1 The following departments will use the system upon completion of validation.

- Corporate
- Clinical

Simons Laboratories, Inc.	**VALIDATION**
E-DocComply Project Plan	**VL 00014-2**

- Drug Discovery
- Development
- Document Control
- Information Technology
- Legal
- Manufacturing
- Marketing
- Quality Assurance
- Quality Control
- Regulatory

6.0 VALIDATION TEAM

6.1 The Validation Team consists of representatives from the following areas.

Department	*Name*	*Title*	*Role*
Corporate	Christian Matthews	Manager of Human Resources	Team member
Clinical	Ginny Paton	Medical Director	Team member
Drug Discovery	Amar Sahai	VP of Drug Discovery	Team member
Development	Harry Fong	Scientist	Team Member
Development	G. Willy Lake	Manager of Neuroscience	Team Member
Document Control	Jan Carol Simons	Director of Technical Communications	Project Lead
Document Control	Linda Tanabe	SOP Coordinator	System Administrator

Simons Laboratories, Inc.	**VALIDATION**
E-DocComply Project Plan	**VL 00014-2**

Department	*Name*	*Title*	*Role*
Document Control	Stella D'Artagnon	Technical Writer	Team Member and Back Up Administrator
Information Technology	Daniel Kukla	Director of IT	Team Member
Legal	John Tessman	Patent Agent	Team member
Manufacturing	David Nettleton	Software Validation Consultant	Validation Manager
Marketing	Astrid Reynolds	Marketing Manager	Team member
Manufacturing	Timothy Osterhaus	Reliability Engineer	Team member
Quality Assurance	Ginger Ogden	Director of QA	Team member
Quality Control	Linda Tanabe	Manager of QC	Team member
Regulatory	Joseph Tamayo	VP Regulatory Affairs	Team member

7.0 PROJECT SCHEDULING

The validation will start on 04 Sep 04. Target date for validation completion is 01 Nov 04.

8.0 BUDGET

8.1 Funding for 20 licensed seats has been allocated by Finance, not individual departments.

Computer Software Validation Deliverables

Here's a list of the key validation documents that the process generates. Note that each of these documents must be reviewed and approved before the next step can occur. Also, in addition to these documents, there will be training records and data collection forms as the process progresses.

1. **User Requirements:** This document identifies the duties the system will perform based on the needs of the end users. What are the needs of the users that led to the implementation of the system? Who will use the system? Who may use the system in the future? How big should the system be? What is the operating environment? What are the hardware requirements?
2. **Validation Plan:** This document tells how the system will undergo validation. Who will validate the system? When will validation take place and what is the estimated schedule? What are the deliverables from the validation? What is the history of the validation (first time validation, revalidation, or partial revalidation)? What are the resources for validation? Which employees comprise the validation team? Who is the validation lead? What are the resources for the validation?
3. **Installation Protocol:** This document gives the plan for installing the software. How is the system to be installed and where will it happen? Which existing systems will be impacted? Who will perform the installation? What are the hardware and software requirements? What are the operating environments that will result from the installation?
4. **Installation Report:** This document assesses the installation. Were there any problems, and if so, how were they resolved? What are the environmental conditions now? Are there any future actions that must occur?
5. **Functional Specifications:** This document establishes the capabilities of the installed system and identifies its functional specifications: application services, interfaces, back up, configuration options, file types, user privileges, licenses, and so forth. This document also tells how the system will be configured and how users can employ each service of the system.
6. **Hazard Analysis:** This document identifies the risks the system poses, in relation to the functional specifications, and provides an analysis of the hazards associated with those risks. It provides a classification of risk — low, moderate, high, and so forth. It tells the highest level of risk and what it relates to and gives the probability for such a risk occurring.

7. **User Testing Protocol:** This is the plan for testing the system in relation to the functional specifications. What is going to be tested and by whom? What is the test environment? What is the plan for change control? How are defects to be managed? How will results be documented? Note that users who will operate the system for testing must undergo documented training on the system.
8. **User Testing Report:** This report assesses the testing. When did it happen? What were the results? Were there any deviations? Who reviewed the results?
9. **System Release Report:** This report releases the system into production. When is the "go live" date? Which documents are complete (validation documents to date and supporting documents such as SOPs)? What are the release instructions? Who requires training and how will that training occur? What is the change control process?
10. **System Review Report:** When a system has been "live" for 30 days — or longer if it is not used frequently — a system review report assesses the system in actual use. Does it work as anticipated? Does it meet the intended needs? Was validation effective? Is the project complete? Does anything else need to happen?

Note: Chapter 5 identifies SOPs and other documents that are necessary to support validation.

Equipment Qualification

Equipment qualification always requires a plan or protocol that tells how qualification will take place, approval of the plan, and then testing and verification of the testing, with final approval. Qualifications typically include both allowances for nonconformances and a plan for requalification should equipment undergo repair or modification.

Some companies prefer to do PQ with process validation, keeping IQ and OQ separate. One school of thought maintains that performance and process testing can be done concurrently. Another school of thought maintains that it's better to include the PQ with the IQ/OQ, because more than one process may require the equipment. However, if equipment is product dedicated, doing the PQ with the process validation can make sense. Determination of the best standard rests with the company and depends very much on the equipment and the process it performs.

Packaging Equipment Qualification Report

Many people responsible for equipment qualification prepare a protocol for testing and secure approval. After testing, they prepare a duplicate report for approval of the results. This method works for many companies, but entails much replication — the uncompleted tests in the preapproval protocol and duplicated documents with entries made during testing in the approval protocol. The following example, however, efficiently handles IQ/OQ/PQ in one document. The following report is ready for the preapproval signatures; after testing, it will receive final approval signatures.

TITLE: Installation, Operation and Performance Qualification	**NUMBER: 2**	Page 1 of 20
SYSTEM: Kaps-All Capper	**MANUFACTURER: Kaps-All Capper**	
DEPARTMENT: Operations/Packaging	**DATE: 19MAR04**	
EQUIPMENT NUMBER: 3413	**SERIAL NUMBER: 8404**	

Installation, Operation & Performance Qualification for the Kaps-All Capper Model A

OBJECTIVE: To verify that the Kaps-All Capper is installed and operates within a validated state of control.

PROTOCOL PREAPPROVAL SIGNATURES				
These signatures indicate that the protocol has been reviewed and been found to be acceptable for the Kaps-All Capper.				
Signatory Name	**Signature**	**Department**	**Title**	**Date**
		Validation	Validation Designee	
		Operations	Department Designee	
		Facilities	Facilities Designee	

QUALIFICATION APPROVAL SIGNATURES				
These signatures indicate that the results have been reviewed and been found to be acceptable for the Kaps-All Capper.				
Signatory Name	**Signature**	**Department**	**Title**	**Date**
		Validation	Validation Designee	
		Operations	Department Designee	
		Facilities	Facilities Designee	

TITLE: Installation, Operation and Performance Qualification	NUMBER: 2	Page 2 of 20
SYSTEM: Kaps-All Capper	MANUFACTURER: Kaps-All Capper	
DEPARTMENT: Operations/Packaging	DATE: 19MAR04	
EQUIPMENT NUMBER: 3413	SERIAL NUMBER: 8404	

TABLE OF CONTENTS

TITLE: Installation, Operation and Performance Qualification	NUMBER: 2	Page 3 of 20
SYSTEM: Kaps-All Capper	MANUFACTURER: Kaps-All Capper	
DEPARTMENT: Operations/Packaging	DATE: 19MAR04	
EQUIPMENT NUMBER: 3413	SERIAL NUMBER: 8404	

1. OBJECTIVE:

The objective of this equipment qualification is to establish documented evidence that the Kaps-All Capper adheres to the following parameters:

1.1 Is acceptably installed per manufacturer recommendations, process requirements, and/or engineering standards. Acceptable installation includes suitable utility connections, components, and critical instruments in current calibration.

1.2 Is fully operational as specified by the protocol. Acceptable operation includes, as applicable, proper control and sequencing functions, recording and reporting functions, and safety and alarm features that meet process requirements and equipment specifications.

1.3 Performs acceptably under production conditions. Acceptable preformance includes consistent operation within specified process parameters under simulated or actual production conditions.

2. SCOPE

The scope of this equipment qualification protocol includes the Kaps-All Capper installation qualification, is limited to the system components and does not include installation of support utilities, other than the connections at the system boundary. Operational testing is limited to demonstrating equipment functionality. Product specific testing is outside the scope of this qualification document, other than where a product is used to demonstrate equipment functionality.

2.1 SYSTEM COMPONENTS

Cap Feeder Bin Assembly
Cap Chute Assembly
Lower Chute and Stabilizer Assembly
Cap Tightening Disc Spindles Assembly
Bottle Gripper Belts and Conveyor Assembly
Electronic Torque Monitor and Printer
Conveyor
Elevator

TITLE: Installation, Operation and Performance Qualification	NUMBER: 2	Page 4 of 20
SYSTEM: Kaps-All Capper	MANUFACTURER: Kaps-All Capper	
DEPARTMENT: Operations/Packaging	DATE: 19MAR04	
EQUIPMENT NUMBER: 3413	SERIAL NUMBER: 8404	

3. **SYSTEM DESCRIPTION**

The Kaps-All Capper accepts filled bottles from the conveyor line and seals them with threaded caps. The capper incorporates a cap feeder, photoelectric cap scanner, conveyor systems for the caps and bottles, motors that power the gripper belts and spindles, speed controllers for the gripper belts and spindles. There is an associated elevator that lifts lids to the Cap Feeder Bin. The gripper belts move the bottles along the conveyor and caps emerge from the cap feeder bin and move down a chute onto the bottles. The torque monitor is used to observe and monitor the torque at which the capper applies caps to bottles. The monitor displays a torque measurement when a bottle passes the right rear spindle.

4. **REFERENCES**

Operator's Manual for Kaps-All Capper, Date: 14APR00 The

Electronic Torgue Monitor Manual. Oct. 03

TITLE: Installation, Operation and Performance Qualification	NUMBER: 2	Page 5 of 20
SYSTEM: Kaps-All Capper	MANUFACTURER: Kaps-All Capper	
DEPARTMENT: Operations/Packaging	DATE: 19MAR04	
EQUIPMENT NUMBER: 3413	SERIAL NUMBER: 8404	

5. TESTING SUMMARY OF INSTALLATION QUALIFICATIONS

INSTALLATION QUALIFICATION SUMMARY	MEETS PROTOCOL SPECIFICATIONS	
	YES	NO
Support Documentation:		
Manuals on File		
SOPs on File		
Component Calibration and Maintenance Program in Place		
Utilities:		
Compressed Air		
Main Power		
Component Specifications:		
Cap Feeder Bin Assembly		
Cap Chute Assembly		
Lower Chute and Stabilizer Assembly		
Cap Tightening Disc Spindles Assembly		
Bottle Gripper Belts and Conveyor Assembly		
Electronic Torque Monitor and Printer		
Conveyor		
Elevator		

Conducted by:	Date:
Reviewed by:	Date:

TITLE: Installation, Operation and Performance Qualification	NUMBER: 2	Page 6 of 20
SYSTEM: Kaps-All Capper	MANUFACTURER: Kaps-All Capper	
DEPARTMENT: Operations/Packaging	DATE: 19MAR04	
EQUIPMENT NUMBER: 3413	SERIAL NUMBER: 8404	

6. INSTALLATION QUALIFICATION

6.1 SYSTEM INFORMATION

SYSTEM INFORMATION
Manufacturer:
Model No.:
Serial No.:
Equipment Number:
Purchase Order No.:
Location:
Additional Data:

Conducted by:	Date:
Reviewed by:	Date:

TITLE: Installation, Operation and Performance Qualification	NUMBER: 2	Page 7 of 20
SYSTEM: Kaps-All Capper	MANUFACTURER: Kaps-All Capper	
DEPARTMENT: Operations/Packaging	DATE: 19MAR04	
EQUIPMENT NUMBER: 3413	SERIAL NUMBER: 8404	

6.2 UTILITIES REQUIREMENTS

UTILITIES	
Required	**As Found**
Compressed Air Pressure: 30 psig-Regulated	Air Pressure:
Voltage: Kaps-All Capper: 220 V AC	Voltage
Amps: 15	Supply Line Amp Rating:
Power Supply Source: Power Panel	Wire Numbers
Supplied voltage is within 10% of	V AC: (Yes/No)
Rated circuit current is ≥ the maximum unit current draw: (Yes/No)	

6.3 CALIBRATED INSTRUMENTATION

Procedure: Review the system and list below instrumentation and gauges that may require calibration. Attach request for calibration forms to the attachments section of this validation report.
1.
2.
3.
4.

Conducted by:	Date:
Reviewed by:	Date:

TITLE: Installation, Operation and Performance Qualification	NUMBER: 2	Page 8 of 20
SYSTEM: Kaps-All Capper	MANUFACTURER: Kaps-All Capper	
DEPARTMENT: Operations/Packaging	DATE: 19MAR04	
EQUIPMENT NUMBER: 3413	SERIAL NUMBER: 8404	

6.4 MATERIALS OF CONSTRUCTION

MATERIALS OF CONSTRUCTION	
List all product contact components and the materials of construction for each.	
Component	**Materials of Construction**
Lubricants:	
List all lubricants used to operate this unit and identify as product or noncontact.	
Lubricant Type	**Product/Nonproduct Contact**

Conducted by:	Date:
Reviewed by:	Date:

TITLE: Installation, Operation and Performance Qualification	NUMBER: 2	Page 9 of 20
SYSTEM: Kaps-All Capper	MANUFACTURER: Kaps-All Capper	
DEPARTMENT: Operations/Packaging	DATE: 19MAR04	
EQUIPMENT NUMBER: 3413	SERIAL NUMBER: 8404	

6.5 MANUALS AND DRAWINGS

MANUALS ON FILE	
Manual Title	Storage Location
DRAWINGS	
Drawing No.:	Revision Date:
Drawing Title:	Storage Location:
Drawing No.:	Revision Date:
Drawing Title:	Storage Location:
Drawing No.:	Revision Date:
Drawing Title:	Storage Location:
Drawing No.:	Revision Date:
Drawing Title:	Storage Location:
Drawing No.:	Revision Date:
Drawing Title:	Storage Location:

Conducted by:	Date:
Reviewed by:	Date:

TITLE: Installation, Operation and Performance Qualification	NUMBER: 2	Page 10 of 20
SYSTEM: Kaps-All Capper	MANUFACTURER: Kaps-All Capper	
DEPARTMENT: Operations/Packaging	DATE: 19MAR04	
EQUIPMENT NUMBER: 3413	SERIAL NUMBER: 8404	

7. TESTING SUMMARY OF OPERATION AND PERFORMANCE QUALIFICATION

Briefly summarize the testing performed, results obtained and any notable observations.

INSTALLATION QUALIFICATION

The unit was installed properly and all components were present. All applicable procedures were in place. The utilities supplied to the unit were inspected and found to be acceptable for proper operation of the unit.

OPERATIONAL QUALIFICATION

All control and operational test function results were acceptable per the protocol test results.

PERFORMANCE QUALIFICATION

All performance testing was acceptable per the protocol test results.

CONCLUSION

The Caps-All Capper has been tested and verified to operate properly according to manufacturer and process specifications. All corrections were noted in the protocol. No Deficiency/Response reports were required.

REQUALIFICATION

Evaluation for requalification will occur if any changes are made to the system following the approval of this report. Change control is specified under VAL-4 Equipment and Facilities Validation Documentation Change Control Policy.

Conducted by:	Date:

TITLE: Installation, Operation and Performance Qualification	NUMBER: 2	Page 11 of 20
SYSTEM: Kaps-All Capper	MANUFACTURER: Kaps-All Capper	
DEPARTMENT: Operations/Packaging	DATE: 19MAR04	
EQUIPMENT NUMBER: 3413	SERIAL NUMBER: 8404	

8. STANDARD OPERATING PROCEDURES

8.1 OBJECTIVE

The objective of this test is to review and list all applicable SOPs for the Kaps-All Capper.

8.2 ACCEPTANCE CRITERIA

SOPs must be in place for the operation and cleaning of this unit. Calibration schedule must be established and a maintenance schedule must be in place.

8.3 PROCEDURE

Review the SOPs that cover the operation and cleaning of this unit. The SOPs may be in draft form; if so, attach to this document. List all applicable SOPs in the following table. Calibration procedures and schedules must be established and in place for all critical instruments. Maintenance procedures and schedules must be in place in the maintenance management system. Review and verify the calibration and maintenance requirements in the following table.

8.4 SUMMARY

STANDARD OPERATING PROCEDURES		
Title	**Control Number**	**Revision Date**
Calibration Verified (Yes/No/Not Applicable):		
Preventative Maintenance Verified (Yes/No/Not Applicable):		

8.5 COMMENTS

Conducted by:	Date:
Reviewed by:	Date:

TITLE: Installation, Operation and Performance Qualification	NUMBER: 2	Page 12 of 20
SYSTEM: Kaps-All Capper	MANUFACTURER: Kaps-All Capper	
DEPARTMENT: Operations/Packaging	DATE: 19MAR04	
EQUIPMENT NUMBER: 3413	SERIAL NUMBER: 8404	

9. OPERATIONAL QUALIFICATION

10. CALIBRATION VERIFICATION

10.1 OBJECTIVE

The objective of this test is to record the instruments on the equipment that do and do not require calibration.

10.2 ACCEPTANCE CRITERIA

Each calibrated instrument must be in calibration at the time of qualification testing.

10.3 PROCEDURE

Review the equipment and record the instruments that do and do not require calibration.

10.4 SUMMARY

Comp.	Model#	Serial#	Cal. ID#	Cal. Reqd (Yes/No)	Cal. Due Date	Manufacturer	Operati ng Range

10.5 COMMENTS

Conducted by:	Date:
Reviewed by:	Date:

TITLE: Installation, Operation and Performance Qualification	NUMBER: 2	Page 13 of 20
SYSTEM: Kaps-All Capper	MANUFACTURER: Kaps-All Capper	
DEPARTMENT: Operations/Packaging	DATE: 19MAR04	
EQUIPMENT NUMBER: 3413	SERIAL NUMBER: 8404	

11. CONTROL FUNCTIONS

11.1 OBJECTIVE

The objective of this test is to verify that all critical controls operate as specified.

11.2 ACCEPTANCE CRITERIA

All critical controls shall operate as designed and specified. Critical controls are to include actions that take place during an out-of-specification occurrence.

11.3 PROCEDURE

Verify that all controls operate as designed and specified. Test all operational controls used during normal operation. If necessary, create conditions that activate an out-of-specification alarm or occurrence. Complete the following summary; include any notable observations in the comments section of the test. AE (As Expected) in the Results column signifies the results are the same as the description.

11.4 SUMMARY

KAPS-ALL CAPPER CONTROL PANEL

Control	Description	Results	Verified By
Start Pushbutton	When the Start Pushbutton is pressed, power is supplied to the equipment.		
Stop Switch	When the Stop Switch is pressed, the power to unit stops.		
Cap Feeder ON/OFF Switch	When the Cap Feeder ON/OFF Switch is pressed, power is supplied to/from the Cap Feeder.		

Conducted by:	Date:
Reviewed by:	Date:

TITLE: Installation, Operation and Performance Qualification	NUMBER: 2	Page 14 of 20
SYSTEM: Kaps-All Capper	MANUFACTURER: Kaps-All Capper	
DEPARTMENT: Operations/Packaging	DATE: 19MAR04	
EQUIPMENT NUMBER: 3413	SERIAL NUMBER: 8404	

11.4 SUMMARY *(continued)*

Control	Description	Results	Verified By
Cap Feeder Speed Control	When the Cap Feeder Speed Control knob is rotated, the Cap Feeder speed increases and decreases.		
Cap Bin ON/OFF Switch	When the Cap Bin ON/OFF Switch is pressed, power is supplied to/from the Cap Feeder Bin		
Cap Bin Speed Control	When the Cap Bin Speed Control knob is rotated, the Cap Feeder Bin speed increases and decreases.		
Spindle Speed Control	When the Spindle Speed Control knob is rotated, the Spindle speed increases and decreases.		
Conveyor/Gripper Speed Controls	When the Conveyor/Gripper Speed Control knob is rotated, the Conveyor and Gripper speed increases.		
Torque Control 1 (Left)	When the Torque Control 1 (Left) knob is rotated, the torque applied to the caps increases and decreases.		

Conducted by:	Date:
Reviewed by:	Date:

TITLE: Installation, Operation and Performance Qualification	NUMBER: 2	Page 15 of 20
SYSTEM: Kaps-All Capper	MANUFACTURER: Kaps-All Capper	
DEPARTMENT: Operations/Packaging	DATE: 19MAR04	
EQUIPMENT NUMBER: 3413	SERIAL NUMBER: 8404	

11.4 SUMMARY *(continued)*

ELECTRONIC TORQUE MONITOR

Control	Description	Results	Verified By
Power ON/OFF Switch (Control Panel)	When the Power ON/OFF switch is pressed, power is supplied to/from the Torque Monitor.		
Power ON/OFF Switch (Printer Operation)	When the Power ON/OFF Switch is pressed, power is supplied to/from the printer.		
Set-Up Switch (Control Panel)	When the Set-Up Switch is pressed, the program scrolls through the set-up procedure.		
Valve-Up Switch (Control Panel)	When the Valve-Up Switch is pressed, the program scrolls through the set-up procedure.		
Valve-Down Switch (Control Panel)	When the Valve-Down Switch is pressed, valves are decreased during set-up.		
Printer ON/OFF Switch (Control Panel)	When the Printer ON/OFF Switch is pressed, every container's torque is printed in KATUs.		
Zero Switch (Control Panel)	When the Zero Switch is pressed, the display zeroes.		
Shift Switch (Control Panel)	When the Shift Switch is pressed, the auxiliary functions are provided.		
Print Stats Switch (Control Panel)	When the Print Stats Switch is pressed, statistical analysis is printed out.		

Conducted by:	Date:
Reviewed by:	Date:

TITLE: Installation, Operation and Performance Qualification	NUMBER: 2	Page 16 of 20
SYSTEM: Kaps-All Capper	MANUFACTURER: Kaps-All Capper	
DEPARTMENT: Operations/Packaging	DATE: 19MAR04	
EQUIPMENT NUMBER: 3413	SERIAL NUMBER: 8404	

12. KAPS-ALL CAPPER SPEED TEST

12.1 OBJECTIVE

The objective of this test is to document the slow and fast speeds of the Elevator, Kaps-All Capper and Conveyor.

12.2 ACCEPTANCE CRITERIA

This test is for baseline information only.

12.3 PROCEDURE

Measure the slow and fast speeds of the Elevator, Kaps-All Capper and Conveyor with a calibrated stopwatch then record the results and the stopwatch used in the summary.

12.4 SUMMARY

Speed	Elevator Ft./Min.	Capper Bottles/Min.	Conveyor/Gripper Ft./Min.	Verified By
Slow				
Fast				

Tachometer Used Indent. No.: ______________ Cal. Due Date: ______________

12.5 COMMENTS

Conducted by:	Date:
Reviewed by:	Date:

TITLE: Installation, Operation and Performance Qualification	NUMBER: 2	Page 17 of 20
SYSTEM: Kaps-All Capper	MANUFACTURER: Kaps-All Capper	
DEPARTMENT: Operations/Packaging	DATE: 19MAR04	
EQUIPMENT NUMBER: 3413	SERIAL NUMBER: 8404	

13. PERFORMANCE QUALIFICATION

14. KAPS-ALL CAPPER ORIENTATION TEST

14.1 OBJECTIVE

The objective of this test is to verify that the Kaps-All Capper is capable of installing caps on small and large bottles consistently over time, and that the equipment operational settings will be recorded.

14.2 ACCEPTANCE CRITERIA

Every bottle must have a cap installed on it. If the caps are not installed properly they will be passed through again until they are.

14.3 PROCEDURE

1. Record the bottle quantity and bottle size for each test set-up, in the summary.
2. Record the Cap Feed Speed Potentiometer Setting in the summary.
3. Record the Cap Bin Speed Potentiometer Setting in the summary.
4. Record the Spindle Speed Potentiometer Setting in the summary.
5. Measure the speed of the Capper (with a calibrated stopwatch) for each test set-up and record the results and the stopwatch used, in the summary.
6. Measure the speed of the Conveyor (with a calibrated stopwatch) for each test set-up and record the results and the stopwatch used, in the summary.
7. With the equipment operating, visually observe 100 (60cc) bottles as they pass through the Capper and verify that 100% of the metal caps are installed on the bottles. If the caps are not installed properly they will be passed through again until they are. Record any additional passes in the comments section, if needed.
8. Remove the caps from the bottles using a calibrated spring torque tester, then record the removal torque value in the summary.

Conducted by:	Date:
Reviewed by:	Date:

TITLE: Installation, Operation and Performance Qualification	NUMBER: 2	Page 18 of 20
SYSTEM: Kaps-All Capper	MANUFACTURER: Kaps-All Capper	
DEPARTMENT: Operations/Packaging	DATE: 19MAR04	
EQUIPMENT NUMBER: 3413	SERIAL NUMBER: 8404	

9. With the equipment operating, visually observe 100 (1300cc) bottles as they pass through the Capper and verify that 100% of the metal caps are installed on the bottles. If the caps are not installed properly they will be passed through again until they are. Record any additional passes in the comments section, if needed.
10. Remove the caps from the bottles using a calibrated spring torque tester, then record the removal torque value in the summary.

14.4 SUMMARY

Bottle Qty	100	100
Caps	Metal	Metal
Bottle Size cc	60	1300
Cap Feeder Speed Potentiometer Setting		
Cap Bin Speed Potentiometer Setting		
Spindle Speed Potentiometer Setting		
Conveyor/Gripper Speed Potentiometer Setting		
Capper/Gripper Speed (Bottles/Min.)		
Conveyor Speed (Ft./Min.)		
# of Caps Installed		
# of Caps Not Installed		
Verified By:		

Tachometer Used Ident. No.:__________________Cal. Due Date:_____________

Conducted by:	Date:
Reviewed by:	Date:

TITLE: Installation, Operation and Performance Qualification	NUMBER: 2	Page 19 of 20
SYSTEM: Kaps-All Capper	MANUFACTURER: Kaps-All Capper	
DEPARTMENT: Operations/Packaging	DATE: 19MAR04	
EQUIPMENT NUMBER: 3413	SERIAL NUMBER: 8404	

14.4 SUMMARY (continued)

Container Size 60 cc			Removal Torque (in-lb)			
1.	16.	31.	46.	61.	76.	91.
2.	17.	32.	47.	62.	77.	92.
3.	18.	33.	48.	63.	78.	93.
4.	19.	34.	49.	64.	79.	94.
5.	20.	35.	50.	65.	80.	95.
6.	21.	36.	51.	66.	81.	96.
7.	22.	37.	52.	67.	82.	97.
8.	23.	38.	53.	68.	83.	98.
9.	24.	39.	54.	69.	84.	99.
10.	25.	40.	55.	70.	85.	100.
11.	26.	41.	56.	71.	86.	
12.	27.	42.	57.	72.	87.	
13.	28.	43.	58.	73.	88.	
14.	29.	44.	59.	74.	89.	
15.	30.	45.	60.	75.	90.	
Average removal torque = in-lb						

Conducted by:	Date:
Reviewed by:	Date:

TITLE: Installation, Operation and Performance Qualification	NUMBER: 2	Page 20 of 20
SYSTEM: Kaps-All Capper	MANUFACTURER: Kaps-All Capper	
DEPARTMENT: Operations/Packaging	DATE: 19MAR04	
EQUIPMENT NUMBER: 3413	SERIAL NUMBER: 8404	

14.4 SUMMARY (continued)

Container Size 1300 cc			Removal Torque (in-lb)			
1.	16.	31.	46.	61.	76.	91.
2.	17.	32.	47.	62.	77.	92.
3.	18.	33.	48.	63.	78.	93.
4.	19.	34.	49.	64.	79.	94.
5.	20.	35.	50.	65.	80.	95.
6.	21.	36.	51.	66.	81.	96.
7.	22.	37.	52.	67.	82.	97.
8.	23.	38.	53.	68.	83.	98.
9.	24.	39.	54.	69.	84.	99.
10.	25.	40.	55.	70.	85.	100.
11.	26.	41.	56.	71.	86.	
12.	27.	42.	57.	72.	87.	
13.	28.	43.	58.	73.	88.	
14.	29.	44.	59.	74.	89.	
15.	30.	45.	60.	75.	90.	
Average removal torque = in-lb						

Torque Tester Used Ident. No.:________________Cal. Due Date:____________

14.4 COMMENTS

Conducted by:	Date:
Reviewed by:	Date:

Courtesy of Phil Cloud

8

Summary Reports

Process reports relate processes, past or future, and draw conclusions. These often serve as the catalyst for larger, more inclusive summary reports that gather large bodies of data and comprehensively draw conclusions. While the distinctions are not always as clear as they are for the purposes of this book, you should think of summary reports as the end result of a project or the completion of a milestone in a larger undertaking.

A Contract Research Organization (CRO), for example, may be working on a study for a therapeutic product developer. The deliverables for such a project are a summary report replete with the study protocol and amendments to the protocols, if any. The end report, the summary of the project, has built from data collection and observation over the course of the study itself, which resides in the study file. The study file for the project may also contain correspondence relative to the project, records of therapeutic component request and receipt, and other support information. So the final document, the summary report, pulls together information in support of the final outcome of the project.

As another example, when a company is ready to begin clinical trials for its developing drug product, it files an Investigational New Drug (IND) application. This is a milestone event, and the IND summarizes information about the drug product, toxicology studies, the plan for the proposed trial, information for the Principal Investigator (PI) in an Investigator's Brochure (IB), and other information pertinent to the clinical study. Then, each year the company must file an IND update, which summarizes the activities from the last IND to the present one. The IB must also be updated, so that the PI has the most current information. Interim amendments may also go to FDA if there are changes to the protocol as the study progresses. It is an interactive webwork of documentation that is summarized in stages.

Other types of documents that summarize are press releases and publications, such as journal articles, abstracts, posters, and slide sets. What they do, in essence, is cull important findings and summarize them; they in turn are supported by the source data and process reports generated along the way.

Organizing Information

The regulations and industry standards help companies determine how to present summary data. While the presentation — the formatting and mechanics of reports — varies from company to company, industry recognizes that presentation is important. Thus, companies develop templates for data presentation in specific documents. Technology has made the ability to create beautiful documents and transfer them securely by electronic means a reality.

Regulators mandate electronic records in many instances. Canada, Japan, and the European Union (EU) must submit dossiers electronically. The electronic Common Technical Document (eCTD) allows companies to organize information in such a way that it is acceptable to many regulators. This is advantageous when companies wish to market their products in many countries. For the therapeutic product developers, the benefits of electronic submissions are enormous. E-submissions mean decreased review time as well as lower costs in transporting huge amounts of paper that comprise a typical submission. (See below.)

The Common Technical Document (CTD)

At present, companies have the option of submitting original NDAs, NDA supplements, and NDA amendments in either the CTD or conventional NDA format. As of July 2003, FDA "highly recommends" that US marketing dossiers be in the CTD format. It will not be a formal requirement, however, until FDA revises 21 CFR Part 314, NDA regulations. CTD is required in EU, Canada, and Japan, so companies planning to market outside the US would be wise to use the CTD format.

Module 1: Administrative Information and Prescribing Information (Specific to country of filing)

Cover letter (if the company decides to submit one)

FDA Form 356h

Comprehensive Table of Contents for the Submission, including Module 1 (includes a complete list of all documents provided in the entire submission; specifies the location of each document with reference to volume numbers and tab identifiers)

Administrative documents (such as field copy certification, debarment certification)

Prescribing Information (copies of all labels and all product labeling)

Annotated Labeling Text

(See the FDA's guidance document Submitting Marketing Applications According the ICH-CTD Format — General Considerations for US submissions.)

Module 2: CTD Summaries

Note: Modules 2 and 3 basically equal the CMC and CMC-detailed information.

2.1 CTD Table of Contents

2.2 CTD Introduction. This section provides a general introduction to the pharmaceutical, including its pharmacologic class, mode of action, and proposed clinical use; it generally should not exceed one page

2.3 *Quality Overall Summary (QOS).* This section follows the scope and outline of the information in Module 3; it needs to include sufficient information for each section of Module 3 to provide the quality reviewer with an overview. This is the place to stress critical key parameters of the product and justify instances of deviations, if any, from the guidances. This summary should also discuss key supporting information from other modules related to quality. All information should be cross-referenced to volume and page numbers in other modules. The QOS should be no more than 40 pages of text (not including tables and figures). Note: Biotech products and products with complex manufacturing processes may require up to 80 pages.

2.4 *Nonclinical Overview.* This overview should present an assessment of the product's pharmacologic, pharmocokinetic, and toxicologic evaluation. This section should offer an interpretation of the nonclinical data, discuss clinical relevance, crosslink with quality aspects of the drug, and discuss the implications of the nonclinical findings for safety and efficacy. It should also explain and provide a rationale for the nonclinical testing strategy. This section should generally not exceed 30 pages.

2.5 *Clinical Overview.* This overview should provide a critical analysis of the clinical data. The focus should be on the conclusions and implications of the data; it should refer to the data available in the comprehensive clinical summary, the clinical study, and other reports. This section should provide a "succinct discussion and interpretation of these findings together with any other relevant information (e.g., pertinent animal data or product quality issues that may have clinical implications)." The clinical overview should present the strengths and limitations of the development program and study results, analyze the benefits and risks of medicinal product in its intended use, and describe how the study results support the critical parts of the prescribing information. (Note: The clinical summary, which follows, gives critical analysis and interpretation.)

2.6 *Nonclinical Written and Tabulated Summaries.* These summaries should provide a factual synopsis of the nonclinical data. This section should include three written summaries: nonclinical pharmacology, pharmacokinetic, and toxicology studies. A breakdown is as follows:

Introduction

Pharmacology Written Summary

Pharmacology Tabulated Summary

Pharmacokinetics Written Summary

Pharmacokinetics Tabulated Summary

Toxicology Written Summary

Toxicology Tabulated Summary

(Examples of tables are in the CTD guidance, *M4S: The CTD — Safety Appendices.*)

2.7 *Clinical Summary.* This section should provide a detailed, factual summary of all clinical information in the CTD, including clinical study reports, meta analyses or cross-study analyses that are represented in full reports included in Module 5, as well as postmarketing data for products marketed in other regions. This summary can typically range from 500 to 400 pages, not including tables. (Note: Length will vary depending on the product.)

Module 3: Quality

3.1 Module 3 Table of Contents

3.2 *Data Section.* This section should include relevant information described in existing ICH guidances. This section of the CTD includes the pharmaceutical development section. For FDA this is a quality or manufacturing-related departure from traditional NDA submissions. More like the development pharmaceutics section in EU marketing dossiers, this section should provide a history of the drug's chemical development and the company's rationale for taking certain paths in development, such as dosage or excipients. This section also includes nonharmonized region-specific information. (For instance, in the US it can include batch records, methods validation, and comparability protocols.)

3.3 Literature References

Module 4: Nonclinical Study Reports

4.1 Module 4 Table of Contents
4.2 Study Reports
4.3 Literature References

Module 5: Clinical Study Reports

5.1 Module 5 Table of Contents
5.2 Tabular Listing of All Clinical Studies
5.3 Clinical Study Reports. This section should contain all the clinical study reports.
5.4 *The Integrated Analysis of Safety Summary (ISS) and Integrated Efficacy Summary (ISE).* While not required elements for the CTD, these should be included in US submissions.
5.5 Literature References

Note: CTD guidance says that "applicants should not modify the overall organization of the CTD," but in the Nonclinical and Clinical Summaries sections applicants can modify individual formats to provide the best possible presentation of the technical information to facilitate understanding and evaluation of results."

Templates are the norm for summary documents, often with boilerplate text built right in. For instance, a company in Phase 2a trials for a medical device will have prepared a protocol based on the protocol for the Phase 1 trial, completed prior to the start of Phase 2a. Once Phase 2a is complete, the company will begin Phase 2b. The common boilerplate information in these documents may be information about the rationale for the study to be conducted and information about the device and the company. Such components as study design and study population may vary somewhat, but there will be common elements since the product is intended to treat certain conditions. Reports emanating from the trial activities can also have boilerplate sections, such as the information about the sponsor and the therapeutic product.

Template design may be according to the conventions established in the company's style guide. (See Chapter 9.) The components may vary depending on the type of report, but the margins, type face, and type size may be standard, as are headings by order of rank. The regulations often dictate the components, so a good document designer, working with managerial staff, can often create aesthetically appealing and functional documents.

Consider the requirements for a study report for a CRO-completed project conducted under current Good Laboratory Practices (GLPs). GLP regulations call for the following elements:

- Name and address of the facility performing the study and the dates on which the study was initiated and completed
- Objectives and procedures stated in the approved protocol, including any changes in the original protocol
- Identification of the statistical methods for analyzing the data
- The test and control articles by name, chemical abstract number, or code number; the strength, purity, and composition or other characteristics
- Stability of the test and control articles under the conditions of administration
- Description of the test system, including, as applicable, the number of animals, sex, body weight range, source of supply, species, strain and substrain, age, and identification procedure
- Description of the dosage, dosage regimen, route of administration and duration
- A description of any circumstances that may have affected the quality or integrity of the data
- The name of the study director and others engaged in the study activities
- Description of the transformations, calculations, or operations performed on the date; a summary and analyses of the data; a statement of the conclusions drawn from the analyses
- Signed and dated reports of each of the individuals engaged in the study
- The locations where all specimens, raw data, and the final report will reside
- Statement prepared and signed by the Quality Assurance Unit (QAU)

The study director must sign and date the study report, and the report must include any corrections or additions to a final report in amendment form.

International Conference on Harmonisation (ICH) and FDA spell out what requires inclusion in large summary reports. Chemistry, Manufacturing, and Controls (CMC) sections of documents, for instance, have clearly defined requirements. So do the guidelines for compiling Drug Master Files (DMFs) and PreMarket Approval Applications (PMAAs) and any number of other documents. Guidance is available on the Worldwide Web.

What summary reports all share is their capability to take discrete components of a larger task and bring them to a conclusion or interim conclusion. Because of their varied applications, elements of these reports are not consistent; rather, the choice of elements depends on the type of information to be relayed.

The following example, a table of contents, identifies the components of an Investigational New Drug (IND) annual report for a product in Phase 2 clinical trials. Each section contains its own summary and is by itself a comprehensive report, replete with supporting data. Such a report in essence tells a regulatory body, in this case, FDA, what the status of the drug testing is from the initial IND approval to date.

2.0 Table of Contents

Section

7.2 In-Progress Activities
7.3 Future Plans
7.4 Summary of Nonclinical Pharmacology Studies
7.4.1 Completed Studies
7.4.2 Ongoing Studies

7.5 Summary of Non-Clinical Biopharmaceutics Studies
7.5.1 Completed Studies
7.5.2 Ongoing Studies

7.6 Summary of Nonclinical Toxicology Studies
7.6.1 Completed Studies
7.6.2 Ongoing Studies

8.0 Foreign Marketing Development

9.0 Log of Outstanding Business

10.0 Appendices

10.1 Appendix A
10.2 Appendix B
10.3 Appendix C
10.4 Appendix D
10.5 Appendix E
10.6 Appendix F
10.7 Appendix G
10.8 Appendix H
10.9 Appendix I
10.10 Appendix J
10.11 Appendix K

Common Report Elements

Beyond complying with the dictates for preparing documents that come from regulators, companies have choices in organizing report components. A document may have an abstract that encapsulates the information at the very beginning of a report. The next element may be an introduction, which restates the abstract in more detail and signals the approaching presentation of supporting data. Reports of this nature may include a project overview or background before presentation of detail. A conclusion then restates the main point.

A report may also include recommendations, as well as a number of appendices that house additional data that readers may need to know, but are not in themselves critical to the report. Other elements that may be included are diagrams, schematics, and glossaries: These elements assist readers in comprehending complex data. (See Chapter Three for information about tables and figures.)

A lengthy summary report may include the following elements.

Executive Summary

This element summarizes what's in the report, usually for an audience that may or may not wish to read the entire report. It is a time saver for busy readers. Reports that typically include executive summaries are industry analyses and project planning reports, although any number of reports can include them. An executive summary may be affixed to a completed report atop the report cover page or may come directly after such prefatory information as required government forms, the table of contents, lists of figures and tables, and acronyms. Again, the regulations often dictate or at least suggest a sequence for elements.

Cover Sheet

This component may include any or all of the following elements: Company logo, Confidentiality statement, Title, Author, Approvals Page, and number. Note: The cover page may include a header that runs throughout the entire report. The header typically includes the title and page numbers (page 1 of 10).

Table of Contents

This is a list of sections and subsections of the report. It is the last component to be written, since it identifies what's in the report by specific page numbers. These are hyperlinked to the actual sections in the report.

List(s) of Figures, Illustrations, Tables

This component identifies the location of specific illustrations within the report. Like the table of contents, it is best prepared after the report is complete and hyperlinked to the tables and illustrations.

Summary or Abstract

This is a short paragraph — sometimes two — that states the key purpose of the report and identifies, but does not elaborate on, major supporting points. Its purpose is to identify, briefly, what is in the report. It saves readers the tedium of poring over reports to find information; with an abstract they will know whether or not a report contains the information they seek. Abstracts are vehicles of convenience for readers.

Introduction

The introduction explains the purpose of the report and cites key information relevant to the findings. Many writers find that it's easier to write the introduction after the main body of the report is complete. Certainly, in reviewing the draft, it may become apparent that key information deserves recognition in this component.

Findings and Recommendations

When a report identifies results or makes recommendations, this element is appropriate. It can come at the beginning of the report, but many companies include it as the last element after a conclusion, but before appendices, if any.

Background

A background section gives a description of the events leading up to the preparation of the final report. This section is often called "History" or "Overview." Sometimes a report will include a history section followed by a background section. This approach works well if, for instance, a generic company is developing a drug product that is already on the market. In that case, a person preparing a summary report of the development of the product may include a history of the brand drug followed by a background of the development of the drug within the generic company.

Supporting Information

The information that offers proof of the premise of the report usually follows logically. Data can be organized in any number of ways. If the document is an analytical development report, for instance, the data may be laboratory methods. If the report assesses equipment that may become additions to a production area, it may include a section discussing each piece of equipment under review. The supporting data, in short, can include any number of sections. The supporting data can also take many forms: charts, graphs, diagrams, schematics — all the useful tools for delivering information.

Conclusion

This component summarizes, again, the gist of the report and confirms the present status of the project.

Glossary

Depending on the audience, writers often opt to include a word list, if certain terms are peculiar to a particular project and readers may misinterpret them. The purpose of a glossary is to prevent misreading. In submissions to the government, a glossary and/or acronym list is usually part of the prefatory information that precedes the actual report.

Appendices

> Any number of formats will work for presenting complex information. The goal is to present it in such a way that it's clear, concise, and consistent.
>
> **David Miller, Quality Manager**

A report may include one or more appendices. Appendices generally contain information relevant to a certain project, but not critical to the report itself. They are vehicles for assembling related information and providing it to readers.

To determine which report elements you need to include, you must first organize the information you have. This is perhaps the most difficult part of writing any complex document. The place to begin is by separating the data into related areas. For instance, a scientist found herself with six months' worth of data that needed to be presented in a summary report for a new generic product, a document that was to be part of an Abbreviated New Drug Application (ANDA) submission. Her data included information about the brand company and product, the product indications, the testing methods, the history of development within the company, and the specifications for the product.

In preparing the report, the author first asked, "What is the purpose of this document?" The answer was to affirm validation of the method for the product, with its supporting data. Here's what she did. She first prepared a short passage to introduce the test procedures and presented the tests sequentially. Then she composed a written discussion about the method and the validation. Next, she described the product and its indications. After that, she sorted through the historical data — the brand product history and the generic product development at her company — and prepared a background section that discussed the history of the brand product and then the history of the generic version to date at her company. She then wrote the introduction and conclusion before finishing the report, with the prefatory information and the cross-referenced appendices holding supporting data. Her finished report followed this order. The steps indicate the sequence in which she prepared the elements.

Cover Page *(Step 10)*

Table of Contents *(Step 9)*

Summary. This component briefly sums up the report. *(Step 7)*

Introduction. This component mentions, but does not explain, the tests and affirms validation of the method. *(Step 5)*

Background. This section gives a brief overview of the history of the development of the product at the company and makes mention of the brand company and product. *(Step 4)*

Properties. This part of the report identifies the product and its properties, gives the structural formula, and presents the specifications. *(Step 3)*

Analytical Procedures. Here the procedures are delineated. *(Step 1)*

Discussion and Methods Validation. This is the "meat" of the report. *(Step 2)*

Conclusion. This sums up yet again the report and its purpose. *(Step 6)*

Appendix. Here is where the writer placed data related to the report and referred to in it. *(Step 8)*

Writing Components of Summary Reports

As we've seen, larger reports don't have to begin at the beginning and go to the end. In fact, when data are accruing for inclusion into a summary report, whether it is for a CRO-conducted study or a regulatory submission, it's a good idea to get the information down as it comes together.

Summaries and Abstracts

Many writers consider the short summary or abstract a difficult component, because it requires extrapolating the key information and condensing it into a short paragraph. It is common for the summary to appear in boldface, alone on a page. However, it can work as well at the beginning of a report on the same page as an introduction. You'll see that approach in some of these examples, because it is efficient.

The following summary is from an analytical development report. The purpose is to establish that testing is complete and that the methods for dissolution have been established for the product.

> Extensive naproxen sodium testing is complete. Assay and related compounds method HPLC/2323/AS1 has been validated. Two dissolution methods, UV/2323/DI1 and LC/2323/DI1, have been validated. LC/2323/DI1 is the method Ronway will employ for dissolution.

Note that many writers might say "This report discusses the validation of naproxen sodium." Logically, however, readers already know that by nature of the report's title. What the summary should include is the most important information in the report itself. It should not talk about the report.

Consider a summary for a product report. Product reports affirm the efficacy of products on a regular basis. They typically present data for specific products in various dosages and include the dates of the review period. An annual product report for a solid dosage product usually includes the following data.

- Table of batches manufactured, with master formula numbers, manufacturing dates, and the disposition of the batch
- Table of granulation/blend information
- Chart of granulation loss on drying (LOD) percentage by batch
- Table of encapsulation information
- Chart of high, low, and average tablet weight by batch
- Table of finished product information
- Table of percentage of final yield by batch

Once all the data are assembled and the report complete, a summary pulls it all together. Here's a summary for a product review for oxazepam capsules:

> Pcatz manufactured twenty batches of oxazepam tablets, 10 mg strength, sixteen batches of oxazepam tablets 15 mg strength, and ten batches of oxazepam tablets 30 mg strength between February 1, 2004, and January 31, 2005. All batches received distribution approval, and all finished product test results were within specification. No customer complaints were received for any of the strengths. All in-process test results and yields were approved.

The Executive Summary

This component is worth examining because it can dramatically help busy readers understand what's in a report. It's particularly useful when readers must evaluate many reports to make decisions. People in project planning, for instance, may be in a position to determine which products the company should pursue and therefore will review a large body of documents to gain the information they need. An executive summary gives more information than a standard abstract because it supplies supporting data as well.

The following passage, an executive summary, explains information from a report that is typically reviewed by executives in a position to make decisions about the direction of the business. If the information in the executive summary is pertinent, they will then read the entire report. If not, they have received a comprehensive overview but haven't devoted too much time to researching material unrelated to their own specific projects.

Executive Summary

Second Generation Dopamine Agonists in the Market for Parkinson's Disease

The second-generation dopamine agonists that entered the market for Parkinson's disease in 1997 are providing early indications that they improve the treatment of the disease. Continuing clinical studies with SmithKline Beecham's Requip (ropinirole) and Pharmacia & Upjohn's Mirapex (pramipexole) demonstrated that the drugs better control Parkinson's disease than levadopa/carbidopa and bromocriptine, standard therapies for the disease. Some studies have found that when ropinirole is used as a single agent during the early stages of the disease, the drug may slow progression of Parkinson's disease more than levadopa does. The studies found that:

The loss of dopaminergic function in the putamen area of the brain was 13% for patients on ropinirole, versus 18% for those on levadopa.

In the worst-affected areas of the caudate region, a 2.6% loss of function was found in ropinirole patients and an 8% loss for patients receiving levadopa.

The loss of function was slower in patients who had symptoms for less than 2 years and who had a significant interaction between treatment and disease progression. For this group, the putamen loss on the worst-affected side was only 1.4% for ropinirole patients but 30.3% for L-dopa patients.

Ropinirole patients preserved function in the worst-affected putamen and caudate areas better than bromocriptine patients. Function in the putamen decreased by 3.6% for ropinirole compared to 13.2% for bromocriptine, and by 2.6% and 9%, respectively, in the caudate area.

Ropinirole may reduce the need for levadopa: 34% of patients receiving ropinirole needed levadopa, compared to 42% of bromocriptine patients.

Both Mirapex and Requip are expected to have a noticeable impact on the Parkinson's disease market beginning in 2000 when their U.S. sales are projected to be about $200 million and $125 million, respectively. The current U.S. market for Parkinson's disease drugs is about $400 million.

An extensive amount of product development is under way for Parkinson's disease. Although many of those products are also second-generation dopamine agonists, several treatments derived from new technologies are also in development. These technologies include glutamate antagonists, COMT inhibitors, neurotrophic factors, and gene and cell therapies. One program that is considered very promising is a collaboration between Guilford Pharmaceuticals and Amgen to develop neuroimmunophilin ligands for 10 neurological disorders, including Parkinson's disease. Neuroimmunophilin ligands are small organic molecules that can cross the blood-brain barrier and bind to immunophilins, which are proteins that are involved in nerve cell growth. Early research data indicate that the molecules promote nerve cell growth and protect nerve cells against damage.

Courtesy of the Genesis Group

Control the Tenses

When summarizing information, be careful to control the tenses. A press release may summarize what was done in a clinical trial using the past tense, but will present the results in the present. A summary of a method validation project might go like this:

The project is complete. — *The status now.* (simple present tense)

We **had** originally **developed** two methods, but **were** unable to validate them. — *Action occurred in the past prior to another action in the past.* (past perfect and simple past)

After careful analysis, we developed a third method. — *Action completed in the more recent past.* (simple past)

We have completed the protocol for this method. — *What has just been completed.* (present perfect)

We **are now validating** this method. — *What is going on right now.* (present progressive)

Upon confirmation of validation, we will do a pilot batch. — *What will happen next.* (simple future tense)

The Background

We **had considered** this project two years ago. — *Occurred in the past before something else in the past.* (past perfect)

Then **we discovered** our production equipment was not capable of producing the product in quantity. — *Occurred after original consideration.* (simple past tense)

Since that time, **we have updated** the equipment **and determined** our capability to manufacture the product. — *Occurred after the equipment deficiency discovery.* (present perfect)

Upon approval of the decision to manufacture the product, **we consulted** the USP, but the **USP does not have** a suitable method for this product. — *The action, and the finding, which is a constant in the USP volume.* (past and present tense)

That **precipitated** the development of our in-house method, which we **have** just **validated.** — *Happened in the past and led to the present stage.* (past tense and present perfect)

And **we are now developing** the product. — The last step in the history of the decision to manufacture the product. (present progressive)

Introductory Elements

Sections of longer reports typically begin with an introductory element that presents an overview of what the section contains. The example on page 275 from the Chemistry, Manufacturing, and Controls section of a submission document for a solid dose drug product for oral administration tells what has changed since the last submission, identifies the sections in the report where detailed supporting data reside, identifies the investigational drug product formulation, briefly discusses risk, and affirms compliance with the regulations.

3.0 Chemistry, Manufacturing, and Controls Summary

WellBeing Incorporated has developed new formulations since the filing of the CMC information amendment for application 11,111 for Welldrug last year. Therefore, we have updated the sections on composition, manufacturing process, specifications, and stability. To achieve a product with a longer shelf life we have developed a new formulation.

Summary of changes to manufacture of Drug Substance (Section 5.1):

- Specifications for the intermediates WC-7-z, WD-2-a, and WF-4-b have been updated.
- Specifications and the analytical method for the Welldrug hydrochloride drug substance have been updated. WD-2-a has been withdrawn from the specifications. WD-2-a is now in the specification for the starting material WB-11-x. We have repeatedly shown that it is not present in the drug substance.
- The stability data for lots 501 and 504 have been updated.

Summary of changes to manufacture of Drug Product (Section 5.2):

- The composition of the capsule has been modified to achieve a more stable formulation.
- Minor modifications of the specifications and analytical method have been made to comply with the changes in the formulation.
- New stability data to support the new formulation.
- Modification in the container closure system to limit the amount of moisture.

This section provides a summary of the chemistry, manufacturing, and controls changes for Welldrug capsules for oral administration. The active pharmaceutical ingredient (drug substance) in the investigational drug product is 2,3,4 Beta-hydrochloride, also referred to as Welldrug hydrochloride or WD-5225. The current formulation contains 1 mg, 3 mg, and 5 mg Welldrug as Welldrug hydrochloride.

In addition to the active pharmaceutical ingredient, Welldrug capsules contain the following well-known pharmaceutical excipients: dibasic calcium phosphate, corn starch, polyethylene glycol 6000, microcrystalline cellulose, sodium starch glycolate, and magnesium stearate.

Table 5.1 presents a summary of the drug substance batches. Table 5.2 summarizes drug product lots. To the best of the sponsor's knowledge, neither the chemistry nor the manufacturing of the drug substance or drug product presents any signals of potential human risk. Completed non-clinical and human safety studies support this statement. The preparation of the drug substance and drug product for the non-clinical GLP studies was performed in accordance with cGMPs.

Courtesy of Kristine Ogozalek

Supporting Information

The example beginning on page 277 is an FDA-requested addition to an IND application for a bioelectronic device that detects HIV in human blood and blood products. This segment presents nonclinical data that was not present in the nonclinical portion of the initial submission. An empirical method (testing serial dilutions of specimens with a known concentration of the target substance) was used to determine the sensitivity data. This additional part to the IND shows that the product has met the goal of 95% detection of HIV-1 with two lots of reagents.

10.1.2.1 Detection of HIV-1 in Dilution Sensitivity Panels

Purpose

The analytical sensitivity of HIV-Gator for the detection of an HIV-1 viral RNA component was measured. Two pilot lots of HIV-Gator reagents were used to test a serially diluted panel of HIV-1 World Health Organization (WHO) standard (97/656). This standard has a known copy level of HIV-1. The analytical sensitivity of HIV-Gator was compared to a commercially available HIV-1 nucleic acid testing kit.

Materials and Methods

An analytical sensitivity panel comprised of HIV-1 WHO international standard (97/656) at 600, 200, 60, 20, 6, and 0 infectious unit (IU)/ml was prepared by serial dilution. HIV-1 panel members were prepared in negative human serum.

Analytical sensitivity panel members frozen in 5 ml single-use aliquots were thawed at room temperature immediately before testing. Panel members were tested using HIV-Gator. Run validity and acceptance criteria for each assay are described in **SOP 000268 and Protocol 0156, "Evaluation of the Analytical Sensitivity of HIV-Gator".**

Four operators tested 10 replicates at each copy level for a total of 40 replicates per target level using 2 lots of reagents, FEE and FEF. The combined results of FEE and FEF were also measured compared to a commercially available qualitative assay for HIV-1 RNA (80 replicates per target). Invalid reactions were not retested or included in the analysis of analytical sensitivity. False positives were measured. If a single replicate of the 10 negative replicates in a given run was reactive, the run was repeated to ensure that any potential contamination encountered during a run would not affect sensitivity analysis.

Results

Tables 10.1.2.1-01 and 10.1.2.1-02-summarizes the results for detection of HIV-1 WHO standard (97/656) with HIV-Gator using Lots FEE and FEF, and HIV-Gator (combined lots) versus the commercially available assay individually and combined. **Tables**

10.1.2.1-03 02 and 10.1.2.1-04 shows estimations of the 50% and 95% detection rates by probit analysis for HIV-Gator with Lots FEE and FEF and HIV-Gator (combined lots) versus the commercially available assayindividually and combined.

Detection of HIV-1 WHO Standard (97/656):

HIV-1 WHO standard detection with HIV-Gator was 100% at 600, 200, and 60 IU/ml for both FEE and FEF **(Table 10.1.2.1-01)**. The detection rates at 20 IU/ml and 6 IU/ml were 71.8% and 67.5% for FEE and 32.5% and 27.5% for FEF respectively. HIV-1 WHO standard detection with HIV-Gator was 100% at 600, 200, and 60 IU/ml for combined lots FEE and FEF versus the commercial assay **(Table 10.1.2.1-0201)**. The detection rates at 20 IU/ml and 6 IU/ml were 68.8% and 30.0% for combined lots FEE and FEF and 66.3% and 33.8% for the commercial assay respectively. There were no significant differences in detection rates observed between the 2 reagent lots or between the 2 assays (data not shown).

Overall Summary of Invalid and False Reactive Rate:

While testing FEE and FEF, no invalid reactions were observed out of a total of 160 80 reactions, or 0%, which met the design goal for of ≤0.5% invalid rate for the assay. No false positive reactions were observed out of 160 80 total negative samples tested, or 0%, which met the design goal for of ≤0.5% false positive rate for the assay.

Probit Analysis

Tables 10.1.2.1-03 02 and 10.1.2.1-04-shows the predicted 50% and 95% detection rates in IU/ml for each target from probit analysis of the results obtained from testing FEE and FEF. From probit analysis, it is estimated that FEE and FEF have a 95% detection rate for HIV-1 WHO standard of 11.5 and 13.5 IU/ml by HIV-Gator. The predicted 95% detection rate for the HIV-Gator for both lots combined was very similar to that of the commercial assay, with predicted 95% detection rates of 12.4 and 14.2 IU/ml for the HIV-1 WHO standard respectively.

The results of the studies performed using FEE and FEF can be found in **Hemetronic Corporation R&D Report D00182, "Evaluation of the Analytical Sensitivity of HIV-1 with HIV-Gator using Lots FEE and FEF" and Report D00187, "Evaluation of the Analytical Sensitivity of HIV-1 with HIV-Gator Versus A Commercially Available Assay".**

Conclusions

Analysis of the analytical sensitivity data indicates that with reagent lots FEE and FEF, HIV-Gator met the goal of ≥95% detection of HIV-1 WHO international standard (97/656) at 200 IU/ml.

Table 10.1.2.1-01 Detection of HIV-1 WHO Standard with HIV-Gator

HIV-1 IU/mL	Lot FEE		Lot FEF		Combined Lots	
	No. Reactive/ No. Tested	Percent Positive	No. Reactive/ No. Tested	Percent Positive	No. Reactive/ No. Tested	Percent Positive
600	40/40	100	40/40	100	80/80	100
200	40/40	100	40/40	100	80/80	100
60	40/40	100	37/40	100	80/80	100
20	28/40	71.8	27/40	67.5	55/80	68.8
6	13/40	32.5	11/40	27.5	24/80	30.0
0	0/40	0	0/40	0	0/40	0

Table 10.1.2.1-02 Probit Analysis of HIV-1 WHO Standard with HIV-Gator

Reagent Lot	Detection Probabilities (IU/ml)	
	50% (95% Fiducial Limits)	95% (95% Fiducial Limits)
FEE	5.7 (4.6 to 7.9)	11.5 (8.9 to 17.5)
FEF	8.0 (6.1 to 9.9)	13.5 (10.0 to 21.5)
Combined Lots	6.0 (4.8 to 7.9)	12.4 (9.8 to 19.7)

Concluding Information

Often interim summary report findings feed into larger projects. Consider the number of development reports that generate for a single product. These can ultimately feed into a submission report for product approval. Each such interim report requires presentation of data and then discussion of findings. This excerpt of a method validation report does that job. The report in total is 55 pages long. It has these components:

- Cover sheets
- Signatures
- Table of Contents
- Summary
- Introduction
- Drug product properties and indications
- Product specifications
- Tests and test results
- Discussion and Methods Validation
- Appendices (28 pages of supporting data cross referenced in the text of the report)

Here's the discussion component of the report.

9 DISCUSSION AND METHODS VALIDATION

We have validated the methods for diflunisal tablets, 250 mg and 500 mg. Appendix contains all figures and tables referred to in the following discussion.

9.1 Appearance

Ronway analysts used the in-house method to determine organoleptic properties. All diflunisal tablets, 250 mg and 500 mg, subjected to organoleptic testing meet specifications.

9.2 Identification

In the HPLC method for identification, the retention time of the diflunisal peak in the chromatogram of the assay preparation corresponds to that of the standard preparation, as Figure 1 shows. In the TLC method the Rf value of the principal spot in the thin-layer chromatogram of the test solution corresponds to that obtained from the standard solution when examined under long wavelength ultraviolet light. Figure 2 shows typical chromatograms.

9.3 Assay

The assay method is based on the assay procedure for diflunisal tablets in the current USP. The concentrations of the Standard and Assay preparations are one-tenth of those called for in the USP procedure because the linearity plot at the USP concentrations gives an unacceptably low coefficient of determination. The diluted solution gives an (R^2) of 0.9999456 and a y-intercept of 1556.7, which are more acceptable values. Tables 1 and 2 and Figures 3 and 4 give linearity data and plots. Figure 1 shows the typical chromatograms of the assay preparation and standard preparations for diflunisal 250 mg tablets.

9.4 Assay Validation

The following tests were used to validate the assay method. The tests verify that the method is rugged and shows no significant variations due to change of system, chemist, or column.

9.4.1 Selectivity of the method from forced degradation compounds.

The selectivity of the method has been demonstrated by showing that the diflunisal peak is resolved from its related substances in solution under stress conditions. A solution of 1 mg/mL was subjected to the following stress conditions and then tested by HPLC.

- **Acid hydrolysis**

1. Prepare a solution of 1 part hydrochloric acid and 10 parts methanol.
2. Then prepare a 50 mg/mL solution of 1 mg/mL diflunisal in 0.1 N methanolic hydrochloride acid.
3. Keep the solution in ambient laboratory conditions for five days.
4. Transfer a 5.0 ml aliquot of the sample to a 50 ml volumetric flask and dilute to volume with the mobile phase.
5. Mix well, filter, and inject the solution into the chromatographic system as listed in the assay section.

- **Base hydrolysis**

1. Prepare a 50 mL solution of 1 mg/mL diflunisal in 0.1 N methanolic sodium hydroxide.
2. Keep the solution in ambient laboratory conditions for 96 hours.
3. Transfer a 5.0 ml aliquot of the sample to a 50 mL volumetric flask and dilute to volume with the mobile phase.
4. Mix well, filter, and inject the solution into the chromatographic system as listed in the assay section.

- **Light**

1. Expose a 1 mg/ml solution of diflunisal in the mobile phase to the intense UV light of a Sun-Lighter at 30¡C for 24 hours. The intense light of the Sun-Lighter is approximately equal to 48 times that of natural sunlight.
2. Cool and transfer a 2.5 mL aliquot of the sample to a 2.5 mL volumetric flask and dilute to volume with the mobile phase.
3. Mix well, filter, and inject the solution into the chromatographic system as listed in the assay section.

- **Results**

Acid hydrolysis

The chromatogram of the acid hydrolysis sample (Figure 5) shows a large peak for diflunisol (99.9% of total peak area) and one small peak at a retention time of 6.78 minutes (0.1% of total peak area).

Base hydrolysis

The chromatogram of the base hydrolysis sample (Figure 6) shows a large peak for diflunisol (99.9% of total peak area) and one small peak at a retention time of 6.97 minutes (0.1% of total peak area).

Light

The chromatogram of the solution exposed to intense UV light (Figure 7) for 24 hours showed a large peak for diflunisal (99.9% of total peak area) and one small peak at a retention time of 6.84 minutes (0.1% of total peak area.

- **Conclusion**

Diflunisal is stable in acidic and basic solutions and to rigorous exposure to UV light with only one small degradation peak evident in each case. The chromatograms show the selectivity of the method by separating this peak from the diflunisal peak.

9.4.2 Linearity

Table 2 lists the peak areas of standard solutions with diflunisal concentrations ranging from 85.0 to 120.0 mcg/ml (representing 8.0% to 120% of the assay concentration). Figure 4 shows the peak area versus concentration plot of the standards solution is a linear curve with a coefficient of determination (R^2) of 0.999893.

9.4.3 System suitability

- **Reproducibility**

  Table 3 shows the peak areas obtained from the chromatograms of five replicate injections of the standard preparation with a relative standard deviation of 0.1%.

- **Tailing factor**
  Figure 8 shows the tailing factor for the diflunisal to be 1.25.

9.4.4 Method reproducibility

Six replicate assays of diflunisal tablets, 250 mg. lot RW69-042A, have demonstrated system reproducibility. Table 4 shows the average result of 250.9 mg tablet with an RSD of 1.6%.

9.4.5 Recovery and precision

Recovery studies were performed by assaying a placebo powder blend for diflunisal 250 mg tablets to which known amounts of diflunisal were added at 80, 100, and 120% levels of the label claim. Table 5 shows an average percent recovery of 100.8% with an RSD of 1.7% from a total of 10 spiked placebo samples at 80%, 100%, and 120% of the label claim for the 250 mg tablets.

9.4.6 Excipient interferences

No interference was observed from the formulation excipients. Figures 9 and 10 show chromatograms of placebo extractions for both diflunisal formulations.

9.5 Uniformity of Dosage Units

The current USP details the procedure for content uniformity. We have validated the uniformity of dosage units by the following tests.

9.5.1 Linearity

Table 6 lists the absorbance values at 550 nm of standard solutions with concentrations of diflunisal ranging from 37.5 to 112.5 mcg/ml (representing 50.12 to 150.6% of the method concentration). Figure 11 shows a plot of absorbance at 550 nm versus concentration of the standard solutions to be a linear curve with a coefficient of determination (R^2) of 0.999937.

9.5.2 Excipient interference

The absorbance of the placebo extract was insignificant (less than 2% of the standard solution) with respect to the absorbance of the content uniformity test solution (A = 0.005 for the 250 mg tablets and A= 0.008 for the 500 mg tablets).

9.6 Related Compounds

9.6.1 HPLC

The related compounds test uses the BP HPLC related substances test procedure for diflunisal tablets except for the test preparation. The quantification of related substances is calculated versus the internal standard of fluoranthene.

The acetonitrile/water extraction (described in the BP monograph for diflunisal bulk drug) was the method of extraction.

9.6.2 TLC

This related compounds test is according to the BP TLC Related Substances test procedure for diflunisal tablets except for the test preparation. The quantification of individual related substances is calculated versus a standard of 4-hydroxybiphenyl. We used the methanol extraction, described in the BP monograph for diflunisal bulk drug, instead of the ether extraction for diflunisal tablets as a matter of convenience.

9.7 Dissolution

Tables 7 to 10 list the dissolution profiles of Ronway diflunisal tablets and the innovator's 250 and 500 mg tablets. Figures 12 and 13 graphically display these dissolution profiles. Five percent USP alcohol is used to dissolve the standard before diluting with dissolution medium because diflunisal has limited solubility in the medium at room temperature. The dissolution procedure has been validated by the following tests.

9.7.1 Linearity

- **250 mg tablets**

Table 11 lists the absorbance of diflunisal standard solutions at concentrations ranging from 0.111 to 0.347 mg/mL (representing 40% to 125% of the test concentration). Figure 14 shows a linear curve with an approximate zero intercept and a coefficient of determination of 0.999988 for the plot of the absorbance versus concentration.

- **500 mg tablets**

Table 12 lists the absorbance of diflunisal standard solutions at concentrations ranging from 0.223 to 0.695 mg/ml (representing 40% to 125% of the assay concentration). Figure 15 shows a linear curve with an approximate zero intercept with a coefficient of determination of 0.999637 for the plot of the absorbance at 306 nm versus concentration.

9.7.2 Placebo interference

The absorbance of the placebo extract was insignificant (less than 4% of the absorbance of the standard solution) with respect to the dissolution test solution (A = 0.018 for the 250 mg tablets and A= 0.021 for the 500 mg tablets).

Analytical Summary Report AD463

CONFIDENTIAL Page 25 of 55

9.8 Moisture

The specification for moisture in diflunisal tablets is not more the 6.0%. The experimental result was 1.8% for diflunisal 250 mg tablets, lot RW69044A. The specification for moisture content was based on the sum of upper limits of moisture allowed for each individual ingredient in the formulation of the diflunisal tablets.

Courtesy of Li Hang Chow, Ph.D.

Publications

Letting industry and investors know what's going on with studies, either *in vivo* or *in vitro*, or more recently in silico (electronically controlled) is a form of summarizing information. Authors prepare journal articles according to the guidelines laid out by the publisher. The *Journal of the American Medical Association* (*JAMA*) for example has explicit requirements, and they are clear cut. Most other journals do as well. Usually there's an article word range and an abstract word limit. You can print out requirements for articles from the web. Conferences also make requirements available electronically, so, for instance, if you are presenting at the American Society for Hematology, you can learn what the specifics are for submitting an abstract or poster. The following manuscript is direct and to the point. Notice how the authors use "we" and the active voice where possible. It includes a table and a figure, which the authors do not position within the text. The publication's staff handle the layout.

Preclinical Antifungal Activity of Chitinase and Glucanse Against *Candida albicans* and *Aspergillus fumigatus*

F. L. Wright, B. S. Bansel*

Mycological Research Institute of America

Research and Development

1735 Everett Avenue

Springtown, CA 94321

*Address correspondence to:

Bushan S. Bansel, Ph.D.

Phone: (723) 650-1234

Fax: (723) 650-1233

E-mail: bbansel@myra.org

Running title: Antifungal Activity of Chitinase and Glucanse against *Candida* and *Aspergillus* spp.

Abstract

We report preliminary *in vitro* and *in vivo* antifungal properties of two cell wall degrading enzymes chitinase and glucanase against the two species of fungi most pathogenic to humans, *Candida albicans* and *Aspergillus fumigatus*. Ninety-nine percent of the cell wall carbohydrate constituents of both these species of fungi contain chitin and glucan. We hypothesized that antifungal activity could be demonstrated by using chitinase and glucanase to disrupt the cell wall. *In vitro* activity was found against the hyphal forms of 5 species of *Candida* and 3 species of *Aspergillus* pathogenic to humans. The minimum inhibitory concentration (MIC) of the enzymes was 1.56 ∞g/ml for *C. albicans* and *A. fumigatus*. There was a synergistic effect using both enzymes together. *In vivo* activity was also prevalent in murine models of candidiasis and invasive aspergillosis with no toxicity.

135 words

Keywords: Antifungal, chitinase, glucanase, chitin, glucan, cell wall, *Candida albicans*, *Aspergillus fumigatus*

Introduction

Candida albicans is a dimorphic fungus that exists in yeast form and mycelial (hyphal filaments) forms. The hyphal form is favored under physiological conditions. It is a commensal organism found in up to 80% of normal humans. However, when alterations occur in cellular immunity (e.g. immunosuppression or AIDS), normal body flora (e.g. loss of normal bacterial flora due to antibiotic or steroid therapy), or normal physiology (e.g. cardiac surgery or indwelling catheters), then *Candida* becomes pathogenic and its growth flourishes. Most frequently this causes skin and mucosal infections, but severe systemic infections can result. Several species of *Candida* have been isolated from humans: *C. albicans*, *C. krusei*, *C. tropicalis*, *C. parapsilosis*, and *C. galbrata*.

Aspergillus fumigatus is a filamentous fungus (mold) normally found in soil, leaf litter, and air. *Aspergillus* spp. are well know to play a role in opportunistic infections, allergic states, and toxicoses in humans. Immunosuppression is the major factor predisposing to development of opportunisitic infections. These localized and disseminated infections are termed aspergillosis. Species of *Aspergillus* isolated from humans include *A. fumigatus*, *A. flavus*, and *A. terreus*.

Among all filamentous fungi, *Aspergillus* is in general the most commonly isolated one in invasive infections. It is the second most commonly recovered fungus in opportunisitic mycoses following *Candida*. Invasive infections by *C. albicans* or *A. fumigatus* are medically serious; they cause significant morbidity and mortality. *Candida* is the most prevalent nosocomial microbial infection (6%) (1). Aspergillosis leaves a mortality rate of 40% despite treatment (2). Like other microbes, *Candida* is acquiring drug resistance to the most commonly prescribed antifungal fluconazole (3). It's necessary to test and report novel antifungals since the rate of immunocompromised patients increases yearly.

We targeted carbohydrate constituents of the cell wall since these sugars are not present in human cells. Human cells do not have cell walls. Breakdown of the cell wall disrupts the integrity of the fungal cell and inhibits growth. The enzymes chitinase and glucanase degrade chitin and glucan. We tested these enzymes separately and together to see if there was *in vitro* activity. This was followed up by testing both enzymes together *in vivo* in murine models of candidiasis and aspergillosis. Preliminary tissue toxicity was tested. This study was conducted to test for antifungal activity of chitinase and glucanase in order to conduct future studies to pinpoint the antifungal effect of chitinase and glucanase and to test for the feasibility of using these purified enzymes as a basis for therapy.

Materials and Methods

Chitinase and glucanase

Chitinase (EC 3.2.1.14) and 1,3-_-glucanase (EC 3.2.1.58) were obtained from the Sigma-Aldrich Corporation (St. Louis, MO). The enzymes were further purified to >99% purity using gel filtration chromatography. Purity was measured by SDS-PAGE. Activities of the enzymes were tested using commerically available *in vitro* assay kits from Sigma-Aldrich. Lyophilized aliquots of both enzymes alone or together (1:1) were dissolved in water for the *in vitro* experiments and in sterile physiological saline for the *in vivo* experiments.

Organisms and cultural conditions

C. albicans was obtained from the American Type Culture Collection (ATCC, Manassas, VA). *C. krusei*, *C. tropicalis*, *C. parapsilosis*, and *C. galbrata* were obtained from the National

Collection of Pathogenic Fungi (Bristol, UK). *A. fumigatus*, *A. flavus*, and *A. terreus* were obtained from ATCC. All fungi were cultured and maintained according to the recommended growth conditions. Only the hyphal forms of the *Candida* spp. were tested.

Antifungal testing *in vitro*

The protocols used to measure the MICs were the National Committee for Clinical Laboratory Standards (NCCLS, Wayne, PA) M27-A2 reference method for broth dilution antifungal susceptibility testing of yeasts; approved standard-second edition and NCCLS M38-A reference method for broth dilution antifungal susceptibility testing of filamentous fungi; approved standard.

Antifungal testing *in vivo*-murine model of candidiasis

After a dose-ranging experiment, the activities of chitinase and glucanase in combination were assessed against infection by *C. albicans* only. A 0.1 ml bolus (4_10^{6} CFU inoculated) of yeast was inoculated into the lateral tail vein of female mice. Two hours after infection, randomized groups of 10 animals were given a test group of chitinase and glucanase dissolved in sterile physiological saline at a dose of 2 mg/kg or a control group of sterile physiological saline only. After 2 days, yeast cell counts were determined for kidney, liver, and spleen.

Antifungal testing in vivo-murine model of invasive aspergillosis

Female mice (n=10/group) were rendered immune-suppressed to *A. fumigatus* infection by treatment with cyclophosphamide at 150 mg/kg via injection into the lateral tail vein 3 days prior to infection with 0.1 ml bolus containing 10^{6} spores of *A. fumigatus*. Treatment was initiated immediately after infection by administration of chitinase-glucanase at 2 mg/kg/day for 7 days. Parameters of efficacy included survival and tissue burden in kidneys, livers, and lungs compared to the negative (untreated) control.

The ethical care and treatment of the animals used in this study were strictly in compliance with all federal guidelines.

Statistical analysis

One-way analysis of variance was and Student's t-test were used to analyze the data. Probability values less than 0.05 were considered significant.

Results

The *in vitro* antifungal activity of preparations of chitinase, glucanase, and chitinase and glucanase together was examined. Table 1 shows the MICs for the chitinase, glucanase, and chitinase-glucanase preparations in 5 species of *Candida*, and 3 species of *Aspergillus*. In all cases and in all fungi tested, the MICs were significantly lower when chitinase and glucanase were tested combined ($P < 0.05$). The MICs ranged from 1.56 ∝g/ml to 10 ∝g/ml for the *Candida* species tested. The MICs ranged from 3.12 ∝g/ml to 10 ∝g/ml for the *Aspergillus* species tested.

The *in vivo* antifungal activity of a chitinase-glucanase preparation was tested in murine models of candidiasis and invasive aspergillosis. Figure 1 shows that the fungal burden was significantly decreased in all particular organs tested in all cases and in both murine models ($P < 0.05$) compared to the untreated controls. *In vivo* antifungal activity was found after a single 2 mg/kg dose of chitinase-glucanase in the murine candidiasis model and after a 2 mg/kg/day dose of chitinase-glucanase for 7 days in the murine invasive aspergillosis model. No post-mortem toxicity was found in any treatment group; no mortality was present.

Discussion

Potent preclinical *in vitro* antifungal activity of chitinase and glucanase has been demonstrated in the most pathogenic species of *Candida* and *Apergillus* (Table 1). Chitinase and glucanase worked synergistically together to produce the *in vitro* fungistatic effect compared to each alone. Further study will have to be made to explore the synergistic effect. Potent antifungal activity has also been demonstrated *in vivo* in murine models of candidiasis and invasive aspergillosis (Figure 1) with no toxicity or adverse effects in the organs examined.

Chitinase and glucanase were postulated to exhibit antifungal activity since chitin and glucan are the major constituents of the cell wall of both fungi. Both enzymes were chosen together since potential broad-spectrum activity against a variety of fungi is desirable. Novel antifungal agents need to be developed since fungi, like bacteria, are capable of developing resistance to current therapies and infection rates continue to increase yearly. The nature of the chitinase-glucanse, substrate interaction needs to be studied informatically.

Acknowledgments

We would like to thank A. Sahai for expert preparation of this manuscript. The research conducted in this study was supported by grant BW-2713293.

References

1. J. R. Walsh *et al.*, *J. Clin. Microbiol.* **40**, 5409 (2002).

2. J. S. Sahai *et al.*, *J. Microbiol.* **299**, 21 (2003).

3. V. Gough, D. Ism, *J. Med. Sci.* **424**, 735 (2003).

Tables and Figures

Table 1. MICs of chitinase, glucanase, and chitinase-glucanase preparations in *Candida* and *Aspergillus* spp.

Species	MIC (∝g/ml)		
	Chitinase	Glucanase	Chitinase-glucanase[a]
C. albicans	3.12	3.12	1.56
C. krusei	6.25	6.25	1.56
C. tropicalis	100	50	10
C. galbrata	50	100	10
C. parapsilosis	12	25	3.12
A. fumigatus	6.25	6.25	1.56
A. flavus	6.25	6.25	1.56
A. terreus	100	50	10

[a] All values were significantly different ($P < 0.05$) than either enzyme alone.

Figure Legends

Figure 1. Fungal tissue burden of the kidney, liver, and spleen/lungs of animals treated with chitinase-glucanase solution in a murine model of candidiasis (2 mg/kg) (A) and in a murine model of invasive aspergillosis (2 mg/kg/day for 7 days) (B) compared to the untreated controls. The symbols represent the means of the fungal burden.

Figure 1.

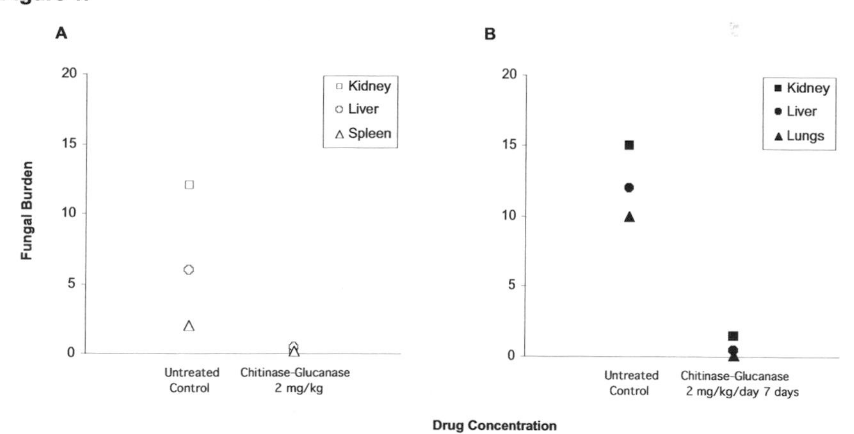

9

Developing a Clear Style

Sometimes people say they have "a way with words." And they do. But often they confuse eloquence with strings of words that lack concrete ideas. Thus, their "way with words" creates gobbledygook that serves no purpose other than to confuse. Consider this excerpt from a technical note:

> Task oriented groups, having had sufficient application of GMP training, will, however, facilitate the achievement of optimal results in the production of product and successfully provide demonstration of an integrated work effort consistent with standards determined to be essential for quality.

The writer here is using an avalanche of words that bury his meaning. "Task oriented groups" indicates that the groups will *work together.* "Having had sufficient application of GMP training" means *training in cGMPs.* "Facilitate the achievement of optimal results in the production of product" means *produce a quality product.* "Provide demonstration of an integrated work effort" means, simply, *show.*

Reworking this message makes the meaning clear.

> Teams trained in GMPs will consistently produce quality products.

> Gobbledygook: writing that is long, pompous, vague, involved, usually with Latinized words.
>
> **A U.S. Senator**

No wonder the original was difficult to understand: It's twice as long as the revised version. Why use 41 words to express an idea that can be clearly stated in eight? When you rid writing of words that add nothing, the information becomes clearer. The purpose of writing, after all, is to present information clearly. If your reader has to wade through words and struggle to find meaning, you risk losing your message. When that happens, you've wasted your time and the reader's.

> A writer expresses himself in words that have been used before because they give his meaning better than he can give it himself, or because they are beautiful or witty, or because he expects them to touch a cord of association in his reader, or because he wishes to show that he is learned and well read. Quotations due to the last motive are invariably ill-advised; the discerning reader detects it and is contemptuous; the undiscerning is perhaps impressed, but even then is at the same time repelled, pretentious quotations being the surest road to tedium.
>
> **Henry W. Fowler**

Gobbledygook uses twenty words where ten will do, as well as strings of words and long terms where clear, direct, short ones will suffice. How can you avoid it? Understand the language you are using and have confidence in your ability to say what you mean. Gobbledygook impresses no one; clean words and structures that communicate clear information do. It doesn't matter who you are or what you do; the ability to make writing clear doesn't rely on the degree of "formal" English background you have. Indeed, language loaded with gobbledygook signals an insecure writer, one more concerned with self than with message and reader.

> Please write in a clear, direct, and active style. Write in the active and use the first person where necessary. Try to avoid long sentences that have several embedded clauses.
>
> ***The British Medical Journal***
> **(http://bmj.com)**

If you multiply the process of constructing wordy, awkward passages hundreds of times every day in hundreds of offices, at clinical sites, and in laboratories, you wind up with reams of useless writing. The process serves to confuse messages, increase workloads, and antagonize workers. There's good news, however. People are becoming aware that clear writing is the best writing.

The Plain Language Movement

"Modern English, especially written English, is full of bad habits which spread by imitation and which can be avoided if one is willing to take the necessary trouble," George Orwell wrote in 1946 in his essay *Politics and the English Language*. It holds equally true today, but the good news is that dictates for clear language have received recognition. In the 1970s, the president of the United states attempted to get rid of "governmentese." And, in 1998, the US president signed an executive memorandum that says all federal documents, regulations, and forms must be written in plain language by 2002. There have been some positive effects to date. The Food and

Drug Administration (FDA) has supported plain language efforts and, also in 1998, issued a Plain Language Action Plan and has offered training to FDA staff. FDA's website has also been hailed as clear and easy to read.

Some regulations that were obtuse have been rewritten to be clear and understandable. Consider the Occupational Safety and Health Administration standard for "egress." It originally said "Ways of exit access and the doors to exits to which they lead shall be so designed and arranged as to be clearly recognizable as such." It now says "An exit door must be free of signs or decorations that obscure its visibility." The Plain Language Movement continues to gain support. In essence, the movement calls for writing that allows readers to find what they need, understand what they find, and act on their understanding.

Tighten Up and Lighten Up

The overwhelming majority of unclear, obtuse communication errs in excessive use of unhealthy patterns. One such pattern is to write loose sentences with words that add nothing to meaning. You can tighten up your messages by making sure every word counts, and then putting your most important words in the core of your sentence. Consider the following sentence:

> The product is a device that delivers antibiotic transdermally.

The core of the sentence, that is the subject and verb group, is "product is." A rewrite gets rid of the "to be" verb that says nothing as well as the subordinate clause where the content lies, and delivers a more direct and meaningful message.

> The device delivers antibiotic transdermally.

In the rewrite, "device delivers" is the subject and verb group and it takes the direct object "antibiotic." Further, it has just five words, while the original has nine.

Keeping sentences on the short side is also helpful in creating readable passages. Shorter sentences result when every word does duty. Sometimes every word is working, but the sentence is long and unwieldy. If a reader has to reread a sentence to understand it, it's probably too long. In such instances it's usually best to make two sentences. Writers who understand this try to have an average of no more than 20 words per sentence.

> Bad writers, and especially scientific, political, and sociological writers are nearly always haunted by the notion that Latin or Greek words are grander than Saxon ones, and unnecessary words like expedite, ameliorate, predict, extraneous, deracinated, clandestine, subaqueous, and hundreds of others constantly gain ground from their Anglo-Saxon numbers.
>
> **George Orwell, Author**

Another way to make writing more readable is to avoid pretentious words. Use sound American English words of few syllables in place of long, usually Latinate root words such as "utilize," "promulgate," and "magnitudinous." You will be better understood if you simply write "use," "make known," and "large."

Free Writing of Unnecessary Words

> Most technical writing is too wordy by 10% to 60%.

Writing all too often buries its force in the sheer number of words. It may be that the strings of words that help in obscuring content are the result of an insecure hand. Often when writers are not sure what they have to say is "impressive" enough, they look for another way to say it. Or possibly they don't think their words have enough power, so they say the same thing twice, and what they wind up writing is nonsense.Most technical writing is too wordy by 10% to 60%.

Most people are familiar with the following expressions, but if you think about it, they say the same thing twice because they add do-nothing words and phrases, and thus make what's written cumbersome and wordy. The following are typical redundancies and wordy expressions, with their more concise alternatives, but there are many more.

Instead of	Opt for
absolutely nothing	nothing
absolutely complete	complete
absolutely essential	essential
actual experience	experience
adequate enough	adequate
advance planning	planning
advance warning	warning
arrive on the scene	arrive
as evidenced by the fact that	because
ask the question	ask
assembled together	assembled
at the present time	now
at the rate of	at

basic fundamentals	fundamentals
but nevertheless	but; nevertheless
brief in duration	brief
bring to a conclusion	conclude
cancel out	cancel
check into	check
circular disk	disk
combined together	combined
come into conflict with	conflict with
complete monopoly	monopoly
completely opposite	opposite
consensus of opinion	consensus
consequent results	results
continue to remain	remain
cooperate together	cooperate
corporately together	corporately
desirable benefits	benefits
disregard altogether	disregard
downward decline	decline
effectuate improvement	improve
exact same	exact; same
few in number	few
filled to capacity	filled
free of charge	free
give consideration to	consider
grateful thanks	thanks; gratitude
has the ability to	can
hold in abeyance	hold
impact on	affect
in an area where	where
in the course of	during
in the event that	if
in the form of a square	square
in this day and age	now
joint cooperation	cooperation
join together	join
job action	strike
just recently	recently
main essential	essential
make application	apply
maximum possible	maximum
mutual cooperation	cooperation
new innovations	innovations
of a similar nature	similar
one particular example	an example
past history	history
plan ahead	plan

postponed until a later date	postponed
qualified expert	expert
red in color	red
repeat the same	repeat
shed light on	clarify
subsequent to	after
totally unique	unique
true facts	facts
ultimate end	end
visible to the eye	visible
visit with	visit

Note, however, that some seemingly redundant statements are not always that. For instance, technical writers have argued that "final product" is a redundancy. Whereas it might well be in some technical arenas, "final product" in pharmaceutical manufacturing may differentiate between the in-process product and the finished product. Judge your words to gain the precise meaning you intend.

Avoid Generalizations and Ambiguities

A generalization is a catchall, a statement meant to encompass an entire concept or idea. It's easy — and human — to use a pat expression that seems to do the job because to be specific involves analyzing what you are saying to make sure you really understand it yourself. You may even say the same thing over and over again, using different generalizations, rather than explain your point. When you find yourself in this pattern, step back and question how well you really understand the point you're trying to make. Often, blanket statements are a signal that the writer doesn't truly understand the subject matter.

It's true that people sometimes deliberately choose to generalize because they don't want to reveal information, and when that happens, generalizations are valid. A writer may say "a suitable bucket with enough capacity to submerge small parts in a solution of potable water and LpHse detergent" in a cleaning SOP, for instance, rather than "a four-gallon bucket," because if he calls for a four-gallon bucket, that's what operators must then use. A generalization here gives leeway and eliminates restriction without compromising the procedure.

Many times, however, generalizations cause problems because they use weak terminology to cover too broad a range of ideas that should be spelled out. Words may carry different associations for different readers, and if that's the case, you run the risk of being misunderstood. What you intend may not be what the reader receives.

Since good writing relays specific information to people who need it, never assume that a reader will interpret a general term or statement correctly. One laboratory supervisor identified a problem as "timely." His manager identified a related problem as "timely" as well. How should the reader interpret this word? In the first instance the supervisor meant the problem was "time-consuming." The manager, on the other hand, was identifying a problem as "current." It's easier to anticipate the variables in interpretation and use words readers cannot misinterpret.

When you've written something, look at it with a fresh eye, always keeping in mind that words like "useful," "plentiful," "imminent," and "tedious" may mean one thing to you and something entirely different to your reader. What do the terms mean? If the "excipients on hand are plentiful," do you mean there's enough to manufacture two batches, or ten, or fifty?

Once you understand what unintentional generalizations are, you'll be able to fix them. You can cure writing of these blights by asking open-ended questions, ones that require more than a yes or no answer. Ask *Who, What, Why, When, Where,* and *How* to elicit precise information. Questions that demand explanation cure much weak writing.

The Generalization	The Query
It was very informative.	What was informative?
	What did you learn?
Production slowdown	What does "slowdown" mean?
is creating problems.	What are the problems?
Something must be done.	What must be done and by whom?
	When and where shall
	the action take place?

Ambiguities

> Writers can usually resolve issues with vagueness through definitions. They can eliminate ambiguity, either semantic or syntactic, through word-choice and arrangement, respectively.
>
> **Scott Simmons, MA**

Closely related to generalizations are ambiguities. Ambiguous statements can be read to mean two different things, depending upon the reader's interpretation. Important here again is control, so the reader won't have to interpret. The classic tongue-in-cheek recommendation that reads, "I cannot recommend this candidate highly enough," is a sound example of ambiguity. Does the writer intend that the candidate receive his highest recommendation? Or does he intend not to recommend the candidate highly at all? It's better to say either, "I give this candidate my highest recommendation," or "I have some reservations about recommending this candidate."

In the documents you produce, a sentence should have one and only one meaning. Choose precise words and structure them so that the meaning is clear. Check carefully to see that sentences can't possibly mean two different things.

Ambiguities do creep into our communication in innocent ways. When people write, *they* know what they mean. But those who read the written account may not. One of the common problem areas is with pronoun references. Usually these glitches require restructuring an entire sentence. Consider the following:

> She asked her manager if she could perform the HPLC testing.

Does the writer mean to ask whether the manager can perform HPLC? Or, is she requesting permission to do the analysis herself? Such structures can usually be corrected by using a question format.

> She asked her manager, "Can you do the HPLC testing?"

Or

> She asked her manager, "May I do the HPLC testing?"

Consider this excerpt from a performance evaluation:

> Mr. Brown likes working with Mr. Harrison because he is receptive to new ideas.

Restructuring ends the ambiguity.

> Because Mr. Harrison is receptive to new ideas, Mr. Brown likes to work with him.

Or

> Because Mr. Brown is receptive to new ideas, he likes to work with Mr. Harrison.

Excising ambiguities from writing requires looking at the piece with the reader's eye. Can the message have different meanings? For instance, how will a statement like "Visiting FDA inspectors can be helpful" be understood? Does "visiting" describe an on-site inspector or the action of an employee going to visit an inspector? When a directive says, "Waste no time in making the departmental adjustments," should the reader make the adjustments quickly or not at all? Grammatically these examples are sound, but their content can be read to mean opposite things.

When you spot ambiguous statements, the best thing to do is rewrite. Don't let a sentence like "The engineering staff will confer on the new plant site" slip by. Determine whether "The engineering staff will meet on the new plant site" or "The engineering staff will meet to discuss the new plant site" is the correct meaning, and rewrite accordingly.

Other places where ambiguity pops up are when words receive misplacement, particularly in phrases. Consider the following sentence:

> In seven to ten days, we will validate the method.

What's ambiguous here is the prepositional phrase "in seven to ten days." Does this sentence mean the validation will take seven to ten days, or that it will begin in seven to ten days? Here a rewrite is in order:

> The validation will take seven to ten days.

Or

> We will begin the validation in seven to ten days.

Constructions that have coordinate elements can also lead to ambiguity. This is particularly the case with the conjunction "or." This conjunction can mean either an alternative or a synonym.

> Silver gauze or wire maintained at 700° C removes halogens, with the exception of fluoride and sulfur products.
>
> Is wire a restatement of silver gauze, or are they alternatives? The sentence is unclear. A rewrite clarifies.
>
> Silver gauze maintained at 700° C removes halogens, with the exception of fluoride and sulfur products. Wire at the same temperature does the same.

Use Metaphor and Simile Carefully

Metaphor calls something what it is not, while simile uses "like" or "as" to draw an image. These two conventions provide effective ways to help readers visualize what the words say. The following sentences present examples of metaphor and simile.

> Much AIDS research has provided the *building blocks* for cancer therapeutics.
>
> After leaving the buffer, the glass proceeds *like a continuous ribbon* under a sequence of magnetron sputtering glass cathodes that deposit the coating layers.
>
> The process works *like a magnet*, drawing the particles from the compound.
>
> The double helix looks *like a twisted ladder.*

Idiomatic Expressions and Cliches

Metaphor and simile provide the underpinnings for creating idiomatic expressions. English is highly idiomatic, and it will continue to be so. Idiomatic expressions are generally well known to native speakers and writers, and they are overused. Many expressions have become clichés. In their original coinage, they carried punch: They presented a fresh image to clarify an idea. Many clichés are now hackneyed and without strength as the result of overuse.

Phrases such as "neat as a pin" or "all our ducks in a row" originally found their way into our language because they were fresh and new — but with overuse they've lost their effect. When Shakespeare coined "a pound of flesh" he drew a mental picture of a heavy and perhaps unjust price. And, as most native born Americans know, the expression has remained in the vernacular. Yet because people hear it so often it has ceased to render the strong image it did when it was first created. So it is with most clichés: They have their place, but as often as not, other phrases serve better.

Think about non-native born readers and writers as well. To a reader not born into the language, an idiomatic expression can entirely confuse because it presents mental imagery totally unrelated to the subject under discussion. On the other hand, many non-native writers often want to learn and use idiomatic expressions because they frequently signal a degree of language mastery. Nevertheless, using clichés presents inherent problems since readers may not understand the nuances of the words, either when reading or writing.

If your readers know what's coming, they may miss the impact of the message. Worse, if they're non-native readers, they may not understand "The ball is in your court" means you are waiting for action on their part. On all counts, then, it makes sense to use straightforward language and spare readers from idioms and clichés. Of course, you don't want to rob a document of its flavor, so use some discretion here.

Excise Excess Nominalizations

Regarded by many writers and readers as the bane of modern communication, nominalizations are everywhere. Good writers avoid them, for they merely serve to confuse. Nominalizations are nouns used where another form of the word would be better. (Nouns name people, places, things, and ideas.) Consider this newsletter headline: "Production Plans Shutdown." Is production actually planning a work stoppage? Or have its plans been shut down? What does the headline mean? It's unclear because all the words are nouns. What is the subject? Is it "production?" Is it "plans?" What is the verb? Is it "plans?" Or is it "shut down?" Look at how a little modification of the word forms clarifies the meaning: "Production's plans shut down" or "Production plans to shut down." In the first version, the noun *production* is made possessive and modifies *plans,* which naturally makes *plans* the subject. In the second, the words *shut down* are put into the infinitive form (by adding "to"), keeping *plans* as the main verb in the sentence.

To avoid this type of gobbledygook, you first need to understand the core of the message and what's happening to it. Then you have to make sure the reader gets the right information. Use the appropriate word forms. Modifiers and verbs clarify. Too many nouns jumble messages.

Sometimes the noun words are strung into a whole sentence. When that happens, the sentence usually needs to be revamped. You can recognize this pattern by looking for the naming words.

> The <u>discovery</u> of the <u>availability</u> of a new raw <u>material supplier</u> in the <u>area</u> had the <u>effect</u> of <u>giving thought</u> to the <u>development</u> of new <u>dosage strengths</u> of <u>product</u>.

This sentence has twelve forms of nouns: *discovery, availability, material, supplier, area, effect, giving,* and *thought, development, dosage, strengths,* and *product*. A much clearer revision has half that many. (Note that *material* and *dosage* are nouns serving as adjectives in that they modify *supplier* and *strengths,* and *giving* is a gerund, a verb acting like a noun.)

> <u>Finding</u> a new raw <u>material supplier</u> led us to consider <u>developing</u> new <u>dosage strengths</u>.

In the revised form, *finding,* a gerund, is the subject — not *discovery* — and *led* is the verb — not *had.* Getting rid of unnecessary nominalizations automatically puts the main information in the main part of the sentence, which is where readers expect to find it.

Weigh your words. Where's the action? What's being talked about? Look at how the following sentences improve once the nominalizations have been excised:

> **Original:** Our expectation was to conduct an investigation of the physical facilities in August. (13 words; four nouns)
>
> **Revised:** We expected to investigate the physical facilities in August. (nine words; two nouns)
>
> **Original:** We have hope that the supervisor's revision will indicate a respect for the content. (14 words; five nouns)
>
> **Revised:** We hope the supervisor will respect the content. (eight words; two nouns)

Pare Prepositional Phrases

Many languages don't have prepositions, so these constructions can cause problems for non-native writers. Even if a person's birth language has prepositions, they don't always translate. Further, native born writers tend to overdo them.

Prepositions indicate relationships between ideas: distance (to, from, beyond); place (behind, within, in, on); identification (from, of); and so forth. Prepositional phrases always consist of a preposition followed by an object of the preposition. Further, the object of the preposition can be any number of elements: nouns, gerunds, pronouns, and noun clauses. Modifiers often come between the preposition and its object, as well. In the following examples, the preposition and its object(s) are underlined:

> He is going to the new lab.
>
> I completed the report on time, without significant interruption or unanticipated problems.
>
> With the help of my peers, I will finish the project by Friday.
>
> I want to work with whoever takes on the project.

Note that in the last example, a noun clause serves as the object of the preposition "with."

Prepositional phrases are plentiful in English. However, too many prepositional phrases can create a sing-song effect, which in turn detracts from the message. The prepositional phrases in the following passage from a status report are underlined:

> This is a report of the status of the laboratory computerization project during the month of July. In the beginning of the project we investigated the possibility of leasing personal computers from the manufacturers of other equipment in our possession, but within a short period of time we realized we should go ahead with the purchase of PCs for laboratory staff from our manufacturers with specialties in laboratory computer packages.

Would you believe 18 prepositional phrases in two sentences? So many in a row sound like chanting, because English places stress on content words, those words that carry primary meaning, and destresses structure words, those components that form the fabric of the language.

You can rid your writing of a sing-song rhythm by eliminating as many prepositional phrases as possible. One way is to combine them and cut unnecessary and repetitive words. Don't say during the month *of July;* say rather, *during July* (one prepositional phrase; two words; same idea). Or you can rephrase by making July adjectival: "This is the July status report of the laboratory computerization project" (one prepositional phrase, down from three in the original sentence).

Do the same with the second sentence. Where a gerund (an "ing" word acting as a noun) comes after the preposition, consider making it an object of the sentence and eliminating some unnecessary words. You can also rephrase to create relative pronoun clauses (*who, which,* and *that* elements that handle information efficiently). Finally, you need to totally eliminate unnecessary and redundant phrases. If, in the first sentence, you've identified the project, you surely don't need to say "of the project" in the second. Look how much easier the edited version of the second sentence is to read:

> At first we investigated leasing personal computers (PCs) from manufacturers who make our other equipment, but we soon realized we should purchase them from manufacturers who specialize in laboratory computer packages.

One handy way to tell if prepositional phrases are running rampant is to count. If you detect more than three prepositional phrases in a row, you should probably restructure. Look at this sentence:

> This guide for the investigation of topical drug product production at sites of various contract manufacturers is for auditors familiar with the provisions of the current Good Manufacturing Practices regulations for

> pharmaceuticals, with guidance on inspecting selected facets of topical drug product production.

It has eleven prepositional phrases, and some redundancy. Rewording eliminates prepositional phrases and makes the sentence more concise, and consequently easier to understand.

> Auditors familiar with current Good Manufacturing Practices can use these guidelines for inspecting selected facets of topical drug product production at various contract manufacturing sites.

The rewrite has four prepositional phrases, an ample amount for any sentence. Eliminating the prepositional phrases and connected verbiage results in a 26-word sentence; the original has 43.

Eliminate Extraneous Expletives

Sometimes writers just can't get started, so they write "There is" or "It is" and in so doing start weakly. "There" doesn't mean place, nor does "it" designate any particular thing. These words are expletives. They're filling in as subjects to get a sentence started. Unfortunately, they're followed by a weak "to be" verb. And the substance, in turn, follows it. Sentences are strongest when they begin with a doer of action. If you find that your writing regularly starts with "it" or "there," rework it so that the key information falls into the main subject and verb spot.

More than a few memos have started poorly. Look at the following opening of an inter-office memo:

> There have been an increasing number of batch record errors and subsequent deviation investigations.

Notice what's important: batch record errors and deviation investigations. Yet that information is buried in a prepositional phrase, and the main part of the sentence, "There have been an increasing number," says nothing. A good way to strengthen this message is to let "batch record errors and subsequent deviation investigations" serve as a compound subject, and the adjective, "increasing," in a verb form, serve as the main verb. The result is much more readable and requires four fewer words:

> Batch record errors and subsequent deviation investigations have been increasing.

The same pattern is working in the following sentence that talks about adverse events (AEs).

> There were two AEs during the Phase 2a trial.

A rewrite gets rid of the expletive and "to be" verb.

> The Phase 2a trial had two AEs.

The logic behind eliminating needless expletives is this: Expletives automatically push the key information into less important elements of the sentence. So instead of saying "There is no expansion or new service included in the plan," which puts "plan" at the end of the sentence in a verb/prepositional phrase structure, start with "the plan" and say, "The plan does not include expansion or new services."

Purge Weak Verbs

When you write and revise, make sure each sentence is strong in its own right. By making short shrift of expletives, you automatically snuff out many "to be" verbs as well. In fact, most writing benefits from a purging of "being." "To be," the most irregular and widely used of all our verbs, can be shelved in many cases. Besides following expletives, "to be" pokes itself into many sentences and weakens by removing action.

Why doesn't the following sentence have sufficient impact?

> We *are* in receipt of your request, and we *will be* in touch as to your report findings.

This compound sentence has two clauses, "we are in receipt of your request" and "we will be in touch." Each clause has a "to be" verb ("are" and "will be") as a main verb. Look how much stronger the sentence becomes when both are eliminated and main verbs denote action:

> We will contact you when we evaluate your report.

In this sentence, the main verb is "will contact," which denotes action. And action is more specific than being.

The following sentences all suffer from too much "to be." Replacing the weak verb with significant action improves the messages.

> **Original:** He is the man who repairs our machinery.
>
> **Revised:** He repairs our machinery.

Original: Since I am to edit the reports, I will be on the lookout for errors.

Revised: As editor, I will look for errors.

Original: As you are aware, we are planning to do the equipment qualification.

Revised: As you are aware, we plan to do the equipment qualification.

Original: There are not very many generic companies that have attempted to develop this product.

Revised: Not many generic companies have attempted to develop this product.

Assess Double Negatives

You have probably heard that using a double negative makes a positive. When you say "We don't have nothing," you are actually saying that nothing is not what you have, so in fact, you have something. Common in speech, most double negatives would benefit from excision in writing. Indeed, even the American Chemical Society, in its guidelines for writers, advises against using double negatives.

Double negatives call for extra words, for one thing, and the more succinct you are, the better. When a writer says "It is not inconceivable," he is using a double negative. "It's possible," serves just as well and uses two words to do the job, not four.

Writers have often argued that there are degrees, and that sometimes the double negative better indicates a degree. For instance,

> It is not uncommon for two generic pharmaceutical firms to be developing the same product.

If you remove the double negative, this is the result:

> It is common for two generic pharmaceutical firms to be developing the same product.

In the first example, "not uncommon" implies a degree that "it is common" does not. Nevertheless, "not uncommon" remains vague. How common is it? The reader doesn't know. It's better to write something like this:

Occasionally, two generic pharmaceutical firms may be developing the same product.

In the second example, one word, "occasionally," serves the purpose of four, and for that purpose alone, the writing is tighter. So when you write a double negative, question whether another construction would serve your message better.

Pare the Passive Where You Can

Using the active voice when writing makes sense because it's more direct than the passive voice and thus easier to understand. Certain information, it's true, does better in the passive, and you don't want to, nor will you be able to, excise passive constructions from your writing entirely. But too often writers choose the passive for no good reason. Probably because it is so prevalent, particularly in technical writing, it has begun to sound like the standard. Yet the passive voice is a weaker construction than its alternative. By its nature it requires an additional verb to hold the tense and relegates the agent of the action to a prepositional phrase or even eliminates it entirely. Thus, it adds extra words and subordinates or omits significant information.

Active voice: Dr. Nolan signed the investigation report.

Subject verb direct object

Passive voice: The investigation report was signed (by Dr. Nolan). ("to be" verb added; prepositional phrase "by Dr. Nolan" optional)

Moreover, reading studies show that people mentally turn a passive construction around to absorb the meaning. It follows then that delivering information by the most direct route, the active construction, improves readability. Nothing is gained in terms of readability in the following passive constructions; the active voice does a better job every time.

Passive: The results are listed in table two.

Active: Table two lists the results.

Passive: The tests results are included in the appendix.

Active: The appendix includes the test results.

Passive: The method has been developed by the analytical group.

Active: The analytical group has developed the method.

The following sentence employs two passive constructions, and they decrease readability considerably.

> The filling line, which has been serviced by an outside contractor, has been requalified by the engineering team.

A rewrite improves readability considerably:

> An outside contractor serviced the filling line, and the engineering team requalified the equipment.

Sometimes, simply eliminating half of the passives makes writing more readable. There will most likely be many instances when you use the passive, but when you do, make it a conscious choice. The passive works well when you don't know who or what initiated an action or when you feel the object of an action is more important than the initiator.

> The laboratory was audited.
>
> It is rumored that our competitor is filing an NDA this week.
>
> The simple formula is well known.
>
> A security leak was discovered.

Variants of passive construction employ verbs other than "to be." Although not as common as the "to be" constructions, you may occasionally see sentences like these:

> He got rewarded for his efforts in reorganizing the warehouse.
>
> Erica has her work evaluated by the supervisor.

The passive voice does not mean past tense, a common misconception. What it does mean is that "to be" or, occasionally, such verbs as "to get" or "to have" hold the tense and that the verb of meaning is always in the past participle form.

Past: The truck was backed up to the holding port.

Present: The truck is backed up to the holding tank port.

Future: The truck will be backed up to the holding tank port.

One final comment is appropriate here. For half a century, writers have viewed the passive construction as the "scientific" voice. The thinking is this: The focus should be on what is taking place or on results, not on who is conducting or has conducted some action. This reasoning holds merit, but a copious amount of unnecessary passive construction has served to make technical information less easy to understand. The trend now is to use the active voice wherever possible. More and more, technical writing reflects the active voice, which results in clearer, easier-to-understand messages.

Apply Common Sense to False Rules

Somewhere along the line, in the latter half of the nineteenth and the first half of the twentieth century, English teachers and writers took it upon themselves to establish some new "rules." You may have heard them as you developed your understanding of the structure of English. Many of these so-called rules make some sense, it's true, yet to be bound by them exclusively is silly: Sometimes they even hinder messages. What follows are some of the rules you may have heard during the course of your language learning experience, with some explanation of them.

Splitting Infinitives

Question anyone who's had a grammar course and chances are you'll learn that the student has been warned about splitting infinitives. In fact, people who know little about English syntax often proudly assert, " I never split an infinitive," as though splitting an infinitive is a bad thing to do. It isn't.

> Writers who insisted that English could be modeled on Latin long ago created the 'rule' that the English infinitive must not be split: *to clearly state* was wrong; one must say *to state clearly.* But the Latin infinitive is one word, and cannot be split, so the 'rule' is not firmly grounded, and treating two English words as one can lead to awkward, stilted sentences.
>
> **The Oxford American Desk Dictionary**

Infinitives in English are the present tense of the verb, preceded, usually, by "to." The thinking has been not to insert an element between the "to" and the verb. This makes sense, because sound English construction places related words as close to each other as possible. Splitting the infinitive often makes sense, however. Words like "over" and "under" have split infinitives so often, that they have become compound elements. Consider "to emphasize." We also say "to overemphasize" and "to underemphasize." Look at the following sentences:

To finally understand what was happening was good.

Our commitment is to really move forward quickly.

Each sentences is clear; and leaving the infinitive intact would create awkward cadence (finally to understand or to understand finally; really to move or to move really). If you are a person who is uncomfortable splitting infinitives, simply don't do it. If you have no qualms about splitting them, however, do guard against weakening your writing by repeatedly separating related elements. Remember, you can always recast the sentence in other words.

Ending a Sentence with a Preposition

Putting a preposition before its object is logical; to do so also adheres to the commonsense principle that calls for putting related ideas as close to each other as possible. In this sense, it's a good idea to watch how you end your sentences. But sometimes people overcorrect — that is, in an effort to keep those wayward prepositions duly attached to their elements, people effect a stilted, and perhaps even laughable, style. We need to remind ourselves that "Never end a sentence with a preposition" is a rule that never existed! It's another shibboleth handed down by our nineteenth- and twentieth-century grammarians.

The following sentences are perfectly fine as they are:

What did you say that for?

Whom are they going with?

That's the training program I'm interested in.

That's something I won't put up with.

Notice that to internalize the preposition creates unnatural and awkward sentences:

For what did you say that?

With whom are they going?

That's the training program in which I'm interested.

That's something up with which I won't put.

Finally, many people, teachers of the language included, mistake verb particles for prepositions. In the sentence, "I *ran across* an old book," *across* serves as a verb particle when the sentence means "I *found* an old book unexpectedly." In the sentence "The truck *ran across* the parking lot too quickly," *across* is a preposition of direction, and "parking lot" is its object. The following combinations are examples of verb-particle unions, but there are many others.

pass out make up take down come upon

If you've been worried about ending sentences with prepositions, don't be. Chances are that, half the time, your prepositions function as verb particles anyway. See Chapter Ten for more information about verb particles.

Beginning a Sentence with a Conjunction

This is another "rule" that never existed. Conjunctions are often clean, effective transitional words that relate sentences to each other. Such an acclaimed writer as Bertrand Russell, who won the Nobel Prize for Literature in 1950, never hesitated to begin sentences or even entire paragraphs with the words "and" or "but." Why then should we?

Beginning a Sentence with "Because"

The same rule that applies to conjunctions applies to "because." Probably the reason teachers tell students never to begin a sentence with "because" is that students sometimes write sentence fragments. "Because" is an adverb, and as such introduces a dependent clause, so the clause can't be a complete sentence in itself. We may answer a question like this:

> Because I said so!

But it's not a complete sentence. However, there's nothing wrong with writing a *whole* sentence that begins with "because."

> Because the testing required a full seven days, I came to the lab on the weekend.

Note here that the independent clause is "I came to the lab on the weekend," and that the "because" clause attaches to it.

Using Contractions

Contractions reflect speech: They meld sounds and assist the flow of language. You may have heard, however, the directive not to use contractions in formal or professional writing, but to do so is not wrong, and often a contraction prevents choppy reading. The danger lies in using contractions that confuse readers, because constructions like "he'd" can mean both "he would" and "he had." Usually the remainder of the sentence tells readers the meaning. To be perfectly clear every time you write, you can opt to limit or eliminate contractions. The choice is yours.

Using the Same Word in a Sentence Twice

Not using the same word in a sentence twice has some merit, provided it's not carried to the extreme. This advice has been directed to content words, not structure words, so in fact, your sentences may have more than one "and," "the," or "of." Writing can suffer when more than one form of the same content word occurs in a sentence, because it's difficult for readers to sort through the meanings. Look at this sentence:

> The interviewer interviewed the interviewee.

Obviously, there's a bit of the ridiculous here. Such constructions are grammatically sound, but the use of the same root in three spots makes comprehension difficult. Rewording does the job — and notice that the sentence has "the" two times.

> The manager interviewed the applicant.

Creating Inanimate Possessives

A popular lesson with many English as a Second Language teachers is that inanimate objects can't possess anything, and so they tell students to use the preposition of possession, "of," instead. This isn't bad advice in and of itself, because you won't be wrong if you say, "The second valve of the autoclave," rather than "The autoclave's second valve." Historically, however, English has no such rule. "The Star Spangled Banner," America's national anthem, contains several examples of inanimate possessives:

> By the dawn's early light…

> By the twilight's last gleaming…

> The rockets' red glare…

So, if you attribute possession to inanimate objects, you'll be in good company. If you opt to say, however,

> The power supply of the equipment is ample.

You won't be wrong. But you can just as easily, and accurately, say,

> The equipment's power supply is ample.

English allows you many acceptable options for delivering information. It's up to you as a writer to select those that will make your writing comprehensible. You can say "by year's end" or "by the end of the year": Both are perfectly correct, and the choice you make helps define your individual writing style.

Consistency in Presentation

People who analyze writing say that words convey 90 percent of a message; other components such as punctuation and white space deliver the remaining 10 percent. Editors have also observed that how a document looks contributes to its readability and overall style. So be aware that you have more tools at your disposal than just words. Technology has made it possible to use a number of options to emphasize key points and to make documents attractive. Typical techniques include boldfacing, underlining, italics, different color ink, or different typefaces and sizes.

Most companies have developed standard templates for different types of documents. These contain typeface and size, automatic numbering, headers and footers, and built-in margins. When you put information into a template, you are assured of consistent layout and appearance. When you are not using a template, certain conventions can help put punch in your writing. Bullets, letters, and numbers help organize information; so does effective use of white space and highlighting.

Bullets, Letters, and Numbers

In technical and business writing, the organization of information is often easier when related elements are set apart, particularly when a document is lengthy. Bullets, letters, and numbers set off information clearly; they signal a series of ideas; and they clarify and encapsulate key points. The admonishment here is for consistency. If you begin itemizing with bullets, don't switch to numbers for corresponding data, for instance. Once more, however, if your company has established standards, and most do, you should adhere to them.

You'll see numbers in varying forms. Some writers use numerals; others spell them out. The key, of course, is consistency. The journalistic style has caught on, and it's a good one: Spell out numbers from one to ten; use numerals for other numbers. Spell out any number that begins a sentence.

Finally, it's a good idea, especially when jotting quick handwritten notes, to enclose the numeral in parentheses after the written-out number. That way, your numbers will be less likely to be misread.

> We expected three (3) candidates, but eight (8) showed up.

> I obtained the same peak in all three (3) chromatograms.

White Space

White space — created by margins, spacing, and indentation — eases the reading task. White space signals an end to one component of a document, whether it is a simple paragraph or a detailed section of a report, and prepares the reader for the next. It also provides a breather, a natural stop in the text that permits the reader to pause, perhaps review, and ultimately assimilate what you've written.

Creating a Style Guide

While English has definitive guidelines for good grammar and usage, some elements have no rules, and a number of conventions are acceptable. That's why many companies develop their own style guides. Two department heads in one company spent an unusual amount of time trying to decide whether standard deviation should be "SD" or "s.d." They finally determined that it should be "s.d.," only to find that yet another department was using SD to mean the same thing. This created an inconsistency in documentation.

This sort of scenario happens in countless companies, because writers bring to their jobs their preferences. Consider the case of a company that prepared Good Laboratory Practice (GLP) study reports containing myriad equations. In some the lower case letter x indicated multiplication; in others the symbols × and • did the same. This was but one example of inconsistency in the reports, because there are so many options available, and all are correct. When different writers choose different options for presenting data, the variations can increase exponentially.

When a company develops a style guide, it spells out the conventions for the writing its employees and contractors produce. To be an effective tool for writers, it must have top down support and effective controls. That means senior management must buy into the concept and issue the directive that the company's style conventions are to be followed. Controlling the style guide means issuing numbers and versions in accordance with the company's document numbering system. Many companies issue individual numbers to each section of a style guide, so that the entire guide does not need to undergo revision each time a change is made to an individual section.

To develop a style guide means looking at the company's existing documentation and identifying the conventions that are common internally. Grouping the conventions comes next. Here's a list of typical sections for a style guide.

1. Acronyms and definitions

This section identifies the industry acronyms that apply to the company as well as all acronyms developed internally. Thus DAD might stand for Discovery and Development in one company, and Deferred Annual Delivery in another. Consider a company developing a bone marrow transplant device. Its guide might include the acronym AML for acute myelogenous leukemia. Such an identification may also include an extended description such as "a form of leukemia characterized by an uncontrolled proliferation of myelopoietic cells in the bone marrow and in extrmedullary sites, and the presence of large numbers of immature and mature granulocytic forms in various tissues and organs as well as in circulating blood." Other entries may be simply definitive. Allogeneic in the same company may bear a description as follows: "The source of the transplanted material comes from a person other than the patient who is receiving the transplant. Solid organ transplants are always considered allogeneic donations."

An acronym and definitions section of a style guide works best in alphabetic sequence, preceded with a directive on how to handle acronyms and terms in writing. These instructions may include how to affix a prefix to an acronym or how to pluralize it. (See also Chapter 14.)

2. Usage

This is the place to indicate preferred usage for vocabulary the company employs regularly. If the company has a lot of writers for whom English is not a first language, certain standard conventions can be included as well, such as how to use "a," "an," and "the," or reflexive pronouns. (See also Chapter 13.)

3. Mechanics

This section is the one writers generally find most useful. Conventions to cover include conventions for writing numbers, measurements, symbols, and abbreviations; directions for capitalizing, hyphenating, and compounding; and advice for punctuating effectively.

For instance, does the company use Arabic numerals for all things that can be counted or measured, or does it prefer the journalistic style of spelling out numbers from one to ten, and using Arabic numerals thereafter? Does the company put a space between a number and the degree sign or not?

Does it put a space between quantities and the symbols for greater than or less than? (See also Chapter 14 for a listing of common symbols and abbreviations.)

How does the company identify states, regions, and countries. What is the convention for identifying academic degrees? What gets italicized and what doesn't? Consider *Escherichia coli* and *in vivo* and *ex vivo*. Determine as well which words receive capitalization — e.g., Androus Dextrose, but the dextrose solution; Down's syndrome; Wilcoxon rank sum test. In writing titles and headings, what needs capitalization and what doesn't? What words should writers hyphenate? How are prefixes affixed? Is it post-dosing, postdosing, or post dosing?

Finally, what are the preferred punctuation marks? Does the company, for instance, use the serial comma or not? When should writers use em-dashes, hyphens, and nonbreaking spaces? These are all conventions that companies must figure out. In essence, they must determine "how we do it here."

4. References

This section should delineate the conventions to which the company adheres in referencing authors, works, and so forth. Subsections could include referencing within a document, such as citing other authors and their work, patents, unpublished work, journal articles, presentations, and other company documents. It should also show how to reference quotations and appendices, as well as how to refer to the company itself.

This part of the style guide should also include how to prepare a reference list, with specific guidelines for books, pages within books, newspaper and magazine articles, regulatory and standards groups, journal articles, corporate authors, anonymous authors, patents, company documents, laboratory notebooks, and other documents and source materials.

5. Formatting

Formatting may include such general conventions as type face, font size, justification, and margins and how to format and position tables and how to number and title them and include legends. Formatting may also include confidentiality statements and where to use them. This section may also show standard templates or select examples with directives of where to get electronic templates.

Other components include such elements as hyperlinking, styles, and keyboard shortcuts.

6. Writing Guidelines

A section such as this may include guidelines for disclosure writing, with reference to a company SOP, if there is one. How does the company reference its products? What are the trademarks?

How should abstracts read? How do authors find specific instructions for writing for publications and presentations? How are posters laid out and what sort of components are standards? If there is standard terminology for products, examples may appear here. Often a company will have a list of key messages that can appear in its publications, so that the company does not disclose what it shouldn't.

Depending on the scope of the guide, this section may also include the "how to" for writing specific documents such as procedures and study reports. An overview of appropriate tenses, sentence constructions, and writing conventions may also reside in this section. Some company style guides also include lists of transitions and power verbs to help writers get their thoughts down on paper without repetition of the same verbs.

7. Editing

It's helpful when editors use the same edit marks across the board. So a section such as this can show the conventional edit marks for transposing elements [TR] or deleting. Showing visuals of the marks is helpful, particularly for administrative staff who often wind up doing the word processing on corrected documents.

Wrapping up

Style guides can include any number of other components. Some include a place for writers to jot down notes to bring to attention when the style guide undergoes revision. A table of contents and an index is also helpful. Style guides work well with tabs or colored paper between sections and housed in three ring binders. The bottom line is this: The guide needs to be a tool that will help writers to write consistently without a lot of debate about how to handle language and conventions.

It's usually best, as well, to reserve certain symbols for visual presentation of data, as in charts and graphs. For example, the symbol for percent, %, works best affixed to a number in a table. In text, it may be better to write "ten percent of the patient population" than "10% of the patient population." Similarly, using lower case letters to indicate "colony forming unit" when there is quantitative data may be preferable as in "10 cfu"; the same is true for plaque forming unit, pfu. Note that there is no rule for using many symbols and abbreviations; however, companies can determine what their style preferences are.

10

Building Strong Sentences

Most people, even those who write for a living, have a few standard patterns for building sentences, and their patterns generally work. The patterns you use when you write reflect your control of the language. What do you think of the following text from a memo?

> We spilled a quantity of excipient during weighing. This requires a deviation investigation. The impact on the batch must be determined and materials must be reconciled.

The information is certainly clear, but what it says about its author is this: He disconnects related ideas and partially repeats the same ones. Further, each idea has equal weight. The same message reworked to connect those ideas reads much more smoothly.

> Because we spilled a quantity of excipient during weighing, we will conduct an investigation to determine the impact, if any, on the batch, and reconcile materials.

The original version has three choppy sentences; the revision has one. Notice that in the second version the cause and effect information is combined logically, and not all information carries equal weight. More than half of the information subordinates to the main idea: "We will conduct an investigation." Isn't the message easier to digest when the ideas relate?

Sometimes people use one sentence pattern repeatedly because it's safe; grammar can be threatening. Surely there are worse things than writing structurally correct sentences that are simplistic and choppy. It shouldn't be a surprise to learn, however, that the most complex element of the English language is the sentence. That fact alone is enough to intimidate even the most courageous writers! Mastery of the sentence nevertheless is at the core of good writing skills.

> Words are symbols of ideas, but they do not begin to "say" anything until we put them together. Sentences, which are syntactical units composed of words, "say" something, partly because of the lexical content (the meaning) of the words and partly because of the grammatical forms that govern words put together in patterns.
>
> **Edward Corbett, Author**

Here's a truth: Native-born Americans learn formal grammar in junior high school, if at all, and many immigrants have learned it as a component of mastering a second language.

In addition, most writers are not consciously aware of what they are doing when they compose sentences. It's a rare person who remembers what a gerund or an adverbial clause is. Yet think of the control you can have over your own writing if you do understand the various sentence elements. You'll be able to present the same information in a host of ways, and you'll make a sound decision about which way is best for what you have to say. A common sense explanation of sentence components and constructions follows.

Sentence Fundamentals

Every sentence has at its core the same principle, although the variations in structure can be infinite. To become a writer with a superior grasp of the language requires understanding the logic behind English construction. Writers build sentences with words, but the words must perform a standard function. That function is the base of every good sentence.

The Base

The basic sentence unit is the clause, which is a group of words containing a subject and verb. A complete sentence *always* has at least one clause and a complete idea. Subjects are what a sentence is talking about, and verbs offer state of being or action about the subject. When a clause has a complete idea, it's a complete sentence, or, in other words, an independent clause.

When it lacks the criterion of a complete idea, it's a subordinate element, or a dependent clause, because it doesn't carry an idea by itself. Always build a sentence on an independent clause that holds the *main idea* in the main subject and verb group. Think in terms of a complete unit of meaning, not a fixed number of words. For instance, "Go!" is actually a complete sentence, because it meets the criteria: It has a subject, which is "you" understood, and a verb, the present tense of "to go." Similarly, a long string of words may not have a complete idea; therefore, that word group does not meet the criteria for being a sentence. Thus, sentence validity is determined not by the quantity of words you use, but rather by the logic behind them.

Sentences may have any number of elements. A variety of phrases, one-word modifiers, and conjoined clauses give the language flavor. In addition, the word order in sentences can vary dramatically, and the constructions writers choose signal their competency with the language. Look at the two examples of clauses that follow. Only the clause with a complete idea is a valid sentence.

Independent Clause

An independent clause is a complete sentence.

> We must replace the bubble point tester on the purified water system.
>
> subject/verb/complete idea

Dependent Clause

A dependent clause by itself constitutes a sentence fragment. The "if" signals that subsequent information should be forthcoming.

> If we replace the bubble point tester on the purified water system,
>
> subject/verb/incomplete idea

Words as Building Blocks

All writing begins with words. The choice of words is the writer's; however, good writers apply logic to the sequence in which they compile their words into sentences.

> English contains about one million words, but only about 20,000 are commonly used today. One-fifth are Anglo-Saxon; three-fifths are borrowed or modified from French, Latin, and Greek; and the rest come from languages around the world.

Words serve two purposes: to deliver meaning and to connect ideas and show relationships. Content words are those that are ripe with meaning.

Structure words, on the other hand, form the fabric of the language; they weave content words into proper word order and sound sentences.

Content Words

Content words are the ones that change the most in the language, those we borrow from other languages, those that shift in meaning, and those we coin when we need new words. They are also the words that build vocabularies.

Content words fall into four categories: nouns, verbs, adjectives, and adverbs. Don't make the mistake, however, of thinking if a word is noun it must be a subject or object, or that a verb must function as the main verb in a sentence and deliver action or being information about a subject. Content words rely on both word order and structure words for impact.

Nouns

Nouns name persons, places, things, events, or concepts, such as "operation" or "occurrence." In short, the label whereby you identify anything is a noun. If a word can take "a," "an," or "the" before it, it is a noun. You can often identify nouns by their endings. Words with endings such as "tion," "ism," "ance," "ence," and "ology" are nouns.

Nouns play many roles in sentences. A noun may serve as a subject in a dependent or independent clause as well as an object for a clause or phrase. A noun also can act as a modifier for other nouns. Nouns can show possession, just like pronouns. Further, multiword constructions can serve as noun elements.

Verbs

Verbs tell about action or being. They also offer tense: past, present, and future and variations of these. Verbs play many roles in sentences. They may position themselves to show action or being in clauses or phrases, or serve as subjects or modifiers. English has over 200 commonly used irregular verbs, which often give both native and non-native users difficulty. (Chapter 11 gives more in-depth explanation of verb usage.)

Action Verbs

Action verbs tell what the subject does, has done, or will do.

> He *performed* the dissolution yesterday.
>
> The SOP *details* how to clean the tablet press.
>
> She *will lead* the project.

Linking Verbs

Linking verbs offer state of being or feeling. Examples of linking verbs are "to be," "to feel," "to seem," "to appear," "to become," and "to look." In general, verbs that essentially mean the same thing as "is," "am," "are," "were," "was," and "will be" are linking verbs. Linking verbs take adjectives, nouns, or pronouns.

I *feel* bad about the failure to validate the process.

She *appears* competent enough.

It *seems* we are subject to the new regulation.

Helping Verbs

Helping verbs modify the meanings of the verbs we choose; they add precision to the meaning of the main verb. Among helping verbs are the words "may," "shall," "will," "could," "ought," and others. To form the future tense in the English language a helping verb is necessary.

We *will consider* your suggestions carefully.

She *might be* willing to help you on that project.

I certainly *shall return* his call.

Infinitives

Infinitives are formed by "to" plus the present tense of the verb, such as "to think." Occasionally an infinitive does not employ "to," but this is the exception. When this is the case, the word order is still consistent. Infinitives can be subjects and objects of clauses and phrases; they are never main verbs.

To finish is our goal.

To prove our hypothesis won't be easy.

We have no way *to establish* a hypothesis.

I come in at 7:30. He lets me *go* home at 3:30. (No "to" marker)

Gerunds

Gerunds are the "ing" form of a verb that serves as a noun ("working," "writing," "sleeping"). For an "ing" word to be a main verb in a sentence requires the addition of a form of "to be," as in "I *am writing* a report." Here "am writing" is a main verb. In the examples that follow, however, the "ing" words are gerunds, serving as subjects or objects.

> *Waiting* for FDA approval can be anxiety producing.
>
> *Writing* poses problems for many scientists.
>
> I hate *running* dissolutions.

Adjectives and Adverbs

These two groups of words are modifiers. That means they alter the meaning of sentence components. Some confusion exists in distinguishing an adverb from an adjective, probably because the language is undergoing changes at present. Linguists have identified one language peculiarity as "overcorrection." In an effort to sound educated, people "overcorrect," changing what's standard to what's nonstandard, using adverbs where an adjective is best, or worse, modifying an adverb by adding an "ly" ending. Using these elements properly, however, is really a matter of understanding the logic behind them.

Adjectives

Adjectives always modify a noun or a word or word group used as a noun. The following examples show some the different ways to use adjectives.

Descriptive Adjectives

These provide some quality of the noun.

> Operating *heavy* machinery requires taking *OSHA-mandated* precautions.
>
> A *bleak fiscal* year looms ahead, I fear.
>
> A *slow but sure* hand will get the job done.
>
> *Three enthusiastic* people joined our department last year.
>
> We had no *early* indications that the *Phase 3* trial for red blood cells would have *so many adverse* events.

Limiting Adjectives

These narrow the scope of a noun. Limiting adjectives are often numericals or the possessive forms of nouns or pronouns.

> *Hal's* job requires concentration.
>
> *Eight* complaints are far too many.
>
> *Several* opportunities came his way.

Demonstrative Adjectives

These modifiers identify nouns specifically.

> *This* copier handles our departmental needs quite well.
>
> *These* requisitions are late.
>
> *That* particular problem continues to plague us.

Interrogative Adjectives

These adjectives demand answers.

> *Whose* job is it anyway?
>
> *Which* chair comes with the desk?
>
> *What* time is the meeting?

Adjective Order

In English, adjectives come before nouns. In other languages they may take other positions. In Arabic and French, for instance, they follow nouns. When adjectives are descriptive, they can often be interchanged in position before the noun as in the following examples.

> We have installed an efficient, easy-to-use laboratory information system.
>
> We have installed an easy-to-use, efficient laboratory information system.

Note that the two nouns "laboratory" and "information" must reside with "system."

For other types of adjectives, there is a more fixed order. Linguists have long pondered adjective ordering in English, and conclude that this pattern of adjective placement is seldom violated:

Determiner	Possession	Measurement	Color	Shape

Here are some examples:

The laboratory's large white board is useful for calculations.

The large Phase 3, heavily-enrolled trial has met its endpoint.

Those blue oval capsules are scored.

A company's main product does not always produce the biggest revenue.

Three large gray Sprague Dawley rats were subjected to the forced rat swim test.

A Final Note About Adjectives

There's a trend to use adverbs as adjectives. But to many readers, misused adverbs can signal a poorly educated writer. The following sentences show the nonstandard use of the adverb "badly" and standard use of the adjective "bad."

Nonstandard: I felt *badly* about the turn of events.

Standard: I felt *bad* about the turn of events.

People unable to distinguish between adverbs and adjectives often opt for the word form with the "ly" ending after a linking verb, thinking it sounds more correct. "Badly" may seem more correct at first glance, but consider applying the same tactic with other adjectives: "I feel gladly" doesn't sound right, and it isn't.

> Forty percent of all French words in English were borrowed between 1250 and 1400. From a flood of more than 10,000 French words inundating Middle English, 75 percent remain in use today.

Adverbs

Adverbs modify verbs, adjectives, other adverbs, or whole sentences. Adverbs offer information about time, place, manner, degree, and condition.

Usually *ly*-ending words are adverbs, although there are a few exceptions. Any one-word modifier that is not used as an adjective or an article is usually an adverb.

> The *heavily* used V-Blender needs regular preventive maintenance.
>
> He is *more* concerned with scale-up.
>
> We validated the method *yesterday.*

Sometimes writers think adding "ly" makes a word an adverb, but that's not always so. The words "fast" and "slow," for instance, can be either adjectives or adverbs, depending on their position in a sentence, while the words "friendly" and "portly" are clearly adjectives.

> **Non-standard:** How fastly can you do it?
>
> **Standard:** How fast can you do it?

Of course, "How quickly can you do it" gets the job done without the dilemma of choosing.

Comparative Adjectives and Adverbs

These adjectives and adverbs differentiate by indicating degrees. Writers show comparison by using the comparative or superlative form of adjectives. The comparative form compares a thing with another thing; the superlative form compares a thing with two or more other things.

> **Comparative:** This project is *larger* than the other one.
>
> **Superlative:** Indeed, it may be the *largest* one the company has undertaken to date.
>
> **Comparative:** Whitworth complained that he was awarded *less* authority.
>
> **Superlative:** In reality, Smythe has assumed the *least* control.
>
> **Comparative:** We argued *more* about how to handle the project than about the project itself.
>
> **Superlative:** What we enjoyed *most* was the presentation entitled "Rx to OTC: Trends in Industry."

You can determine whether to use adjectives ending in "er" or "est" or the words "more" and "most" by this rule of thumb. When the length of the word or its sound requires it, use "more" and "most" instead of the endings "er" and "est." For downward comparisons, all adjectives and adverbs use "less" and "least." For upward comparisons, most one-syllable adjectives and adverbs and many two-syllable adjectives take the endings "er" and "est." "More" and "most" form the comparative and superlative for most adverbs of two or more syllables — including nearly all that end in "ly." "More" and "most" form the comparative and superlative for all adjectives of three or more syllables.

> Simonsen's proposal is much *more interesting* than Von Duc's.
>
> That's the excuse we hear *most* often around here.

Structure Words

The good news about structure words is that you can learn them, and once having learned them you won't need to worry about learning new ones. Structure words, those elements that help impose order on content words, rarely change, and when they do, the process is generally evolutionary. For instance, "Ms.," as a form of salutation, is a pronoun that has found its way into the language within the last half of the twentieth century. Structure words may change as society does over time, but the process is slow, whereas content words enter the language with rapidity. Structure words include pronouns, expletives, articles (three little words that can wreak havoc), prepositions, conjunctions, and words that indicate the negative.

An intensifier is a linguistic element used to give emphasis, color, or additional strength to another word or statement. Intensifiers come from several parts of speech and grammatical categories. Here are a few:

> Where *the dickens* is my report? (Article and noun)
>
> It was *very* hot in the lab. (Adverb, the largest class of intensifiers.)
>
> It's a *complete* failure. (Adjective)
>
> She is a stark, *raving* mad scientist. (Participle as adjective)
>
> Where *in heaven's* name is the auditor? (Prepositional phrase)
> He *himself* designed the modified release facility. (Pronoun)

Pronouns

Pronouns usually take the place of a noun. The words "him," "she," "it," "they," and "them" are pronouns. Thousands of nouns operate in the English language, but there are a limited number of pronouns. Pronouns usually substitute for specific nouns, which are grammatically known as antecedents.

> When Ruiz prepares the standard response line, *he* averages the diameters of the standard reference concentration and the diameters of the standard response line concentration tested for each set of three plates.

The pronoun "he" refers to the antecedent "Ruiz." Pronouns save you from repeating the noun.

Pronouns can also function as adjectives modifying nouns. In these two examples they serve as possessive elements.

> *Her* promotion to chief chemist was the direct result of *her* years of dedicated work.

> The new project may well be *my* downfall.

Pronouns are either nominative, objective, reflexive, or possessive in case. Use the nominative case as subjects and after "to be" or other linking verbs. Use the objective case after action verbs and prepositions. Use the reflexive to intensify, and use the possessive case to show ownership. Pronouns don't use apostrophes to show possession, because case is built in.

Of late, a change seems to be occurring in the application of pronouns within the American language. People are doing a form of over correction. That means they are using the nominative form where the objective form has been the norm. It is not unusual to hear, "Edgar dropped Vitthal and I off at the contract laboratory." The "I" here is considered nonstandard by many purists. Look logically at pronoun use: No one "drops I off," but certainly, someone can "drop me off." All too common are constructions such as these: "Would you like to go with John and I?" or "Can you tell Edna and I how to get there?" You may employ these constructions in speech, but when you use them in your writing you run the risk of having it judged ungrammatical. In addition, the written language is not as variable as spoken English, so it's usually best to use the currently acceptable standard.

Here are the standard pronouns in their nominative, objective, reflexive, and possessive forms:

Nominative	Objective	Reflexive	Possessive
I	me	myself	my, mine
you	you	yourself	your, yours
he	him	himself	his
she	he	herself	her, hers
it	it	itself	its
we	us	ourselves	our, ours
you	you	yourselves	your, yours
they	them	themselves	their, theirs

Nominative Case

The current standard is to use the nominative case as a subject of a clause or with linking verbs. Linking verbs are verbs of nonaction. They serve to link a subject to information about the subject itself. In the following sentences the "to be" verb form dictates "she" and "he," rather than "her" and "him."

> If there's anyone who's to be praised, it's *she.*

> It appears to be *he* who's to receive credit for the ingenious proposal.

Use the nominative case when pronouns serve as subjects, as in the following sentences.

> *You and he* must work collaboratively.

> *She and I* have volunteered to take minutes.

Objective Case

Use the objective case after prepositions.

> Strictly speaking, between *you and me*, no margin for error is permissible.

> My work was waiting for *me,* but for Ella and *him,* no work was available.

> If you would like to come along with Alexander and *me,* we'd be delighted.

Finally, use the objective case when pronouns receive action.

> He gave *me* the formula.

> Give *her* the recognition she deserves.

Reflexive Pronouns

Think of a reflexive pronoun as an intensifier for nouns and other pronouns. A tendency now is toward using the reflexive as you would the objective case, as in "Call the stability supervisor or myself for the results." Language mavens, however, generally decry this usage. Writers who employ the reflexive pronouns to add intensity to other elements produce strong sentences.

> It was common in the 17th century to interchange the nominative and objective forms of the second person plural pronoun, 'ye' and 'you,' with the result that eventually the objective 'you' replaced 'ye' altogether. Interchange of nominative and objective forms of other pronouns was also anything but rare.
>
> **Webster's Dictionary of English Usage**

Eli is directing the project *himself.* (Eli himself)

The loss of the second supplier is *itself* a serious concern. (loss itself)

Writers have flexibility in their sentence constructions here: Reflexives can assume many positions within a sentence.

Eli *himself* is directing the project.

The loss of the second supplier *itself* is a serious concern.

You may also opt to use the reflexive as a subject, but many purists find that practice unacceptable. Nevertheless, many writers are shelving the "I" pronoun in favor of "myself," perhaps to be self-effacing, as in the following: "Hernandez and myself plan to redo the experiment tomorrow." Logically, however, "myself" can't plan anything. "Myself" should intensify "I" or "me."

The following sentences use reflexives appropriately to intensify the message.

As for *me,* I'll answer only for *myself.*

He, himself, was responsible, and no one else.

Possessive Pronouns

The possessive pronouns signal ownership. However, they take no punctuation. Even though we would indicate ownership for a noun in this way — "John's book" — we must remember that pronouns signal possession inherently, as in "his book." A common error is to include an apostrophe with "it" when the pronoun is used as a possessive, but "it's" always means "it is." Think of "its" as working the same way as "his."

This computer seems to have a mind of *its* own.

Surely *his* and *your* reports counter the previous findings.

To do *their* best was *their* goal.

Possessive pronouns may refer to an object in a previous sentence without identifying the object again.

Is that your book? No, it's his.

That one is not yours. It's mine.

You can always trade her for yours.

They bought the house next to ours.

I don't want to take our car. Let's go in theirs.

Other Pronouns

The English language uses pronouns to replace nouns in many ways. Following are major groupings of standard pronouns:

Reciprocal Pronouns

"Each other" and "one another" are the reciprocal pronouns in the English language. They act either as adjectivals or objects, always referring to previously cited nouns or nominative pronouns. "Each other" refers to two nouns; "one another" usually refers to three or more.

Solveig and Carolyn trust *each other.*

They share *each other's* project findings.

All the people in our division help *one another* with research projects.

Demonstrative Pronouns

Demonstrative pronouns indicate number and proximity. They specifically point to a noun, as in the following examples:

Those findings were inconclusive.

That's the information we need.

This copier isn't working.

Relative Pronouns

Relative pronouns introduce subordinate clauses. You'll find details about them in the section about subordinate clauses in this chapter.

Interrogative Pronouns

These words ask questions.

> *What* did you accomplish today?
>
> *How* many lots were rejected?
>
> *Which* line clearances did packaging perform?

Note that these pronouns are followed by the verb.

Indefinite Pronouns

Words such as "all," "some," "many," "much," "any," "plenty," and "enough" are indefinite pronouns. So are "each," "every," "neither," and "either." These pronouns reference nouns preceding them. If you pose the question, "Which of these two would you like?" An acceptable answer is "I'll take either." (Note that when "either" is used with "or," and "neither" with "nor," these words are serving as joining, or conjunctive, words.)

Quick Tips: Indefinite Pronouns

Indefinite pronouns function as noun substitutes, and you should treat them the same way you do nouns. They usually refer to something that has been identified before. Here are some common noun substitutes.

enough — I have had enough.

plenty — We have plenty.

other — Would you prefer the other?

another — He needs another.

any other — Do you have any others?

one another — Be kind to one another.

each other — Good chemists help each other.

many — There are so many!

much — We can't do much about the situation.

several — When we tested for impurities, we found several.

some — You asked if I'd like cake. I would very much like some.

none — I know of none.
more — You always want more.
most — I believe most feel that way.
much — Heaven knows, I don't do much.
either — You can have either. The choice is yours.
neither — Neither is acceptable.
all — You can't have all.
both — She took both for herself, and left me with none.
each one — Do a study of each one.
each — We ran a test for each.
few — Many are called, but few are chosen.
less — I actually have less than the others do.

Expletives

"It" and "there" are the most frequently used expletives in the language. When these words are used as expletives, they don't refer to anything specifically, nor do they indicate place. They simply serve as fillers to start a sentence or clause.

It's going to be tough to comply with the new regulations.

There's a new inspector on the swing shift.

It's going to be a busy day.

Articles

Three little words in current English are articles: "a," "an," and "the." Native speakers and writers of the language rarely even think about articles. They know when to insert "the" and when not to. It's not that simple, however, for those who are foreign-born, because many languages simply don't have comparable words, or, if they do, they're structured differently. "Hunden," for instance means "the dog" in Norwegian, which, like English, is a member of the Germanic group of languages. However, the "en" in "Hunden" is the article "the." Thus, directly translated, "hunden" means "dog the," and in English, word order is different. If articles spell out trouble for people who know the Germanic languages, they present dilemmas for people whose native tongue comes from a different family of languages. Many of the world's languages do not have articles, and the concept of marking a noun is illogical to speakers and writers of these languages.

Articles are noun markers. When you see an article, you can be sure that a noun will follow. Sometimes a group of words follows an article; rest assured: the last one is the noun. This is helpful, since English often employs nouns as adjectives. Consider the following sentence:

She is wearing a white cotton lab coat.

The word "cotton" is a noun, but here it's used as an adjective to modify lab coat. The article in front of the phrase "white cotton lab coat" lets you know that coat is the noun initially marked by "a."

In addition, "a" or "an" signal a noun that's one of several or many. "The" signals a noun that's a specific one or group. For instance, "Use a desiccator" means use one of a number of desiccators. Were the sentence to say, "Use the desiccator," it would indicate there is but one desiccator. When a noun is plural, however, use "the," since "a" or "an" literally means "one." "The pipettes need washing" (specific pipettes).

"A" and "an" can spell trouble, too, because there's confusion as to which precedes a vowel and which precedes a consonant. Here's a simple solution: stop thinking about the letters and listen for the sound. Use "a" in front of consonant sounds, and "an" in front of vowel sounds: an extra-special event; a union contract; a once-in-a-lifetime opportunity.

> Articles are usually the last elements non-native writers master because their omission or misuse doesn't interfere with meaning.
>
> **Ellen Measday, ESL Professor**

Use articles when you refer to specific things that are set apart from a general group. When you refer to general nouns that you cannot count specifically, do not use an article. The following categories of nouns *do not take articles* because they cannot be counted specifically.

1. Names of substances that you can obtain in various forms: food, bread, meat, butter, vegetables, chemicals, groceries, paper goods

   Potato chips may contain saturated fat.

   But: The potato chips are stale. (specific potato chips)

2. Names of materials that can change shape, depending on the product: wood, metal, grass, cement, plastic, rubber, cotton

   Cotton is a natural fiber.

   But: The cotton has turned yellow. (specific cotton)

3. Names of liquids that can change shape, depending on the container: milk, oil, coffee, water

   Coffee is a natural diuretic.

   But: The coffee tastes funny. (specific coffee)

4. Names of natural substances that can change shape, depending on natural laws: steam, smoke, fog, fire, ice

   Fire can be deadly.

   But: The fire raged for two hours. (specific fire)

5. Names of substances with many small parts: rice, sand, sugar, beans

   Rice is the staple food of many cultures.

   But: The rice was overcooked. (specific rice)

6. Names of groups of things that have different sizes and shapes: clothes, furniture, luggage, automobiles, pipets

   Cars are a source of pollution.

   But: The car is fuel-efficient. (specific car)

7. Names of languages: Hindi, Chinese, Russian

   Russian can be difficult to learn.

8. Names of abstract concepts, often ending in "ness" "ance," "ence," "ity," "ment," "ology": nuance, permanence, contentment, ambiguity, biology

   Curiosity often drives the research scientist.

   But: The curiosity he felt then was due in large part to the unexpected unknown. (specific curiosity)

9. Most "ing" forms (gerunds): learning, working, eating

   Learning is a lifelong process.

Examples of Working Articles

Business has been poor lately. (business in general)

Becoming successful in the pharmaceutical industry takes hard work. (a specific business sector)

Let's make a decision now. (specific decision)

Decision is the first step in reaching resolution. (the general decision-making process)

Give me a glass of milk. (one glass)

The milk has spoiled. (specific milk)

Milk contains calcium. (substance in general)
I gave the doctor a complete history. (specific history)
History repeats itself. (history in general)
They are disciplined people. (specific people)
Japanese has over three hundred symbols. (the language)
She has a pain in her shoulder. (specific pain)
Pain is difficult to live with. (general condition)
She heard a noise. (one noise)
Noise can be stressful. (all noise in general)
Did you read the paper? (specific paper)
Paper can be flimsy or heavy. (general substance)
A large rock marks the corner where you will turn. (one rock)
Geologists can date rock. (general substance)
The war in Vietnam brought controversy in many forms. (one war)
War is an old as mankind. (war in general)
I can't find a space to put the beaker. (one beaker)
Beakers come in all shapes and sizes. (all beakers)
Florida has a hot and humid climate. (specific climate)
Climate varies dramatically from state to state. (general climate)
The computer had a head crash. (specific computer)
Computers have changed communication worldwide. (general technology)
I got stuck in a traffic jam. (specific situation)
Traffic in New York continues to create problems. (general situation)
He has a long commute. (specific commute)
Commuting long distances is exhausting. (general action)

Prepositions

Prepositions show relationships. When used in phrases, they function as adjectivals, adverbials, and occasionally as nouns. Prepositions also serve as verb particles; that is, they form part of a verb phrase and in so doing alter the meaning of the verb to which they are attached. Often, the way a word is used determines whether or not it is a preposition. For instance, the simple conjunction "but" can be a preposition when used like this: "No one *but Jack* was available."

Sometimes prepositions are phrasal; that is, they combine to make a new meaning. These are common prepositions:

about	below	from	save
above	beneath	in	since
across	beside	inside	through
after	besides	into	throughout
against	between	like	till
along	beyond	near	to
among	but	of	toward(s)
around	by	off	under
at	down	on	until
atop	during	outside	with
before	except	over	within
behind	for	past	without

according to	in spite of
along with	off of
but for	out of
in connection with	on top of
in front of	up to
in place of	with regards to

Commonly Confused Prepositions

into — This means going from outside to inside.
I'm going into the house now.

in — This means already inside and doing something.
He is working in the lab.

beside — This means next to.

besides — This means in addition to.

Prepositions start a phrase. The end word is the object, as in "to the laboratory," where "to" is the preposition and "laboratory" functions as the object.

Verb Particles

It's common in American schools for teachers to admonish students about ending sentences with prepositions. However, the student who writes "I looked it up," is ending the sentence with a preposition that's acting as a verb particle. In this sentence, the verb is "look up." In fact, "up" as a particle is very common. We acknowledge when things are looking up; we put up with adversity or poor working conditions or difficult people; we hang up curtains; we use up groceries; we get hung up in details; we take up a hobby. The list of "ups" is seemingly endless.

Verb particles are not limited to "up." "Off" is a common particle. "To knock off" was probably coined to mean "stop." It can take a direct object, as in "knock it off," which means, colloquially, "stop it." It also means to copy: "He knocked off the product." In other words, he made a copy of it. "At Ron-way Labs, we knock off brand products when they come off patent." The term "knock off," then, can shift from being a verb to a noun: "Our product is a knock-off of Alacrity Laboratories' topical zinc ointment."

When a preposition becomes part of a verb or shifts to being part of a noun, it adds meaning, and thus it becomes a content element of the language. You can "check in" to a hotel; you can "check out" again. You can "turn on the TV" or "turn the TV on." Often, with verb particles, you can insert the object between the verb and its particle, which is a good way to tell if the element is working as a preposition in a phrase or as part of the verb, as in "Try it out."

Verb particles do not change when the tense changes. For instance, you can look something up; you can be looking something up; or you can have looked something up. In addition, often a direct object is inserted between the verb and the particle, as in "I tied it up."

Common Verb Particles

Off

We *drained off* the excess.

We *rounded off* the numbers.

The chemical can *leach off* into the ground.

Turn off the equipment when you are done.

Don't *put* it *off*.

Up

We *cleaned up* the spill according to the approved procedure.

We have a computer *back up* SOP.

We need a *follow up* plan.

Pick up the compound in the pharmacy.

We will *look up* the method in the USP.

Down

Always *write down* the results.

Take down this number.

He *fell down*.

The project is *winding down*.

Keep the pressure *down*.

Out

Cross out the error and initial and date it.

Check out the new book on software validation.

Fill out every line on the form.

They *handed out* coffee cups on the convention floor.

The equipment is *worn out*.

On/upon

I *came upon* the solution yesterday.

The supervisor *looks on* during QC testing.

We *are up on* the new regulations.

We can't *take on* another project.

We need to *bring on* a consultant.

Around

Turn it *around.*

The HR manager *is taking* the new hire *around.*

That condition *is going around.*

He *hangs around* in the breakroom.

I was *searching around* for some information.

Over

We will have *to do* the test *over.*

Go over the results one more time.

Think it *over.*

When it's raining hard, it's best *to pull over.*

When traveling on business, try *to stay over* on a Saturday night.

Into/in

We need *to bring* him *in*.

That feature *is built in*.

I *ran into* an old professor.

All hourly personnel must *sign in*.

He is ready to *turn in*.

Across

He *came across* well during the presentation.

She wants *to get* this idea *across*.

I ran across an old friend yesterday.

By

get us *by* until the end of the month.

Drive by the cash machine.

Through

The mixture should *pass through* a number 10 sieve.

We are working it through.

Of

We are *out of* excipient.

I *can't think of* anything to say.

Away

Take it *away.*

Get away from the door.

We want *to move away.*

Keep the flammable materials *away* from the building.

Negatives

Elements indicating the negative are standard in English. "No" and "not" are the most common. Writers use them to indicate the opposite meaning of their content words:

I did not complete the qualification.

They are not going.

I received no warning.

Negative are commonly contracted. So the previous sentences can also appear like this:

I didn't complete the qualification.

They aren't going.

Conjunctions

English has lots of joining elements. Writers use them to connect parts of sentences and whole sentences. The simple conjunctions are "and," "or," "for," "so," "nor," and "yet."

Li and Lily are methods development scientists.

I am ordering pipettes, flasks, and a sampling thief.

I came in late, *but* I worked overtime.

She got an OOS result, *so* she started an investigation.

Correlatives

English has several pairs of words that serve as conjunctive words. They show relationships between complete sentences and parts of sentences. The common correlatives are "not only, but also"; "either, or"; "neither, nor"; "both, and"; "not, but"; "not, nor"; "never, nor"; and "whether, or."

The certificate of analysis must include *not only* the lot number *but also* the product code.

Neither method AD002 *nor* AD003 has been validated.

They will *either* reconcile the batch record *or* conduct an investigation.

I can do *both* dissolution *an*d chromatography.

It's not just the process *but* also the result that's important.

We *never* willingly deviate from a process without investigating the impact *nor* do we institute change without tight control.

Whether to continue developing this product *or* not depends on the FDA's ruling on the patent extension.

Conjunctive Adverbs

Conjunctive adverbs connect ideas and show relationships. Here are the most common conjunctive adverbs.

Transitional phrases often work like conjunctive adverbs, too. Refer to the section on transitions in Chapter 3 for a list. See Chapter Twelve for punctuating conjunctive adverbs.

also
anyway
besides
consequently
finally
further
furthermore
hence
however
incidentally
indeed
likewise
meanwhile
moreover
nevertheless
next
nonetheless
otherwise
still
then
therefore
thus

Types of Sentences

We use content and structure words to build our sentences — each of which has a distinct function. These are the standard types of sentences.

Declarative Sentences

Declarative sentences state information. In short, they declare. Most of the sentences we write for documentation are declarative.

Interrogative Sentences

Interrogative sentences ask questions. "Would you consider working overtime this weekend?" "How many will attend the in-house documentation briefing?"

Emphatic Sentences

These sentences are distinguished by the use of "do" with the verb. They make emphatic statements.

She does exude enthusiasm.

Harrison does check the results carefully.

They do audit for cGMP compliance.

Imperative Sentences

Imperative sentences give commands, and the subject "you" is understood.

First, investigate what went wrong.

Always ask for direction if you're not sure how to proceed.

The imperative sentence is a particularly effective way to convey instructions and works well in master formulas, laboratory methods, standard operating procedures (SOPs), instructions, manuals, and other documents that tell "how to."

The Sentence Core

As we've already established, each sentence has at its core an independent clause, and a clause is a unit containing a subject and verb. Five standard patterns form the basis of most clauses in English, whether they are independent or dependent. These patterns can incorporate any number of modifying elements as well. Nevertheless, the core constructions are fairly constant. Occasionally writers transpose elements to give their writing zest; doing so, however, does not alter the basic patterns inherent in the language. The following sentences are all in the active voice.

Pattern 1 — Subject and intransitive verb. The verb does not take an object; nothing receives the action of the verb.

Yelena laughed.

Pattern 2 — Subject, transitive verb, and direct object. Something receives the action of the verb.

Carlos conducted the experiment.

Pattern 3 — Subject, transitive verb, indirect object, and direct object. Something receives the object, which is a result of the action.

Pankaj gave me the report.

Note that indirect objects receive the action of the verb indirectly. In the following sentences, the indirect objects are underlined.

The director offered Dr. Kovalerchik a promotion to senior chemist.

The grandfather clause guaranteed the early retirees medical benefits.

Prepositional phrases can also show these relationships.

The director offered the newly created senior chemist position to Dr. Kovalerchik.

The grandfather clause guaranteed medical benefits to the retirees.

Pattern 4 — Subject, transitive verb, direct object, and object complement. An object complement serves to add information about the direct object.

Suresh called the results a breakthrough.

Pattern 5 — Subject, linking verb*, and noun, pronoun, or adjective. Linking verbs serve as a bridge between subject and what follows.

Jerry Hsu became a senior chemist.

It is she.

Margarita is inquisitive.

*Remember, linking verbs show state of being. The most common linking verbs are "be," "feel," "seem," "become," and "appear."

The Passive Voice

Sentence patterns 2 and 3 contain elements that make the passive construction possible. In the passive voice, the direct object takes the subject position, and the agent of action moves into a prepositional phrase or disappears altogether. The passive has its uses, particularly when the focus is on the result of an action.

Active: We found two unknowns.

Passive: Two unknowns were found. ("We" has disappeared.)

English teachers have often admonished writers to use the active voice, and that's generally good advice, because the active voice is more direct. Sometimes, however, no clear alternative to the passive exists. The section on verbs in this chapter and Chapters Nine and Eleven give more details about the passive voice.

Writing Questions

Questions generally transpose elements to elicit a response. Often the helping verbs "can" or "do" play active roles in formulating questions.

Does this mean the project is finished?

Look at this sentence, which follows pattern 2.

He can perform dissolutions.

In question form, "can" precedes the subject.

Can he perform dissolutions?

This sentence is also a statement of fact.

He gave the assignment to Enrico.

In question form, "to do" holds the tense and the order is transposed.

Did he give the assignment to Enrico?

In speech, questions take many forms. A question may be posed orally by simply raising the voice at the end of a sentence. Or a sentence may be "tagged," with an interrogative construction, such as "The results are within specification, aren't they?" In writing, however, questions are usually stylistic devices to signal a point that's coming.

What does this mean for Ronway Laboratories?

With this wording, the writer is anticipating the readers' question or pointing out to the reader what to question. This technique allows succinct delivery of information that provides the answers.

English poses two types of questions: direct and indirect. The indirect question does not invert the word order. Here's an example of an indirect question.

We asked whether the method is suitable for scale-up.

Here's the same information in a direct question.

Is the method suitable for scale-up?

Questions in writing are useful in that they offer variation in the general, declarative delivery of information. They trigger where text is going.

Building the Basic Clause

The structural variety of sentences is endless. As a writer, you can choose how to compose your sentences so they work best for you. Begin with the basic subject-verb pair, make each word count, and vary your sentences to give your writing interest. Follow these guidelines for framing each clause.

Choose Subjects Wisely

When you write, you can deliver your message with impact if you select your subject carefully. Look at the variations in the following examples, and consider the strength of each.

Noun as Subject

Fernando exceeded his goals for the quarter.

Noun Clause as Subject

That you have finished the project on time is excellent!

Pronoun as Subject

He exceeded his goals for the quarter.

Infinitive as Subject

To exceed your quarterly goals is commendable, Fernando.

Gerund as Subject

Exceeding your quarterly goals is commendable, Fernando.

Noun Clause as Subject

A noun clause is a subordinate clause element. A noun clause makes a strong subject, but can also function as other parts of sentences.

> *That Fernando exceeded his quarterly goals* is commendable.

See page 357 for more information about noun clauses.

Select Precise Verbs

Whenever you can, choose an action verb over a being verb.

Look at the different messages you can convey when you change the verb, and consider the strength of each.

Kakashan is in the lab.

Kakashan works in the lab.

Kakashan labors in the lab.

Kakashan performs in the lab.

Kakashan excels in the lab.

Agreement

Probably the most important rule in English is that subjects and verbs must agree. People learning English have often pointed out that the conventions for achieving agreement are not always logical. To a degree, it's true, but English, like other living languages, has developed over time as the result of usage preferences of native speakers. In most cases, subject-verb agreement in the present tense calls for using the s-ending for third person singular constructions and omitting it for first person singular and plural constructions. The illogic here is that English takes an s-ending to form the plural, but uses it on the present tense verb in the third person singular, as the previous examples show. Thus, "Kakashan excels in the lab" is correctly the third person singular, but "We excel in the lab," is the correctly the first person plural. Here are some more examples:

Third person singular:

Joan wants to perform the dissolution.

The train pulls up to the loading dock.

The water bath heats slowly.

First person and plurals:

I record patient data.

You enter the data, too.

We write protocols.

The technicians install the new equipment.

Measure Words

English has a number of measure words that make mass nouns countable. These words are helpful to indicate quantity by container or portion. The preposition "of" usually follows them. Thus we can say "a can of isopropyl alcohol," "a lecture bottle of liquid waste," "two pieces of glassine paper," or "a unit of blood." Note that there are some exceptions to this standard: "a thousand people" not "a thousand of people," and "a dozen patients" not "a dozen of patients." However, English does call for the "of" when a number is used as the unit word preceding a plural noun or pronoun, as in "a hundred of the patient volunteers" or "fifty of them."

Collective Nouns

Collective nouns are nouns that are inclusive; that is they refer to more than one unit. In American English collective nouns usually take a singular verb, but depending on context, they can take a plural verb and the construction will be grammatical. The key is to determine whether the collective noun refers to the whole unit or to the individual members or components that comprise the unit. Thus it is acceptable to write the following:

The trial staff record data daily. (plural)

The staff is courteous. (singular)

The validation team are planning the revalidation. (plural)

The validation team is in place.

The committee choose their new chair in January. (plural)

The committee meets on Wednesday. (singular)

Adding to the Base

Once you've decided what's most important and have positioned your key subject and verb unit holding a complete idea, you can augment the information with dependent clauses, phrases, and words. You can compound your sentence by adding other independent clauses. Be careful, however, because overly long sentences can be cumbersome. Yet choppy, short, and abrupt sentence structures don't do messages justice either.

Dependent Clauses

Dependent clauses, also called subordinate clauses, contain subjects *and* verbs, but not a complete idea. Dependent clauses can include some components of independent clauses. English has three types of dependent clauses.

Adverbial Clauses

Adverbial clauses offer time, manner, condition, degree, or reason to support a main idea. The following are examples of adverbial clauses.

Because I completed the assay

When Matthew comes back from vacation

If tomorrow comes

So that we can succeed

Relative Pronoun Clauses

Relative pronoun clauses contain a relative pronoun. The relative pronouns are "who," "which," "that," "whom," "whoever," "whichever," "whomever". Relative pronoun clauses add information about nouns, words acting as nouns, or pronouns. The following are examples of relative pronoun clauses.

who arrived from Australia yesterday

which is in the reception area

that controls the data

whom we all know and love

whoever wants it

whichever one you want

whomever you wish

Noun Clauses

Noun clauses take the place of nouns. As we've seen, they can function as subjects. They can also work in other parts of sentences where nouns do. The following are examples of noun clauses. To determine if an element is a noun clause, see if you can replace the clause with "it." If you can, and you still have an intact statement, the clause is functioning as a noun. Note that when a noun clause beginning with "that" is in an objective position, "that" can be omitted when the clarity of the sentence is unhindered.

> That we are having trouble with the formulation is obvious. (It is obvious.)
>
> But: It's obvious we are having trouble with the formulation. (That understood)
>
> Whether you go or stay depends on you. (It depends on you.)
>
> I see you are prepared for the meeting. (I see it.)
>
> She wouldn't tell me what she was doing. (She wouldn't tell me it.)

Constructing Sentences with Clauses

You can combine your ideas effectively by using clauses. However, too many clauses muddle messages, so use some discretion here. If you build strings of subordinate clauses, you probably need to reevaluate your information and create separate sentences. Avoid the temptation to put unrelated information in the same sentence. The following examples present key information effectively. Remember, too, that length does not determine what type of sentence you've written. The subjects and verbs in each clause are in italics.

Simple Sentence: One Independent Clause

> *I hold* you responsible.
>
> Reviewing the report, *he discovered* his errors.
>
> Afterwards, *Dick Luscher delivered* the stunning news.

Complex Sentence: One Independent Clause and One or More Dependent Clauses

> *Norton, who has* contacted us more than once, *deserves* a response.

If *you don't succeed, try* again. (Here, the second subject *you* is understood.)

Whenever *I feel* overwhelmed, *I take* a walk through the plant.

Jean, who supports our position, *voted* favorably.

Compound Sentence: Two or More Independent Clauses

I came; I saw; I conquered.

You will be pleased with the computer, but *you'll* surely *have* a few questions.

We regret our action; however, *we've learned* from it.

Compound-Complex Sentence: Two or More Independent Clauses and One or More Dependent Clauses

Wong, who has been diligent, finally *quit*; as a result, *we lack* a technician.

Because *downsizing* has become a reality, our staff now works too hard; furthermore, one employee, who is pregnant, will go on maternity leave in July.

The *computer, which is* down more often than not, *could use* servicing, and the *copier delivers* inadequate results.

Phrases

Phrases differ from clauses in that they have either a subject or a verb, but not both. They can never be complete sentences, and must connect to another element. You can often tighten your writing by rephrasing a dependent clause as a phrase. Here are the standard types of phrases in English.

Prepositional Phrases

Prepositional phrases are very useful. However, you should guard against overusing them. Remember, also, that prepositions always take the objective case noun or pronoun. The current tendency to use the nominative case alarms language purists and offends many readers. The following are examples of efficient prepositional phrases.

The instructions that came *with the machine* are now posted *on the wall behind it.*

He gave clear instruction *about the new procedure to Michael and me.*

(See Chapter 9 for more information about using prepositional phrases.)

Infinitive Phrases

An infinitive is the present tense of a verb, usually preceded by "to." Infinitive phrases consist of an infinitive and its object, as well as any modifiers. Infinitives function as nouns, adjectivals, and adverbials.

To have knowledge is more valuable than to have wealth.

"To be or not *to be,"* Hamlet's memorable question, might well be our own.

I rarely want *to go* to work in the morning.

Often infinitive phrases incorporate other phrases, particularly prepositionals, which is the case in the last example.

Participial Phrases

Participial phrases are formed by using the present or past participles of verbs. The present participle is the "ing" form of the verb; the past participle is usually the "ed" form of the verb, unless the verb is irregular. For some reason, many writers avoid participial phrases, although they can be useful. The following examples illustrate how participles efficiently add information.

The technician *compiling the data* is reliable.

Having already completed his report, he was frustrated to learn of the new and relevant information.

Asked to contribute to the project, she felt she had no alternative but to do so.

Gerund Phrases

Gerund phrases are present participial phrases (starting with "ing" words) that function as nouns. The following sentences contain gerund phrases.

Missing the opening remarks hindered his comprehension. (subject)

Taking time off rejuvenates you. (subject)

Isn't it important for us to be *working as a team?* (object of the verb *to be)*

Appositives

Appositives are efficient modifiers. Their function is to repeat a noun. Knowing how to use appositives will help you cull unnecessary words from your writing. As often as not, appositives are tidy little phrases, but they can also be one word.

Elizabeth Liv, *an experienced engineer,* will design the new production facility.

Thomas Wanat, *the founder of Network Concepts, Inc.,* will speak at the next luncheon.

Our plant engineer, Hector Gamboa, directs all maintenance activities.

Single-Word Sentence Modifiers

Not all modifying information comes in the form of a phrase or clause. Often a single word will modify a complete sentence. These modifiers are adverbs.

Tomorrow, we begin project Svarteper.

Finally, we secured permission to purchase the autoclave.

Guidelines for Composing Good Sentences

You've already discovered that putting your principal information in the main clause and building from there is a good idea and that succinct phrasing makes your writing stronger. Here are a few guidelines that build on these basics.

Put Elements as Close to the Words They Modify as Possible

Did you have an English teacher who scolded you for misplacing modifiers? Few students escaped this criticism. You may recall the rebukes, but do you remember what misplaced modifiers actually are? For most writers, that's the problem. They remember the criticism, but not what it means. Misplaced modifiers, to refresh your memory, are elements incorrectly positioned in a sentence. Consider the following:

> I saw the new dissolution equipment walking through the laboratory.

It's a ludicrous image, isn't it? But a sentence like this is easy to write because the writer knows what he means. The writer, here, probably meant this:

> Walking through the laboratory, I saw the new dissolution equipment.

The culprit in the original sentence is the participial phrase, "Walking through the laboratory." It belongs by the subject, "I," the word it modifies. Misplaced elements aren't limited to participial phrases, however. Often a single word wends its way into the wrong position. Consider this sentence:

> He told us to only do one assay.

In this example, "only" modifies "do," yet it should rightfully modify "assay." This sentence should be reworked to flow logically.

> He told us to do only one assay.

When you use modifiers such as "only," "even," "almost," "nearly," and "just," put them in front of the exact word or words they modify.

Also, position other phrases and clauses so that your readers can see at a glance what these elements modify. The following sentence risks baffling readers:

> The compliance specialist returned to the U. S. and drafted his audit report on the 10:30 flight.

The wily phrase here is prepositional — "on the 10:30 flight." What this phrase modifies is "returned to the U. S." So, logically, the phrase explains the mode of travel. The way the above sentence reads leads readers to believe the specialist audited the flight.

> The compliance specialist returned to the U. S. on the 10:30 flight and drafted his audit report.

Don't Dangle Your Modifiers

Dangling modifiers are elements just like those writers misplace, except that they don't have anything to attach themselves to. You can't apply the rule to put elements as close to the word they modify, because it isn't there. The following sentence is nonsensical, even though readers probably understand what the writer means.

> To reward the scientists who worked overtime, they were given a pizza party.

Who initiated the reward is lacking here, so the phrase "to reward the scientists who worked overtime" has nothing to connect to. Illogically, it links to "they," who are the scientists, and they certainly didn't reward themselves! When we attach the infinitive phrase that begins this sentence to a logical doer of the action, we have a sentence that makes sense.

> To reward the scientists who worked overtime, the manager gave them a pizza party.

The following sentences flaunt dangling modifiers, and they demand revision.

> **Wrong:** Participating in casual gossip, confidential information can be inadvertently released. (Can confidential information participate in casual gossip?)
>
> **Right:** By participating in casual office gossip, people can inadvertently release confidential information.
>
> **Wrong:** The tension was growing, awaiting FDA approval. (Can tension wait for approval?)
>
> **Right:** Our tension grew as we waited for FDA approval.

Make Elements in Sentences Parallel

Mixing constructions that should be parallel is fairly common in English. However, doing so interrupts the flow of sentences and makes them more difficult to read. Sentences like this give foggy messages:

> We had thought *to review* the report, *evaluating* the validity of its ideas, and *presenting* it for approval.

What are out of kilter here are the verbs. "To review" is an infinitive; "evaluating" and "presenting" are participles. When the verbs are parallel the sentence flows smoothly, and readers understand it easily. Note that the "to" of the first infinitive carries the other verbs as well.

We had thought *to review* the report, *evaluate* the validity of its ideas, and *present* it for approval.

It is not only with verbs, however, that writers violate principles of parallelism. Other elements can stumble as well. See how the following sentences benefit from revision to parallel structure.

Original: The new clerk is punctual, energetic, and you can depend on him.

Revision: The new clerk is punctual, energetic, and dependable.

Original: In his new position, Nyugen felt intimidated and without support.

Revision: In his new position, Nyugen felt intimidated and unsupported.

Original: She reads *The Pink Sheet* every month, as well as skimming the *Drug GMP Report* and other publications.

Revision: Every month she reads *The Pink Sheet* and skims the *Drug GMP Report* and other publications.

Create Logical Sentences

Writing reasonable sentences requires a combination of many things, but certain illogical patterns surface in writing at times, and it helps to be able to spot them. These patterns mix apples and oranges; they often talk about an idea or concept and call it a person, for instance. Examine the following sentence. Can you see what's wrong?

A chemist in our heavily regulated state is a demanding discipline.

This sentence in effect says, "a chemist is a discipline," and that doesn't make sense. Such a sentence needs reworking.

Chemistry in our heavily regulated state is a demanding field.

Or

A chemist in our heavily regulated state works in a demanding field.

To revise illogical sentences requires a sharp eye. Look especially for the explanations of the subjects and objects.

> Three things concerned us: the job description, the starting date, and the candidate herself.

> Logic dictates that a *candidate* can't be a *thing,* and we need to revise accordingly.

> We had three concerns: the job description, the starting date, and the candidate's ability.

The same illogic is at work in the next sentence.

> The clerk offered two suggestions: a person to handle complaints and another computer.

What's wrong with this sentence is that a *person* and a *computer* have been identified as *suggestions.* This sentence, too, deserves revision.

> The clerk offered two suggestions: that we hire a person to handle complaints; and that we lease another computer.

The next sentence is faulty as well.

> Harassment is when a person is subject to any sort of abusive treatment.

Analysis of this sentence reveals that "harassment" is "when." And that's illogical. "Harassment" can't possibly be time. Again, revision is in order.

> Harassment subjects a person to abusive treatment.

Make Pronouns Agree with Their Antecedents

Another illogical pattern has crept into our language, and it's so widespread that it's just about acceptable. But if you think about it, this pattern mismatches numbers. Professionals concerned about writing logical sentences guard against it. Consider the following sentence:

> Everyone needs refresher cGMP training to do our jobs well.

Awry in this excerpt are the pronouns: They don't mesh. How can *everyone* be explained with *our?* Here's a more logical statement:

We all need refresher GMP training to do our jobs well.

Writers probably mismate nouns and pronouns because they have become concerned about sexism in language. It's true that it sounds stilted to write "his or her" or "his/her," or even "one's." To say, "Each needs their own work area," is to violate the tenets of sound English construction, because "each" is singular, and "their" is plural.

The next examples show illogical constructions and their revisions.

Wrong: If an employee finds an out-of-specification result, notify the supervisor immediately.

Right: If an employee finds an out-of-specification result, he must notify the supervisor immediately.

Or

Employees who find out-of-specification results must notify the supervisor immediately.

Don't Mix Your Sentence Patterns

When you speak, you may take the liberty to shift gears in mid-sentence, but you surely shouldn't attempt the same when you write. Sentences like this:

Although I think Mr. Stief is an excellent chemist, but he shouldn't be required to train new hires.

work better like this:

Although I think Mr. Stief is an excellent chemist, he shouldn't be required to train new hires.

or like this:

Mr. Stief, an excellent chemist, shouldn't be required to train new hires.

The following sentences also profit from logical revision.

Original: Depending on the number of copies we need, the possibility of using the copier, and the rest of our workload will determine whether the project gets done today or not.

Revision: The number of copies we need, the possibility of using the copier, and the rest of our workload will determine whether the project gets done today or not.

Original: Looking out on Elmora Avenue is far more preferable than the macadam of the rear parking lot.

Revision: Looking out on Elmora Avenue is preferable to viewing the macadam in the rear parking lot.

11

Managing Verbs in English

Verbs wield power in the English language, and your choice of verbs can make significant differences in your messages. Consider the verb "show." It is akin to "demonstrate," "delineate," and "prove." Which verb you choose to use depends on the precision of your message and your role as a writer. Further, how you use verbs also impacts messages because verbs can take many roles. All verbs provide action or identify a state of being, regardless of their function in a sentence. When you build sentences, each clause must have a main subject and verb, and that coupling should be the most important part of the construction. Verbs can control subordinate clauses, serve as adjectives or adverbs, act as nouns, or steer supporting phrases.

A primary role of verbs is to put information into a timeframe, and to do so verbs have tenses. When you use verbs to indicate time, you have many options. Verbs indicate past, present, and future, although there is no true future tense in English. The future in English always requires a helping verb affixed to a present tense of another verb. Verbs offer flexibility in that they help you place activities in the past before other activities in the past; they allow you to speak to completed actions in the future and ongoing activities that are habitual; and they provide a vehicle for showing activities in progress.

When verbs present action, there is often a "doer of action" followed by a verb, but that's not cast in stone. Consider the following sentence:

The scientist dissolved the tablets.

In this sentence, the scientist performed an action on the tablets, "dissolved." But consider this sentence:

The tablets dissolved.

In this sentence, in the simple past tense, the tablets did not perform an action, but the verb determines that the tablets are no longer intact, so in effect, the verb indicates that something has happened to the tablets. You can change the tense from the past to the present and have the same working relationships:

The scientist dissolves the tablets in the acid bath.

The tablets dissolve in the acid bath.

Thus, word order determines meaning in English, with verbs acting as potent controllers. The verbs you employ when you write dictate the interaction of sentence elements. Often, too, verbs work with other words to indicate chronology. Words such as "just," "soon," "usually," and "nearly" make time more specific. Notice how the sentence

He has finished.

becomes more specific when "just" is added. The adverb "just" places the time in the immediate past, while the above sentence makes a statement of fact with an indefinite time reference.

He has just finished.

Auxiliary words and phrases can serve to alter chronology. For instance, when you affix "will" or "shall" to a verb, you put the action into the future. But you can also use the present tense to convey the same meaning, because "next week" indicates the future. Consider these two sentences:

Andrea will begin writing the protocol next week.

Next week Andrea begins writing the protocol.

You have many choices, equally good, for conveying information, and the choices are yours. When you make your choices as a writer, it's helpful to understand the options. While the study of the function of verbs in English can fill volumes, some basic tenets apply. The writing task is less difficult when you, as a writer, both understand what drives meaning in English and master verb usage.

Verb Functions

Action verbs indicate activity, while a few show state of being. These are often called linking verbs because they link what follows to the subject that precedes the verb. The most common verbs of being are "to be," "appear," "seem," "feel, and "become." Whether a verb is action or being depends mostly on its context, however, and many verbs we consider state of being verbs often serve as action verbs in certain situations.

Take the verb "feel," for example. In the sentence "I feel ill," the writer is, in essence, saying he is sick. However, if a glass washer says "I always feel the edges of the beakers to make sure there are no nicks or chips," he is using "feel" as an action verb.

To determine whether a verb is action or being, a simple test is to replace it with "to be" and see if the meaning is still clear.

She seems efficient.

She is efficient.

He became the manager.

He was the manager.

> The English word "tense" has its origins in the Latin "tempus," meaning time.

Verbs, either action or being, have four principal parts: an infinitive, a present participle, a past tense, and a past participle. The infinitive is the dictionary form of the verb; you can recognize it because it can be preceded by "to" — "to go," "to eat," "to read." Writers use the infinitive when action is presently occurring and change its form a little bit depending on who the subject is. For instance, if the subject is "he," "goes" — rather than "go" — is the correct form. The present participle indicates ongoing action. The present participle incorporates "ing" as in "We are working." The past tense indicates action that has already occurred.

Writers form regular past tenses by adding "ed" to the infinitive, such as "banged" from "bang." The past participle indicates action that has occurred and been completed in a time context. Whenever you use "have," "has," or "had" before a verb, you should use the past participle. Note, also, that the passive construction, which uses "to be" as a helping verb, requires the past participle. The past participle in regular verbs is the same as the past tense, so regular verbs don't present trouble, nor do irregular verbs that have the same form for the past and past participle. Since there are, however, over two hundred irregular verbs in the English language, it's a good idea to familiarize yourself with the ones you need to use most frequently.

Here's an example of the principle parts of a few regular verbs. Note that when verbs end in "y," in the past tense, the "y" changes to "i." And, when a verb ends in a silent "e," the "e" drops. In other verbs that end in a single consonant following a vowel, the final consonant usually doubles; this is to make sure the vowel preceding it remains short. Two consonants keep the preceding vowel sound short.

Stop + ed = stopped

Infinitive	Present Participle	Past	Past Participle
to learn	learning	learned	learned
to study	studying	studied	studied
to examine	examining	examined	examined
to decide	deciding	decided	decided
to hurry	hurrying	hurried	hurried
to work	working	worked	worked
to seem	seeming	seemed	seamed
to stop	stopping	stopped	stopped

This consistency is lacking in irregular verbs.

to speak	speaking	spoke	spoken
to write	writing	wrote	written

(See the list of most frequently used irregular verbs at the end of this chapter.)

Verb Tenses

English has these tense groups: simple, progressive, perfect, and perfect progressive. Each of these tenses allows for the past, present, and future. In addition, these tenses can be in the active or passive voices.

The Simple Tenses

The Simple Present Tense

If you listen to the way people use this tense you will notice that they tend to use it to indicate an action or activity that they do usually or habitually. For this very reason, the present tense is also called the habitual tense. And because it's known as the habitual tense, it's the preferred tense for many documents such as procedures and instructions. In addition, the present tense indicates a condition or state of being that is always true or always exists.

Margarita *starts* chromatography testing in the morning.

The formula *is* not perfect yet.

Vicki Li *prepares* her own summary reports using the document template.

He likes to read his mail before beginning the day's activities.

In the above sentences, the verbs indicate habitual action or status. No actual action is going on. In a sense, the present tense here indicates that the action or activity is not taking place at any particular or specific time.

Using the Present Tense

When you use this tense, keep the base or simplest part of the verb. The only change you ever need make is in the third person singular (he, she, it, Jack, Jane, the laboratory). With "he," "she" or "it," add the "s" or "es" to the base. In addition, change verbs ending in "y" to "i" and add "es" for the third person.

I *like* to come to work.

He *likes* to come to work.

I *go* to work.

He *goes* to work.

I try hard.

He tries hard.

Sentences often contain more than one clause (a subject-verb group). In an independent clause (main clause), the verb controls time. A dependent clause relies on the independent clause. The dependent clause explains, modifies, or adds information to the main clause. The time or tense of this clause does not necessarily agree with the tense of the main clause changes.

I work very hard to make sure that our error will not happen again.

The main clause, "I work very hard to make sure," is present tense; "that our error will not happen again" is future tense.

Jerry Hsu anticipates positive results from the study we have contracted out.

Here again, the main clause is in the present tense, and the clause "(that) we have contracted out is in the past perfect tense.

About "to be" in the Present Tense

The verb "to be," the most frequently used verb in English, is very useful, but also highly irregular. Be careful when you use it as a main verb in a clause.

Here is the present tense of "to be":

I am

he is

she is

it is

you are

we are

they are

you are

> It's somewhat illogical, when you think of it, that we use "s" to show plural constructions — as in two test tubes — but affix it to verbs to make them singular.
>
> **Ginger Ogden, Technical Writer**

The two more frequently used forms of "to be" are "is" and "are." They often appear in contractions (he's, they're). You'll see these forms working with other verb tenses and as part of auxiliary verbs.

The Simple Past Tense

Remember, regular verbs indicate the past by adding "ed" to the base.

love + ed = loved

walk + ed = walked

Occasionally verbs do not change to become the past tense. The following is an example of an irregular verb that does not change in the past:

I set the specifications as needed. (present/habitual)

I set the specifications yesterday. (past)

Unlike the present tense, the past tense does not change in the plural or singular for the third or first person. It is always the same.

I talked.	It talked.
You talked.	We talked.
He talked.	You talked.
She talked.	They talked.

There is one notable exception: The verb "to be" is affected by number, just as it is in the present tense.

About "to be" in the Past Tense

I was

he was

she was

you were

we were

they were

Use "was" with a singular subject, but use "were" with a plural subject.

The past tense often works in conjunction with other words the make time more specific.

Yesterday, the company announced a merger.

In July, there were two adverse events (AEs).

The CRO planned to start the study by May 17.

The Simple Future Tense

There is no inherent verb conjugation for the future tense in English. The future tense relies on the word "will" and occasionally "shall" coupled with the base form of the verb.

The company will always comply with FDA regulations.

The verb "to be," which takes the form "is/are" in the present and "was/were" in the past, uses "be" in the future.

> The director will be in his office all afternoon.

The future tense is not affected by a change in the subject from the singular to the plural. You need not be concerned about subject-verb agreement when your sentence is in the future tense. However, because dependent clauses sometimes use different tenses, always make sure that there is subject-verb agreement.

Many writers use other forms for the future tense.

> Frida is going to test the substance once again.

> He is about to receive the lab results.

> Edgar can leave the packaging line soon.

When indicating action performed in a series, only one "will" is necessary.

> The new scientist will study and learn the formula by heart.

"Study" and "learn" are both dependent on the same "will."

Although "will" generally comes immediately before the verb, the negative "not" comes between "will" and the verb, and so do words such as "always," "sometimes," "never," "probably," and "also."

> Manufacturing personnel will not monitor the deionized water systems.

The Progressive Tenses

When you wish to indicate that an action is ongoing or continuous, you can use the progressive form of the verb. If the action or activity is continuous right now (in the present), you can use the present progressive tense. If the action or activity was continuous in the past, you can use the past progressive, and when the activity will be continuous, you can use the future progressive. The progressive form of the verb is the "ing" form, also called the present participle.

The progressive uses a form of "to be," which holds the tense, and the "ing" form of the verb. When you add "ing" to a verb, you may have to drop a letter from the base. If the verb ends in a silent "e," drop the "e" before attaching the "ing" suffix.

shake + ing = shaking

take + ing = taking

When a verb ends in a single consonant following a vowel, double the consonant to keep the vowel sound short.

sit + ing = sitting

ban + ing = banning

When the sentence already contains "is" or "are," or "was" or "were," you must use the verb "to be" specifically in its progressive form, "being." "Being" is the present participle of "to be."

Dependent clauses do not necessarily have the same tense as the independent or main clause.

I am telling you that the results were inconclusive.

The Present Progressive

To form the present progressive use "am," "is," or "are" plus the verb stem and "ing."

Charla is reading the file.

Keith is conducting the experiment.

Shannon is examining the results.

I am working as a CRA.

They are doing an audit of the site now.

Note: Some people use "getting" instead of "being." The word will be understood, but its use is informal. Be careful with this word, for it cannot always replace "being."

He is being consistent in his work habits these days.

The new pilot development staff are being honored tomorrow.

"Getting" does not serve well in any of these constructions.

The Past Progressive

The helping verb in the past progressive is the past tense of the verb "to be." It takes "was" in the singular and "were" in the plural. "You" takes "were" in both the singular and plural.

I was trying to save time.

We were investigating the spill.

You were not mixing the compound correctly.

The Future Progressive

Future time also contains a progressive. It is used to express some activity that will be ongoing or in progress in the future. The future progressive is formed with "will" + "be" + the "ing" form of the verb.

In September, we will be completing the research.

The Perfect Tenses

In addition to the simple and progressive tenses, you can use the perfect tenses to indicate time. The perfect tenses indicate completed action, whether in the past, present, or future.

Started in the past and completed in the past:
I had performed that experiment.

Completed in the recent past or at an indefinite time:
I have performed that experiment.

To be competed in the future:
I will have performed that experiment hundreds of times by then.

To form the perfect tenses, use "to have" and the past participle of the verb. Even though the perfect tenses take the past participle, the time may be past, present, or future. The tense is indicated by "to have." Regular verbs will not give you difficulty when you use the perfect tenses, because the past tense and the past participle are identical. Irregular verbs, however, may have completely different forms.

The verbs "to do" and "to eat" are irregular:

Infinitive	Present	Past	Present Perfect	Past Perfect
to go	He goes.	He went.	He has gone.	He had gone.
to eat	She eats.	She ate.	She has eaten.	She had eaten.

The differences between simple and perfect tenses are sometimes very subtle. One interesting distinction is the way in which the perfect can be used for description.

> Harry works in research and development. (This is a statement of fact: He works there now.)
>
> Harry has worked in research and development. (This refers to an indefinite time in the past.)

The Present Perfect

The present perfect can express an action that occurred at some indefinite time in the past, continues into the present, and may even extend into the future. Generally, it means that the action is complete or "perfect" in the present, even though it also occurred in the past.

> The Director of Compliance has spoken with the FDA inspector.
>
> I have found these indicators to be inconsistent.
>
> Kui has wanted to improve his presentation skills for a long time.

The Past Perfect

The past perfect is used when an event in the past occurred before some other event in the past. The past perfect distinguishes between these two past times. The simple past indicates an event in the past. The past perfect moves even further back, to an event that was completed before another event.

> The production scientists had encountered difficulties with the formulation before they changed the excipients. (This indicates that the difficulties occurred in the past before something else in the past.)

The Future Perfect

> Documentation is the way we verify what we did, what we are doing, and what we will do. It is the window whereby we can view complete operating activities from start to finish.

The future perfect indicates action that will be completed in the future. Information about time and place accompany this tense, because to be a completed action, these elements must be present.

By this time next year, I will have completed the requirements for my Ph.D. degree in chemistry.

Tony will have served as company spokesperson for three consecutive years by the time he retires.

As in the other applications of the perfect tense, you need to use the auxiliary "will have" only once. Consider this:

I will have read and I will have summarized the entire report by Friday.

The sentence can be written more effectively like this:

I will have read and summarized the entire report by Friday.

The Perfect Progressive Tenses

These tenses combine the perfect and progressive forms. In the perfect progressive, like the other perfect constructions, time is indicated by "to have," and "to be" is in the past participle form, followed by the progressive (ing) form of the verb.

Present Perfect Progressive Tense

This tense indicates action started at an unidentified time in the past that is now ongoing and will most likely continue.

The regulatory associate has been reviewing FDA regulations carefully.

Margaret has been walking to work for six years.

Mike Yang has been working on stability software since the 1980s.

Frances and Kathy have been getting excellent results lately.

The Principal Investigator has been reviewing the adverse events.

Past Perfect Progressive Tense

This tense indicates action begun in the past that was ongoing and then stopped.

I had been experiencing some doubts about what we would find, but was pleasantly surprised that the results were excellent.

Jones had been working on a major confidential project when he left our employ.

The division chief had been planning to permit flex hours for the duration of the project.

Andersen had been looking into the disparities in yield, when he realized the change in granulation was the cause.

Future Perfect Progressive Tense

This construction usually relies on additional qualifiers for clarity. It indicates ongoing action that will have been completed by a specific time in the future. Like the future perfect, it takes information about time and often place.

Thor will have been working here for two years next month.

I will have been learning about the new computer applications for a record length of time, I'm sure.

To Have and to Be

To Have

"Has" and "have" are sometimes difficult to notice because they "disappear" into contractions. Yet it is important to remember that they are integral parts of the tense.

They've bought some new laboratory equipment. = They have bought some new laboratory equipment.

When you are forming the past perfect of the verb "to have" you still use "to have" as the auxiliary verb.

Pietro has had many positions at Ronway. (present perfect)

Pietro had had some experience before he joined Ronway. (past perfect)

Pietro will have had six years of service by next June. (future perfect)

To Be

Linguists have long recognized that the "to be" verb is not a typical verb. In fact, they've given it its own name: the copula. Because "to be" has had so many uses over the course of history, linguists feel that "to be" is best kept separate from other verbs. And in fact, they may be right. "To be" is dramatically different in form from other verbs. It often helps other verbs, and it doesn't follow standard patterns of irregular verbs. In short, it's not easily categorized. Therefore, to understand "to be" all by itself is best.

I am	I was	I have been
you are	you were	you have been
he is	he was	he has been
she is	she was	she has been
we are	we were	we have been
you are	you were	you have been
they are	they were	they have been

Note that the past participle for all subjects is "been," and that the tense is held by the verb "to have" as it is in all uses of the past participle.

We have been successful with Liqui-Powder products.

We had been confident that a strong market existed.

We will have been a publicly owned corporation for a quarter of acentury by the year 2010.

Voice

The Passive Voice

The passive construction places emphasis not on an agent of the action but rather on the results of an action. Thus, an active sentence such as this:

> Yale Jen writes monthly forecast reports for the pharmaceutical industry.

can be made passive like this:

> Monthly forecast reports for the pharmaceutical industry are written.

Note that by using the passive you can eliminate the agent (Yale Jen).

The passive voice can also take the perfect tenses. When this construction is used, "to have" carries the time, and "to be" appears in the past participle form.

> A burn ointment has been developed. (present perfect, passive voice)

> A burn ointment had been developed. (past perfect, passive voice)

> A burn ointment will have been developed. (future perfect, passive voice)

(Chapters 9 and 10 include more information about the passive voice.)

Time Line — the regular verb "to work" and the irregular verb "to go"

Tenses	Past	Present	Future
Simple	indefinite time in past I worked. I went.	habitual action I work. I go.	will happen in the future I will work. I will go.
Progressive	ongoing in past I was working. I was going.	ongoing in the present (e.g., on a project) I am working. I am going.	will be ongoing or happening in future time I will be working. I will be going.
Perfect	in the past before something else in the past I had worked. I had gone.	completed in present time I have worked. I have gone.	will be completed at some point in the future, with time and place (e.g., at the company for two years next August) I will have worked. I will have gone.
Perfect Progressive	ongoing in the past and completed in the past I had been working. I had been going.	ongoing in the present (on a project) I have been working. I have been going.	will continue to be ongoing in the future, with time and place (e.g., at the company for two years next August) I will have been working I will have been going.

The Emphatic Voice

To form the emphatic, use the verb "to do" and the root form of the verb. As with "to have" and "to be," "to do" holds the tense. The emphatic places emphasis on the verb itself; it increases the significance of the action. The future tense cannot take the emphatic.

Present Emphatic

I do appreciate the group's efforts.

He does attempt to be more precise.

The seminar presenters do anticipate some positive responses.

They do investigate what the competition is doing.

Past Emphatic

I did do the report.

We did conduct that particular test seven times.

Peter and Andrea did indeed take responsibility for the overrun.

They did not follow through as thoroughly as they should have.

The Emphatic in Negative Constructions and Questions

The emphatic plays a strong role in posing questions and creating negative constructions. Using a form of "to do" allows the inverted verb construction of questions.

Do you think we will be able to validate the method? (present tense)

Did you validate the method? (past tense)

"Do" and "did" are commonly used in responses to questions.

Who reviews the CMC section of the submission? The vice presidents of chemistry, development, quality, and regulatory do.

Who entered the data into the system? I did.

Did QA calibrate the scale? They did.

To add "not" to negate a statement requires a form of "to do" as well. Note also that "not" is often contracted with the verb.

Didn't the investigation show analyst error to be the cause of the deviation?

He doesn't think the vendor is suitable.

We don't plan to revalidate the system any time soon.

The Imperative Voice

The imperative voice uses "you" as the subject, and the subject is understood. This voice is used to give directives. Structurally, an imperative sentence gives a message such as "(You) measure 50 ml of solution." In writing, the "you" is omitted but understood:

> Measure 50 ml of solution.

The imperative construction is the only place in English where a sentence does not have a written subject. But it does have a subject: it is "you," and it is understood.

Helping Verbs

Helping, or auxiliary, verbs alter the meaning of main verbs, in both the active or passive voice. Often, they hold the tense in a sentence as well. The following are the helping verbs in English.

May

Permission

> You may leave the office.

Possibility

> That solution may be volatile.
>
> The acid bath may be prepared an hour in advance. (passive construction)

Benediction (or malediction)

> May your new job bring you satisfaction. (Note: In formal English, "may" replaces "can.")

Can

Ability

> Our department can easily perform that experiment.
>
> That trial run can easily be duplicated. (passive construction)

Permission

You can park your car in the back lot.

Possibility

Even expert chemists can record results erroneously.

Must

Obligation or Compulsion

You must be back from lunch in time for the meeting.

Logical Necessity

He must be working late again.

I must be dreaming.

Obligation Imposed by the Listener

Must we have lost it at the station? It could be anywhere. (This expression is infrequently used.)

Have To

The meanings of "have to" correspond to those of "must."

Note: "Got to" is colloquial and has the same meaning as "have to." Occasionally "have got to" does appear, but it is not used frequently in American English.

Obligation or Compulsion

I have to be going now.

We have to comply with FDA requirements.

The method has to be validated. (passive construction)

Logical Necessity

There has to be some logical reason why this test produced different results than the previous one did.

Future Time — "have to" combines with "shall" or "will"

We'll have to meet next week.

Past Time

Someone had to find a logical conclusion to the problem.

Will

Future Time

I will be arriving in Australia tomorrow.

Willingness

Who will help me improve my speech?

I will do anything for money.

Will you open the door for me?

Insistence

He will always disagree with anything you say.

Intention

I will write it, for sure.

The cause will be ascertained. (passive construction)

Predictability

By now he will be completing the audit.

Shall

Willingness

He shall be rewarded if he is patient. (passive construction)

Casey shall work overtime if he has to.

Insistence

You shall do as I say! (I insist.)

Requirements (often legal)

There shall be written procedures, and such procedures shall be followed.

Intention

I shall definitely announce my candidacy tonight.

The new organization plan shall be presented at 8 p.m. (passive construction)

Need

"Need" as a helping verb may be considered the negative and interrogative counter to "must" in the sense of "compulsion." Need I have a passport? (Do I have to have a passport?) Need anyone be lying here?

Obligation

You need to get a haircut. (Less categorical than "must")

Need Be

This construction is very similar to "need."

Need he be disturbed? If need be, we will perform an assay.

The last example uses the expression "if need be," a relatively common idiomatic device meaning "if there's a need."

Might

Condition

We might find the results to be entirely different from what we have expected.

He might change his mind.

The results might be proved inaccurate. (passive construction)

Past Time — *"Might" is the past form of "may."*

Might he go, he asked.

Would

Past Time/Predictability

He would spend whole weekends in the lab.

They would be given their lab assignments on Monday. (passive construction)

Request

Would you consider helping me?

Desire — *In this context "would" usually accompanies "rather."*

I would rather be on vacation.

Condition — *Often the conditional uses "if" to complete the idea.*

If you would complete the form, I'd appreciate it.

The method would have been validated by now had the dissolution results been acceptable.

Should

***Past Time/Obligation** — This construction is often used with the perfect tenses.*

The report should have been done by now. (passive construction)

Obligation

You should devote more energy to your work.

Should can usually take the place of "ought to" and "had better."

Ought To

"Ought to" has the same meaning as "must," — expressing not confidence, but rather lack of confidence. It signifies also that it is used like "should."

You ought to consider looking for another job.

Had Better

This construction is not as categorical as "must," and it signifies exhortation or strong recommendation, rather than compulsion.

You had better be quick about that. (I urge you to be quick.) "Had better" is like "ought to," in that these word groups mean pretty much the same thing.

We had better be going.

Could

Past Time/Condition

He could work for hours on end with no respite.

It could have happened on the second shift. (passive construction)

Condition

Could you help me?

Am, Is, Are To

This construction is followed by the infinitive verb and is similar to "to have to" and "ought to." It can frequently substitute for either of those forms.

> Vince is to return to Norway tomorrow. (He has received instructions to return.)

Sometimes this construction is used to indicate a command.

> You are to report back to me by 12 noon.

> They are to be promoted in June.

> The meeting is to take place tomorrow.

Past Tense

> We were to report back about the meeting immediately. (implies explicit instruction)

Compounding Verbs

Verbs function well in compound constructions. When compounding verbs it's not necessary to restate the helping verb or the "to be" or "to have" forms that constitute the passive and perfect constructions. In the following sentence "has to" is the helper, and investigate and write compounded with "and."

> The company has to investigate the out-of-spec result and write a report.

In the passive construction, "to be" needs only to appear once when verbs are compounded.

> The facility was cleaned and painted.

The same is true for the perfect tenses.

> We have held a concurrence meeting and agreed to the changes.

Contractions

These combinations of verbs are often contracted. However, in formal writing, it's best to spell out the words.

have — I should've attended the conference. We should've prepared a slide set.

had — I'd often thought about that idea. He'd no way of knowing.

would — He'd often remark on the uniformity of the tablets. I'd like to see the process.

will not — We won't have the results of the audit for another week.

Common Irregular Verbs

Regular verbs form the past tense with "ed." In regular verbs there is no distinction between the past tense and the past participle. The irregular verbs, however, can pose difficulties even for accomplished writers because no "rule" governs them; they are truly irregularities in the language. If you familiarize yourself with the forms of the more common irregular verbs, you will spare yourself the embarrassment of saying or writing "You should have went," which is nonstandard for "You should have gone." Here is a list of the most commonly used irregular verbs in English.

Infinitive	Present Progressive	Past Tense	Past Participle
arise	arising	arose	arisen
awake	awaking	awoke, awaked	awoken
become	becoming	became	become
begin	beginning	began	begun
bid	bidding	bid	bid
bite	biting	bit	bitten, bit
blow	blowing	blew	blown
break	breaking	broke	broken
bring	bringing	brought	brought
buy	buying	bought	bought
catch	catching	caught	caught
choose	choosing	chose	chosen
come	coming	came	come
cut	cutting	cut	cut
dive	diving	dived, dove	dived
do	doing	did	done
draw	drawing	drew	drawn
dream	dreaming	dreamed, dreamt	dreamed, dreamt
drink	drinking	drank	drunk
drive	driving	drove	driven
eat	eating	ate	eaten
fall	falling	fell	fallen
feed	feeding	fed	fed
feel	feeling	felt	felt
fight	fighting	fought	fought
find	finding	found	found
flee	fleeing	fled	fled
fly	flying	flew	flown
forget	forgetting	forgot	forgotten
forgive	forgiving	forgave	forgiven
freeze	freezing	froze	frozen
get	getting	got	gotten
give	giving	gave	given
go	going	went	gone
grow	growing	grew	grown
hang	hanging	hung, hanged	hung, hanged
have	having	had	had
hear	hearing	heard	heard
hide	hiding	hid	hidden
hold	holding	held	held
keep	keeping	kept	kept
know	knowing	knew	known
lay*	laying	laid	laid
lead	leading	led	led
lend	lending	lent	lent
let	letting	let	let
lie*	lying	lay	lain

light	lighting	lighted, lit	lighted, lit
lose	losing	lost	lost
make	making	made	made
pay	paying	paid	paid
prove	proving	proved	proven
ride	riding	rode	ridden
ring	ringing	rang	rung
rise	rising	rose	risen
run	running	ran	run
say	saying	said	said
see	seeing	saw	seen
set	setting	set	set
sit	sitting	sat	sat
sing	singing	sang	sung
sink	sinking	sank	sunk
sit	sitting	sat	sat
sleep	sleeping	slept	slept
slide	sliding	slid	slid
speak	speaking	spoke	spoken
stand	standing	stood	stood
steal	stealing	stole	stolen
strike	striking	struck	struck/stricken
sweep	sweeping	swept	swept
swim	swimming	swam	swum
swing	swinging	swung	swung
take	taking	took	taken
teach	teaching	taught	taught
tear	tearing	tore	torn
tell	telling	told	told
think	thinking	thought	thought
throw	throwing	threw	thrown
wake	waking	woke	woken
wear	wearing	wore	worn
win	winning	won	won
wind	winding	wound	wound
write	writing	wrote	written

*See following section.

Note: Lie and Lay

These two verbs are often confused — and it's easy to understand why. The past tense of "lie" is "lay," and therein stands the reason for the confusion. Just remember that people "lie down," but we "lay" objects down. People *may have lain* in bed during an illness, but books *can have laid* on a table for months.

Examples of Irregular Verbs in Sentences

Note that when the verb "have" is present, the past participle often follows. Unfortunately, in the common vernacular, the distinction between past tense and the past participle is diminishing. However, educated people do make the distinction. To literate people, sentences like "I would have fell" sound jarring to the ear. Listen for the "have" — and recognize it as a helping verb that indicates the past particle is to follow. It's easy to miss the "have" when using contractions that sound like — but aren't — "of," such as "would've" and "should've." It's usually best to avoid this type of contraction in writing, anyway.

In addition, "have" holds the tense — "have," "had," "will have" — and can be accompanied by other "helping" verbs such as "should," "would," "may," and so forth.

> I *swam* three laps. I *have* often *swum* more.
>
> I *went* to the bank. I *should have gone* before today.
>
> We *froze* at the skating show. The ice *had frozen* unevenly.
>
> He *gave* at the office. He *has* always *given* enough.
>
> I *fell* down. I *have* never *fallen* so hard before.
>
> By the time it's over, I will *have had* more than enough. I *had* plenty when the project started.
>
> I *lay* on the sofa all day. I *had* never *lain* there for so long a time.

Verbs as Subjects and Objects

> Lie and lay: these two verbs cause more confusion and trouble than almost any other words in the English language.
>
> **Edie Schwager, Writer and Editor**

Chapter 10 explains how to use verbs in subject and object roles. When verbs take these positions in sentences, they are either the infinitive or the "ing" form of the verb, which grammarians also call a gerund. For native born writers of English these constructions pose few problems. People born into the English language generally know which form to use, and usage tends to be idiomatic. The preferred form has more to do with usage over time than grammatical convention.

Using the correct form of the verb in verb phrases is often problematic for writers for whom English is not a first language. If you say, for example, "I stopped to talk to her," where "to talk" is the infinitive, it means that you stopped at a time and place to chat. However, if you say "I stopped talking to her," you mean that you no longer speak with her. How does one know when to use the infinitive or when to use the gerund (the "ing" form)? Linguists have theorized at great length. A view is that the infinitive often expresses something hypothetical, future, or unfulfilled, while the gerund expresses something real, vivid, or fulfilled, or an ongoing activity.

For instance, you can say, "I want to go to the clinical site." Or you can say, "The statistician is going to see if there is a trend." Or "We are hoping to complete the project by next week." In all these instances, you would be talking about something future or something that has not happened yet. These sentences all take the infinitive.

On the other hand, if you want to talk about something that is ongoing and real, the gerund is usually the best verb choice. "I like running the tablet press equipment." Or "The system keeps recording data." These sentences take the gerund, which is really the present participial or progressive form, which indicates ongoing activity. This may be an oversimplification, however, and usage is often idiomatic.

Some verbs can take both forms but the meaning differs.

I stopped to watch the film.

I stopped watching the film.

Some verbs do best with just the infinitive:

I want to transfer to Research and Discovery.

I expect to get the results by Friday.

I hope to finish the testing today.

I have decided to come in on Saturday.

I refuse to consider that option.

I plan to start the next phase of the study shortly.

I asked to see the data.

I have chosen to go to this seminar.

These verbs take the gerund, or "ing" form of the verb:

I enjoy listening to classical music.

I always avoid discussing the results of the trials.

Do not risk disclosing patent information.

He admits causing the spill.

I finished going to classes.

He denies having seen the report.

She defends giving the assignment to the technician.

Some verbs take either the infinitive or gerund with little change in meaning.

I like to work late.

I like working late.

The trainer begins to present the data after the introduction.

The trainer begins presenting the data after the introduction.

I start to test the first thing in the morning.

I start testing the first thing in the morning.

I continue to test for at least three hours.

I continue testing for at least three hours.

12

Punctuating Effectively

Most people recognize that the mechanics of writing — commas, quotation marks, periods, semicolons, and the like — are second in importance to the message. Yet a misplaced or omitted punctuation mark can alter a writer's message. Consider this line from a requisition:

> Please send us the ten cubic inch containers by Friday, March 25th.

At first glance the request seems clear enough. Certainly it's grammatically sound. But this request for drug product packing containers produced the wrong results! The writer intended to order containers that measured 10 cubic inches each. Yet he received ten containers, each one cubic inch. This error proved costly. He could have clarified his message by using a hyphen in the appropriate spot.

> Please send us the ten-cubic-inch containers by Friday, March 25th.

The hyphens in the revised version connect "ten," "cubic," and "inch," so there's no doubt that ten is a measure of size, not an amount.

Clarity is the goal of punctuation. The mechanics are tools for delivering information. Most people harbor some reservations about their ability to punctuate properly, usually because they try to apply rules, and the rules don't always work. Commas, for instance, can be appropriate in some instances, but in nearly duplicate constructions unnecessary: It depends on the meaning of the message.

> The earliest punctuation mark was a dot used by Greek scribes in the third century BC to mark a break in a sentence. It is the ancestor of the period, comma, colon, and semicolon.

In regulated industries, the need for precise punctuation is acute because abbreviations and symbols permeate much of the writing. Abbreviations and symbols indicate precise quantities and relationships. Writing carefully means that you indicate gram as g and not gr or g. because g is the standard. It takes no punctuation. Applying your own version of such an abbreviation can confuse messages.

Knowing what to punctuate and what not to requires an understanding of how punctuation works.

Punctuation marks are like road signs. They provide signals for working the reader's way through a network of ideas. They draw attention to points of interest and move the reader through complex construction areas so he doesn't lose his way. What follows are some simple guidelines to help you handle standard punctuation so your writing delivers the message you want it to.

Periods

Use Periods to End Statements, Indirect Questions, and Mild Commands

The period is the easiest way to punctuate independent clauses and is the endmark writers use most frequently. Periods provide a strong break between complete ideas. Periods also give readers time to absorb and understand what's been said before going on.

> The testing is slated to begin May 25. The results will be in your hands by June 5.
>
> He asked whether or not the new USP contained an approved method for the product.
>
> The sample is in the stability chamber.

Use Periods with Most Abbreviations

Abbreviations usually take periods, but acronyms generally don't unless they spell out a word unrelated to the meaning of the acronym. In addition, most acronyms are in capital letters, even though a few companies prefer to write acronyms in upper and lower case. Just remember to be consistent with the style dictated by the firm to avoid confusion.

Mrs.	Dr.	Ave.	Blvd.
a.m.	b.c.	Ph.D.	NBA
OSHA	FDA	SOP	IPA

(See Chapters 2 and 14 for more information about acronyms and abbreviations.)

Exclamation Points

Exclamation points function like periods, but they're more emphatic. The problem with exclamation points is this: Some writers use them overabundantly, and in so doing rob their writing of the punch it could have were the emphasis selective. When you write, don't exclaim too much.

Sometimes, however rarely, you may use an exclamation point after a word or two, usually in an informal context.

> Well done, Jack!
>
> We have six responses to our inquiry!

Question Marks

Question marks take their place at the ends of direct questions. Most writers don't misuse these marks too often, but occasionally they may be tempted to drop one at the end of an indirect question, and that's nonstandard.

Direct questions require question marks.

> Do you plan to work overtime?
>
> Who will be the new director of QC?

Indirect questions take periods.

> I want to know if you plan to work overtime.
>
> I wonder who the new director will be.

Question marks sometimes appear appropriately after single words in a sentence:

> The journalists' questions Who? What? When? Where? and How? help us find the details we need to write effectively.

Commas

It shouldn't surprise you to learn that commas are the most misused of all punctuation marks. What often happens is that writers become comma-crazy and drop them into text whenever they "hear" a slight pause. Many journalists follow this bit of advice for comma use: When in doubt, leave it out. Insert commas with caution: Use them to delineate your message. The meaning of the text generally determines whether or not commas are in order. The following are standard conventions for commas, but they are not cast in stone.

Use a Comma Before a Conjunction Joining Two Independent Clauses

To avoid the choppy rhythm you can sometimes create by having all your independent clauses stand alone, you can join them by using a comma and a conjunction. (An independent clause is a complete sentence, containing a subject and verb and cohesive idea. See Chapter 10 for more information about sentences.) These are the simple conjunctions in English:

> and, but, for, so, nor, or, yet

Notice that in the following sentences, the comma offers a softer pause than a period does between the two ideas, and the conjunction shows their relationship.

> The testing is slated to begin on August 3, and the results will be in your hands by September 4.
>
> The testing is slated to begin on August 3. The results will be in your hands by September 4.
>
> Our research team developed the product, but the marketing staff remains uncertain of its appeal.
>
> Our research team developed the product. The marketing staff remains uncertain of its appeal.
>
> The product should be successful, for it meets a need no other product on the market does.
>
> The product should be successful. It meets a need no other product on the market does.

Joining two independent clauses with a comma and no conjunction is generally considered a poor way to join two complete ideas, unless they are short and parallel in construction, as in the following example:

> I entered, I looked, I left.

Note: Constructions like this are uncommon. Even when sentences are short and parallel in construction, periods serve well.

> I entered. I looked. I left.

Use a Comma to Set Off Introductory Clauses and Phrases Unless the Elements Are Short

> Because the task proved to be so difficult, the director allotted more time to its completion.
>
> If the procedure is routinely performed under more than one condition, make sure cGMP training covers all variables.
>
> By year's end we will be able to report our findings.

Use a Comma to Set Off Elements that Give Additional Information

You can tell if the information is purely additional by omitting it. If the sentence makes sense without it, the information is not essential, and commas set it off. The thing to remember is this: Use either two commas or no commas, unless the information comes at the beginning or end of a sentence and the comma is balanced with either a capital letter or an endmark.

> The new EPA requirements, which were issued just last month, affect how we handle disposition of drug products.
>
> Myhre, who has already written a number of validations, is developing a new format the whole company can adopt.
>
> Lathia, after evaluating the USP guidelines, decided to develop an in-house method.
>
> It was an old facility even then, forty-five years ago, and by today's standards it is obsolete.
>
> The senior scientist's assistant, John, has been working on the dissolution.

In the last example, "John" is appositive to "assistant," and in this instance provides additional information. What the commas indicate is this: The senior scientist has only one assistant. But consider the following:

> The senior scientist's assistant John has been working on the dissolution.

Here the lack of punctuation makes it understood that there is more than one assistant. The trend in this type of construction is to omit commas, however, or to restructure.

> John, who is one of the senior scientist's assistants, has been working on the dissolution.

Do not Use a Comma if the Information Is Essential to the Meaning of the Passage

The following sentences make no sense without the explanatory information. Logically, commas are not in order.

> The Right-to-Know Act that guarantees workers access to information about the chemicals they work with was designed to ensure safe working conditions.

> The programmer who designed our database will modify the system and then oversee the revalidation.

Use a Comma to Set Off Asides, Interrupters, and Transitions

Sometimes writers insert words for cadence, and sometimes they include them to move from one idea to another. When they do that, they usually set them off in commas.

> Your last report, for example, was letter perfect.

> Any manager, I suppose, makes controversial judgment calls.

Use a Comma to Set Off "Yes," "No," or a Name in Direct Address or Interjection

These constructions generally appear in less formal, direct communication.

> Ellen, scale-up for naproxen sodium should proceed smoothly.

> Yes, I understand.

No, you can't.

Use a Comma to Set Off Contrasting Ideas

The problem lay with the register, not the roller.

Document the follow-up investigation and corrective action, if required, in the report.

Be Consistent in Using Commas between Words and Phrases Forming a Series

Many writers, when composing serial elements, omit the last comma before the conjunction. Others include it. Either style is acceptable, and to omit the comma is becoming the standard. In fact, the argument for omission is logical: Commas themselves can mean "and." So, if you use a comma followed by "and," you are effectively saying "and and." Nevertheless, this book employs the serial comma, because, in certain instances, meaning may not be clear without it. In delivering technical information its inclusion can dispel all doubt as to meaning.

All container labels must show compound identity, strength, purity, expiration date, and destination.

As the safety director, he was systematically quick to assess, address, and resolve causes of repeat accidents.

The large pharma evaluated the biotech company for its pipeline, financial soundness, and potential for partnering.

Use a Comma between Coordinate but not Cumulative Adjectives

Not all adjectives do equal duty. The way you position and punctuate them depends on what you want to say. When adjectives are coordinate, commas are in order; when adjectives are cumulative, they're not. That is, adjectives that modify a noun or pronoun with equal weight are coordinate. They can change positions and not mar the message. Cumulative adjectives, on the other hand, must be in a specific order to make sense and require no commas.

> I always use a comma before the final conjunction in a series because it prevents misreading. For instance, were I to say the components are a, b, c and d, I could be indicating that there are four components equal to a quarter each, or that there are three, one third being a, the second third being b, and the final third having two parts, c and d, equal to a sixth each. The comma eliminates all doubt.
>
> **Li Hang Chow, Ph.D,**
> **Research Chemist**

Coordinate Adjectives

> Informative, clear, easy-to-read documentation should be the standard.

> Easy-to-read, clear, informative documentation should be the standard.

Note here that *and* can replace the commas between coordinate adjectives. This is not the case with cumulative adjectives.

> Informative, clear and easy-to-read documentation should be the standard.

Cumulative Adjectives

> Three scented liquid compounds go into the formulation.

Note that when we interchange these adjectives, the sentence becomes ridiculous.

> Liquid scented three compounds go into the formulation.

Because each cumulative adjective modifies the one that follows it, each takes its logical place in the sentence and requires no comma. Adjectives of number and color are cumulative, with number beginning the series, followed by size and color. The adjectives in the following sentences are cumulative and require no commas.

> Every interested employee should attend.

> A fantastic new proposal has reached our desk.

> Six recurring problems continue to frustrate us.

> Four tall green trees lined the drive.

A cumulative adjective can be followed by two or more coordinate adjectives. When this happens, only the coordinate adjectives take commas or are joined by "and."

> We have a choice of four innovative, exciting packaging designs.

> We have a choice of four exciting and innovative packaging designs.

Use a Comma to Prevent Misreading

Occasionally you may need to insert commas to prevent misreading. And you should do so, even if you violate the comma "rules." Consider the following statement:

> Soon after we adjourned.

What does it mean? Is it an adverbial clause of time or a sentence stating a fact? Insertion of a comma in the right place clarifies.

> Soon after, we adjourned.

Use a Comma after Geographical Locations

> The lab was located in Butte, Montana, for five years.

> We looked for plant sites in Florence, South Carolina, and Galveston, Texas.

It's noteworthy to add, however, that the geographical comma seems to be losing its significance. More and more, writers are leaving it out.

Semicolons

Many writers dread the semicolon and use it incorrectly or not at all. Nevertheless, it's a handy mark to have at your command. It signals a harder break than a comma, but a gentler one than a period. What you need to know about semicolons is that they work well *between* independent clauses or in lists. To use them where a comma rightfully belongs is non-standard.

Use a Semicolon and Conjunctive Adverb between Two Independent Clauses

The conjunctive adverbs also define relationships between ideas. These are some of the most common conjunctive adverbs:

accordingly	however	nonetheless
also	in addition	otherwise
besides	indeed	still
consequently	instead	then
furthermore	moreover	therefore
hence	nevertheless	thus

You can recognize a conjunctive adverb because it adds adverbial information to a sentence. (Remember, adverbs offer information about time, place, manner, condition, or degree.) Another quality they possess is their ability to take different places in sentences they link. When it occurs in the middle or at the end of a sentence, the conjunctive adverb is set off in commas.

> The technician was unable to obtain the correct part; therefore, the job remains uncompleted.

> The technician was unable to obtain the correct part; the job, therefore, remains uncompleted.

Use a Semicolon Alone between Two Independent Clauses if the Relationship between the Ideas Is Clear

> The labor shortage is reaching epidemic proportions; the attrition rate has doubled.

Use a Semicolon to Separate Independent Clauses Joined by Coordinating Conjunctions When the Clauses Are Long and Complex, or Have Internal Punctuation

> Somewhere along the way I discovered you could go through life and avoid the dreaded semicolon altogether. I usually choose to write sentences separated by periods and capital letters.
>
> **Constance McGoff**

Exercise common sense here. When compound-complex sentences become unwieldy and difficult to follow, you may need a stronger break between the independent clauses, purely to point out the break between the main ideas. If, however, a sentence contains so much information as to be difficult to read, you should probably write two separate sentences.

> Record the results of the experiment; in addition, note the equipment you used.

Or

> Record the results of the experiment. In addition, note the equipment you used.

Use a Semicolon to Separate Items in a List that Contains Internal Commas

Although commas neatly separate most serial elements, semicolons help avoid confusion in lists with internal commas.

> We delineated the following objectives: to pinpoint the precise problems; to design, install, and monitor a corrective plan; and to address any new trouble spots if they develop.

> The new director of marketing has lived in Des Moines, Iowa; Trenton, New Jersey; Chicago, Illinois; and San Francisco, California.

Colons

Colons are useful marks that let readers know there's more to come. Colons are divisive: They set off elements and direct readers to them.

Use a Colon to Introduce a Series

What follows a colon explains. When the explanation is a series, the colon works well.

> The investigation revealed the following: samples were put up for stability testing according to schedule; they were routinely checked at the test stations; and at the two-month station, the information was incorrectly recorded as being the three-month results.

Use a Colon to Introduce a Second Independent Clause that Explains the First

This construction is handy, because the colon indicates that the *second* sentence is an explanation of the first. Whether you capitalize the second independent clause or not is up to you. If you do, however, do so consistently.

Our evaluation of the USP method tells us our course of action: We must develop an in-house method.

Use a Colon to Introduce a Final Appositive

Appositives rename nouns or pronouns.

Sorensen proposed a new function: documentation management.

In this example, "documentation management" explains "new function."

Use a Colon to Introduce an Example that's Set Off from the Rest of the Text

You'll see colons used this way frequently. Again, remember to place them at the ends of complete sentences. This usage is handy when you're employing bullets or numbers.

When we explore which products to develop, a primary resource is a bi-monthly publication: The Genesis Report/RX.

Note: It's becoming more commonplace to use colons to set off a list even though they do not follow a complete sentence.

They are:

They include:

Purists still find these constructions jarring, however. An easy way to avoid criticism is to complete the sentence before the colon.

They are the following:

They include six options:

Use a Colon to Separate Subtitles and Titles, Subdivisions of Time, Parts of Biblical Citations, and City and Publisher in Bibliographies

12:02 p.m. 1:45 p.m. 9:00 a.m.

Summary Report: Diclofenac Modified Release Tablets

CGMPS and You: The Jackson Pharmaceutical Co., Inc. Training Program

Boston: Little Brown and Company

Use a Colon at the End of a Formal Salutation

Dear Dr. Tavares:

Dashes

Dashes emphasize a sentence element or indicate a break in tone or thought. Don't use them indiscriminately. They shouldn't habitually take the place of commas, semicolons, or periods.

Use a Dash to Emphasize Appositives and Parenthetical Elements

The arbitrator ruled that the company — previously cited for EPA violations — was responsible.

The art of negotiation — some might call it manipulation — requires confidence in your position and a willingness to concede some issues.

Use a Dash to Separate a Series that Comes at the Beginning of a Sentence

Attention to detail, patience in directing others, and a willingness to listen — these are attributes of a good lab manager.

Parentheses

Parentheses set off incidental information, either within a sentence or after it. They work pretty much the way commas and dashes do, although each mark gives different emphasis to the inserted information. It's usually best to work information into the text rather than drop it in as an aside, which is the effect parentheses create. The following examples are all grammatical, but the first wields the most strength, because the relative pronoun clause is more strongly connected to the main clause.

> We have finally identified the unknown, which the chromatogram showed to be significant.
>
> We have finally identified the unknown — which the chromatogram showed to be significant.
>
> We have finally identified the unknown (which the chromatogram showed to be significant).

Use Parentheses to Enclose Descriptions or Explanations

When you place a parenthetical sentence between two sentences, start it with a capital letter and use end punctuation *inside* the parentheses. If you place a parenthetical phrase or clause inside a sentence, you don't need any punctuation other than the parentheses. If your parenthetical element is a sentence in its own right and you've inserted it in another sentence, punctuate it inside the parentheses. The latter form appears rarely, however.

> The corporate headquarters (with no less than six separate buildings) has only one cafeteria.
>
> The corporate headquarters is big (Would you believe six separate buildings?), but it has only one cafeteria.

Use Parentheses to Enclose Letters or Figures that Identify a Sequence of Items in a Series

> This is the vacation policy: (1) All employees must request vacation time a month in advance; (2) no employee may carry vacation days into the following year; and (3) employees who wish to buy out vacation time must do so before the end of November.

Use Parentheses to Enclose Acronyms in the First Citation

> Our new Standard Operating Procedure (SOP) system is in place.

Note that if you identify an acronym initially, all you need to do in subsequent text is use the acronym.

> When you write an SOP, follow the guidelines set forth in our manual.

Use Parentheses to Adhere to Select Referencing Systems

Certain referencing systems, such as the Modern Language Association (MLA) Style, call for parenthetical citation in the text.

Depression levels in patients dosed with 25 mg product were significantly lower than in those who received placebo (Lily et al 2004).

Brackets

Use brackets to enclose your own comments or corrections when quoting someone else, or when indicating with *sic* that text is in error. The term *sic* means, in Latin, "thus it is," and although we should generally stay away from latinate expressions, we have no English equivalent, so *sic* does the job.

He said, "I [have] been working on the formulation."

"I wroted [sic] down the procedure," he said.

Hyphens

Hyphens are little lines that join words or parts of words. They're useful for making text fit a line or for compounding words. What they do is facilitate reading.

Use a Hyphen to Compound Words

English, like other Germanic languages, joins words to make new words. These are compound words. Historically, hyphens have combined two existing words to make a new word, and after a while, the hyphen drops. For example, the two words, *book* and *keeper* were originally joined by a hyphen *(book-keeper)*. But gradually, the idea of a *bookkeeper*, or "one who keeps the books," gained acceptance as a concept in its own right, and the hyphen disappeared. What this means is that a hyphenated word is probably a word in transition: It started as two separate words and is on its way to being an unhyphenated word.

His brother-in-law works in production.

A cross-reference would be a big help.

It's usually wise to consult a dictionary, however, simply to make sure the compound word takes a hyphen; most don't. Note that many companies address hyphenation in their style guides because their use can be quite arbitrary.

Use Hyphens to Join Two or More Words Functioning as a Single Descriptive Modifier Before a Noun

We can hyphenate words that serve as a single modifier before a noun.

Over-the-counter products are our specialty.

We also make some exceptions here. If a combination is common, we generally don't hyphenate it.

Use a Hyphen to Spell Out Compound Numbers and Fractions

Seventy-nine compliance specialists share their expertise in the new book, *Doing It Right the First Time Around.*

We expect that *one-half* to *three-quarters* of our employees will choose the alternate medical plan.

Use a Hyphen to Break a Word at the End of a Single Line of Text

To avoid wide gaps of white space at the end of a line of text, you can break a long word with a hyphen and continue it on the next line. Just make sure to break the word between syllables. It is best not to separate a short beginning or end syllable from the rest of a word.

Note that many word processing systems allow you the option of hyphenating or not, and if you elect to include hyphenation, the program will do it for you as the need arises. Some corporate style guides dictate that words not be broken at the ends of lines, however, because breaking words by syllable, or simply arbitrarily, may contribute to difficulty in reading.

Apostrophes

Use an Apostrophe to Denote Possession

Use apostrophes after nouns and indefinite pronouns but not after pronouns, which already denote possession by their very nature. Add the apostrophe before an *s* in a singular citation, and after the *s* in a plural.

> Li's experiences, however unpleasant, taught her skills that were invaluable in her new position.

> Who will be appointed to the Board of Directors is anybody's guess.

> The chemists' conference will be held next month.

When a singular noun ends in *s*, you can add *'s*, if you choose. Most writers, however, omit it.

> James' decision to earn an MBA was a good one.

Or

> James's decision to earn an M.B.A. was a good one.

When two or more nouns show collective possession, add *s* to the last one.

> Carrie, Amy, and Lori's new work station is adequate for all three since they work different shifts.

When two or more nouns show individual possession, add *s* to each

> Carrie's, Amy's, and Lori's work stations are comparable in size.

Add an apostrophe to the last word in a compound structure to form the possessive.

> His father-in-law's firm handled the lawsuit.

Note: Don't confuse this usage with the plural in compounds of this nature.

> The event was to honor all fathers-in-law.

Use an Apostrophe to Form a Contraction

It's a wonderful position in a promising biotech.

Don't think twice about it.

We'll schedule the meeting for one o'clock.

Use an Apostrophe and "S" to Form the Plurals of Letters, Numbers, and Certain Words Used as Nouns

The speaker revealed a decided lisp when he pronounced his r's.

The 1980's triggered increased government scrutiny of the pharmaceutical industry.

His talk was laced with well's, you know's, and see's.

Quotation Marks

Quotation marks are used mainly to assist readers by indicating the beginning and end of a direct quotation.

At the last committee meeting, she specifically said, "I take vehement exception to the latest proposal."

When a quotation takes up more than three or four lines of text, it's best to offset it.

The Genesis Report/Rx offers this prediction about the dermatology market.

> The therapeutic technologies—steroids, topical tars, and immunosuppressives—that currently dominate the psoriasis market will remain the leading technologies for at least five years. Signal transduction technology is not expected to make major incursions against the established technologies. The primary new development will be cell adhesion molecule and antisense technologies. (Genesis 9)

Note also that when longer quotations are indented, the quotation marks are omitted.

A quotation within a quotation is set off with single, not double, marks.

> "Perhaps," she said, "I should read your report, 'Achieving Excellence in the Throes of Downsizing.'"

Commas and periods come *inside* the marks, whether or not they're part of the quotation; colons and semicolons come *outside*. And dashes, questions marks, and exclamation points come inside *if* they belong to the quotation; otherwise, they come outside.

> "It's been a stressful year. We've lost two major accounts, and therefore," Smith paused, overwhelmed himself by the news he was about to deliver, then continued, "we must cut the budget significantly."

> "Should I ask for a raise or hope that one is forthcoming all by itself?" she queried her coworker.

Ellipses

The ellipsis is useful when excerpting pertinent information from longer passages, such as regulations, when you wish to include just information that applies to your message, and not the whole passage. Ellipses are three periods in a row that signal an omission. When an ellipsis comes inside a sentence, use three periods. When it ends a sentence, include the fourth period, the endmark for the sentence itself. When it begins a sentence, use three periods followed by a capital letter.

> Sixty-two healthy men participated in the study. Thirty-one received a placebo and thirty-one received the active drug. None of the volunteers knew whether the tablet he was taking was the active drug or placebo. Study results show the active drug ...was well-tolerated.

The above excerpt indicates that additional information exists about the active drug but is not included here.

Capitalization

It used to be that writers would capitalize words that they felt were significant. Over the years, however, English has become more standardized and consistent. Now writers capitalize less frequently, but for greater impact. Company style guides usually address capitalization, and the conventions of capitalization may vary from company to company. The issue, of course, is that they achieve consistency.

Capitalize the First Word of Every Sentence

How did the discussion go?

Sanchez plans to retire shortly.

Capitalize the Key Words in Titles; Don't Capitalize Coordinating Conjunctions, Articles, and Prepositions

This guideline generally holds true. But when a title includes a colon, the first word after the colon always takes a capital. In addition, hyphenated words all take capitals if each is as important as the others.

The report "Safety Practices and Safety Records" should be required reading for all who manage.

"Training the Trainer" is the video we'll see.

Before our next meeting, please review this month's article, "Management by Committee: An Argument."

He recommended the book *The Dressed-for-Success Professional* to the middle-level managers.

Generally, Capitalize Proper Nouns

It's usually pretty clear which nouns are proper and which are common, but if you're in doubt, consult a dictionary. If the noun is capitalized, the word will start with a capital letter in the dictionary.

Proper Nouns

Names of People: Su Chow, Pankaj Parikh

Titles: Dr. Jones, Rabbi Davis

Note that when a title word refers to a general class rather than as a term of address, it is not capitalized.

Titles of Family Members: Aunt Astrid is funny.

Don't capitalize words like aunt and uncle, however, when you make a general reference: Avuncular means characteristic of an uncle.

Names of Languages: English, Swedish

Names of National and Ethnic Groups: the Swiss, Hispanics

Names of Companies and Organizations: The Simons Group, the League of Women Voters

Names of Regions and Continents: the Northeast, Asia

Note that points of the compass are not capitalized unless they refer to a specific region: "We travel south every winter." "The West has a great amount of desert land."

Names of Countries: Norway, Poland

Names of Cities and Towns: Newark, New York

Names of Neighborhoods, Parks, Roads, and Streets: Brookdale Park, Valley Road

Names of Bodies of Water: Lake Hopatcong, the Atlantic Ocean

Names of Specific Vehicles: Challenger III, Amtrak

Names of Religions and Religious Groups: Judaism, Protestants

Historical Periods, Events, and Documents: Ice Age, the Battle of Gettysburg, the Constitution

Names of Political Organizations and Bodies: the Republican Party, the New Jersey State Senate

Names of Months, Days, Holidays, Holy Days, and Holy Seasons: September, Monday, Christmas, Ash Wednesday, Lent

Note that the seasons are not capitalized: spring, summer, winter, fall.

13

Working on Words

When a comedic television character calls a captured burglar "a bungler," we laugh, because we know what he intends, and recognize that "bungler" is somehow apt, as well. But slips like this in the workplace don't always have such a happy effect. What they may accomplish is to make the writer look ridiculous, or worse, alter the message. Look at the following line, written by a quality assurance supervisor.

FDA's comments were dully noted.

This writer's audience may be reduced to laughter, if they understand she means "duly," as "in due course." But it may be just as likely the audience will think "dully" is what she intended, in which case, her message has taken on a different meaning. Either way, the error has damaged the message. If the audience realizes "dully" is a misspelling of "duly," the secondary message is that this supervisor can't spell; worse, she and her staff can't catch simple errors. But, if the message is taken at face value, pity the department, which has indicated their lack of interest in FDA comments.

> The shape of our language is not rigid; in questions of usage we have no law-giver whose word is final.
>
> **E. B. White, Author**

Correcting errors like this in your writing can be difficult because these glitches are usually caused by confusing two sound-alike words or an innate inability to spell. A computer spell checker can help you find misspellings, but it won't assist you in locating inappropriate words. If you use "it's" for "its," for instance, a computer won't tell you because both are viable. Further, while the "rules" for spelling will help in some cases, in others they won't at all. In addition, English, as a living language, is in a state of constant change. We form new words all the time. That's the process of a living language. We borrow words; we invent words; we create new forms of existing words.

The "rules of good diction" can stifle even the best of writers. What word is acceptable? Many words, like "hopefully," mean "full of hope." However, while modern "experts" say "hopefully" should only be used in that sense, they're not correct in their assertion. A word like "hopefully" is a disjunct, a word that permits the writer to directly comment on the sentence. It is no different than "clearly," "frankly," "unfortunately," and a spate of other words. Unfortunately, readers may not accept "hopefully" the way they would "unfortunately" in this very sentence. Your consideration, then, is whether or not to use a word like this, knowing your usage is correct, and risk the criticism of pedantic readers. This choice is yours.

The following glossary lists words that can be problematical for various reasons. Bear in mind, too, that the average person has a vocabulary that may be as large as 18,000 to 20,000 words, so the list may not answer all your questions about word choice. It's a good idea, therefore, to ask yourself three simple questions:

1. Am I entirely sure of the word's meaning?
2. If not, is there another word I can use that I *know* conveys exactly what I mean?
3. Is the word as specific and concise as it can be ?

> We all have three vocabularies: reading, which is the largest, because we recognize many words we don't actually use ourselves; speaking, the second largest, because we can be spontaneous in speech and experiment with words; and writing, the smallest, because writing requires more control and precision, and we are unlikely to risk the inappropriate or inaccurate word.

Glossary of Usage

a, an Use a before consonant sounds, and an before vowels sounds. It is not the preceding letter that determines the correct article; rather, it is the sound.

An FDA preapproval inspection

An SOP

A union contract

absorbance, absorptance, absorptivity Absorbance refers to the ability of a layer of a substance to absorb radiation, expressed mathematically as the negative common logarithm of transmittance. Absorptance and absorptivity refer to the ratio of energy a body absorbs to the energy striking it.

The success of radiation therapy depends on the degree absorbance in the subject.

Absorptivity varies from patient to patient.

absorbent, adsorbent An absorbent is a substance that can take in and incorporate a gas, liquid, light ray, or heat. An adsorbent can attract and hold to its surface a gas, liquid, or substance in a solution or in suspension.

The device absorbs the drug.

The design provides an immobilized adsorbent bead for reducing the concentrations of residual drug substance.

accept, except Accept is a verb meaning receive. Except is usually a preposition or conjunction meaning but for or other than.

She could not accept the results since they contradicted earlier findings.

All worked on the special project, except for Rogers, who was out of town.

accuracy, precision Accuracy is the degree of correctness of a measurement or statement, while precision refers to the degree of refinement of a measurement or statement.

Regular monitoring confirmed the accuracy of the refrigerator temperature.

The results of the first study were 76 percent, but subsequent studies showed greater precision with readings of 76.345.

access, excess Access is a noun meaning a way in. Excess is a noun meaning too much.

He sought access to the building.

His boss was disturbed by the excess charges.

adverse, averse Adverse is an adjective that means unfavorable or harmful. Averse means reluctant or opposed.

The bio-study looks for adverse drug reactions.

I am really averse to accepting that proposal.

advice, advise Advice is a noun. Advise is a verb.

Take my advice; do as your supervisor advises.

affect/effect Usually affect is a verb meaning to influence, and effect is a noun meaning results. Sometimes, however, effect is a verb meaning to bring about, and affect is a noun meaning facial expression.

The change in the blood bank procedure affected the results.

The study showed the added step had a negative effect on the process.

A change in project management effected positive change.

The patient showed no visible affect.

after, following After means later than a time or event. It can also imply cause and effect. While following is synonymous with after, it works best as an indicator of position unrelated to time.

One patient suffered an adverse event after taking the study drug.

The following authors are cited: Cimino, Lin, and Wollowitz.

aggravate, exasperate, irritate Some critics hold that aggravate only means make worse, but it is frequently used to mean irritate or annoy. Exasperate certainly means annoy and frustrate. And irritate means annoy. In writing, you may choose not to use aggravate to mean "irritate" or "exasperate" if you wish to avoid criticism.

The CEO was irritated by his managers' stubbornness, because he feared a delay in making a decision would aggravate the already poor company morale.

His reaction exasperated the committee.

aliquot, sample Aliquot is a noun meaning a measured portion of a whole. Use it as a verb only when you wish to indicate distribution of measured portions. A sample is a part taken as a representative of source material.

Two aliquots were drawn. (Not "Two samples were aliquoted.")

but

The mixture was aliquoted onto microtiter plates.

The quality assurance inspector sampled the raw material prior to release.

all ready, already All ready means completely prepared. Already means by now or before now.

We were all ready to go to the meeting, but it had already started.

all right All right is always two words. Alright is a common misspelling.

Whatever you decide is all right with me.

all together, altogether All together means in unison, or in one place. Altogether means entirely.

It's not altogether true that we never work all together.

allude, elude Allude is a verb meaning to refer. Elude is a verb meaning to avoid.

She alludes to his past errors constantly.

We would prefer to elude the issue.

a lot A lot is always two words. Alot is a common misspelling.

The restructuring of the stability group has generated a lot of speculation.

alternate, alternative, option These three are often confused. Alternate means to pass back and forth successively from one state, action, or place to another. Alternate also means a replacement or backup. Alternative means one of two choices. Historically, however, it has been used to mean a choice between two or more choices. Option can mean any of a number of other choices or additional features.

The alternate states are ice and water.

We arrived by an alternate route.

We must have an alternative plan.

The computer system came with a number of options.

The document control system offers many options.

a.m., p.m. These abbreviations should be used with numerals. Don't use them as substitutes for morning and evening. It's considered nonstandard to say "He worked until late in the p.m." Use lowercase letters in general correspondence. (See also the section on mechanics in Chapter 14 for using abbreviations.)

Charles left the office at 2:00 p.m.

We find that cGMP training is best conducted during the morning hours.

ambiguous, ambivalent Although these words don't sound alike, they're still frequently confused. Ambiguous is an adjective meaning unclear. Ambivalent is an adjective meaning having mixed feelings.

Documentation must be free of ambiguous statements.

My feelings about the corporate changes are ambivalent.

amiable, amicable Amiable, an adjective, means having a pleasant and friendly disposition, or being good natured. Amicable, which is also an adjective, means peaceable or showing good will.

The new employee proved to be an amiable fellow.

We hope the negotiations will continue to be amicable.

among, between Among is used to refer to three or more people or things. Between is generally used to refer to two people or things.

The workload was divided among the staff.

Between you and me, I think she'll quit.

amount, number Amount refers to a quantity of something that cannot be counted. Number refers to countable items.

There's a small amount of residue remaining, despite the number of times we cleaned the equipment.

anatomy, morphology, structure Anatomy, a scientific discipline, studies and describes structure of living things. Morpohology, also a scientific discipline, studies and describes the origins, shape, and structure of living things. It also refers to the system of word-forming elements and processes in language. Structure refers to parts of a whole, in living or nonliving things, and how they relate to each other.

Dieticians must study anatomy.

With an advanced degree in morphology, he devoted his life to the study of primates.

What is the chemical structure of the compound?

How is the argument structured?

and etc. And etc. is a redundant expression, since etc. (etc.) means and the rest. Generally it's best to avoid Latinate expressions when they serve no purpose.

Once the chemist finishes the testing, the supervisor will review the lab notebooks, monthly reports, and other documents.

anesthesia, anesthetic Anesthesia means the loss of sensation with or without consciousness. Anesthetic, on the other hand, refers to a substance that induces anesthesia.

Which anesthetic did they use to induce anesthesia?

ante, anti- Ante, as a prefix, means before. Anti means against.

Pronouns should refer to their antecedents.

The drug contains an extract of the antigenic protein from the pollen of plants.

anxious, eager Anxious, since the beginning of the twentieth century, has been dictated to mean nervous or worried and is usually followed by "about," although historically it has often been used to mean eager. Eager means looking forward and is usually followed by "to."

I'm eager to start the experiments, but I'm anxious about their complexity.

anymore Anymore, an adverb, means any longer. Use it in negative contexts. You may invoke criticism if you use it when you mean "nowadays" or "now."

The union demands don't surprise me anymore.

anyone, any one Anyone means any person at all. Any one means a person or thing in a specific group. The same distinction applies to "everyone" and "every one" and "someone" and "some one."

Anyone who meets the criteria can join the association.

Any one of these options will suffice.

appraise, apprise Apprise means to inform, and appraise means to evaluate. Problems occur when writers use "appraise" to mean "apprise."

Please keep us apprised of the ongoing dialog between regulatory and FDA.

The Principal Investigator will appraise the adverse events.

archaic, obsolete These two adjectives are often used interchangeably. However, they really do have different meanings. Archaic means ancient and seldom used; obsolete means no longer in use. Do not use obsolete as a verb.

"Thou" is the archaic form of "you."

Version A of the document is obsolete.

as, for, because, since As is an adverb meaning while, or a comparative term. It is often used to mean "because," but some purists rail against that usage, and there is some merit, because using as may create ambiguity. Does it mean time or reason? Consider a sentence like this: "As the laboratory was locked, he went home." For this reason you may opt to use it only in the sense of time. For, a preposition, means in place of or destined. Since, an adverb, means after or because. Because also shows cause.

As I was reviewing the data, I discovered an unaddressed problem.

Exchange the software program for the updated version.

I have three status reports for you.

Since our last meeting, I've investigated the issue.

Because the method failed dissolution, we had to develop another one.

The company was forced to comply with the new regulations, since we were in danger of being cited.

as, like In formal speech and writing, as may be either a preposition or a conjunction; like functions as a preposition. Generally, use as if what follows is a full clause (a group of words having both a subject and verb).

The fourth chromatogram shows well separated peaks, as we expected. The third, like the first two, doesn't.

assumption, conjecture, law, theory, hypothesis Assumption is a belief presented at the start of an argument as the basis for deduction and inference. A conjecture is a speculation but not one proposed for testing. A theory is a broad concept based on extensive observation. A hypothesis is narrower in concept than a theory, and typically offers a potential explanation for phenomena that can undergo testing by experiment and observation. A hypothesis can form a conclusion at the end of an argument or draw an inference based on data.

The assumption was that the trial would require 100 patients.

That's purely conjecture; there's nothing behind it.

From these data we can postulate this hypothesis.

assure, ensure, insure Assure has the connotation of promise. This is the context of quality assurance, the promise of quality. Ensure has the connotation of make certain. Insure is often used interchangeably with ensure, but purists maintain it should refer only in reference to legal and financial protection.

Let me assure you we will launch the product in March of next year.

Cline allowed extra planning time to ensure excellent results.

It's expensive to insure the plant against flooding.

as to As to, meaning about or concerning, is considered stilted by many writers. A sentence like this is probably better rewritten: "I will notify you as to my decision." About can usually replace as to.

I will notify you about my decision within the week.

at present, presently This pair invokes much controversy, although historically they have both been used to mean now. If you wish to avoid debate about your word choice, use at present as a prepositional phrase to mean now. Reserve presently to use as an adverb meaning in the immediate future.

At present, we have an opening for a lab technician.

We will address that issue presently.

average, typical In scientific writing reserve average for use as a synonym for statistical mean. Typical is a workable word to use in place of a nonstatistical average.

The typical patient is a 30-year-old male.

averse See adverse.

a while, awhile Awhile is an adverb. A while is an article and a noun. Awhile can modify a verb, but cannot act as the object of the preposition.

I will be gone awhile.

I will be gone for a while.

bad, badly Bad is an adjective. Badly is an adverb. Use bad to modify nouns and after linking verbs. "I feel badly" is a common misusage, what experts call "overcorrection," although popular usage may make it standard in time. "I feel bad" is always correct because "feel" is a linking verb. Further, consider saying "I feel sadly" or "I feel gladly." They're equally incorrect and sound so to our ears. But, "badly" takes its rightful place after an action verb.

I feel bad about the changes I'm going to have to make.

He performed badly on the aptitude test.

because of/due to Substitute because of or as a result of for due to when possible. See also as.

being as, being that These phrases are seldom used in formal print and are considered colloquial for "because." Avoid using them in professional writing, opt for because.

Because our new product received FDA approval, the CEO commended the regulatory group for the efficient ANDA submission.

benchmark, criterion, standard Benchmark is often used as a synonym for criterion or for standard. Criterion, however, is often restricted in meaning to be comparable to measure, but with possibly different specifications for judgment. Standard is the term that best describes a criterion assigned a specific value against which a judgment can be based.

The company benchmarks salaries against industry standards.

What's the criterion for measurement?

We compared the HPLC chromatogram to the standard.

beside/besides Beside is a preposition meaning next to. Besides is a preposition meaning except as well as an adverb meaning in addition.

The director sat beside her assistant.

No one, besides Sue, can compile the data we need so quickly.

Besides Netgo, six other companies are working on anti-depressants.

between See among.

bi, semi Bi, as a prefix, traditionally has meant every other, while semi has meant twice in. Most people don't have a problem with semi, but bi is a bugbear. Consider bimonthly. In the publishing world, bimonthly means "every two months." However, in educational circles, it usually means "twice a month." To avoid confusion, spell out what you mean.

The board of directors meets every other month.

We sanitize the equipment twice a month.

bring, take Bring means to carry something from a farther place to a nearer one. Take means to carry something from a nearer place to a farther one.

Take these reports to be copied, and bring them back as quickly as you can.

but, hardly, scarcely These words are negative in their own right. Using "not" with any of them to indicate negation is redundant.

I had but an hour.

Kathleen hardly had time to complete the task.

I could scarcely make out the words.

can, may Our language is losing the distinction between can and may, but diligent writers still observe it. Can is traditionally reserved for ability; may denotes permission.

How well can he perform liquid chromatography?

You may assist either Patel or Kovalerchik.

capital, capitol Capital means chief or principal, and derives from the Latin "caput," meaning "head." Capital is often used in reference to worth. A capitol is the building or buildings in which legislative bodies meet.

How much capital does the company have?

We visited the capitol building in Sacramento.

case, patient, subject Case describes an instance or episode of disease, disorder, or dysfunction. Patient refers to a person affected by the disease, disorder, or dysfunction. In clinical studies, many writers distinguish subjects as healthy volunteers.

Twenty-two cases of fever were reported during the study.

The patient went into anaphylactic shock.

Of 100 patient volunteers, five had type II diabetes.

center around Center on and center in are more logical than center around, which is contradictory.

His talk centered on joint ventures.

The company is firmly centered in providing services.

chord, cord Chord means three or more musical tones sounded simultaneously. It also means a straight line joining two points on a curve. Reserve cord for all other uses.

We tied it with cord.

He had a spinal cord injury.

circadian, diurnal Circadian means a cycle of approximately 24 hours. Diurnal means each day in the daytime and comes from the Latin word "diurnus."

Homosapiens function on a circadian rhythm.

While some animals are nocturnal, humans are diurnal.

compare to, compare with, versus Compare to means to regard in a similar manner. Compare with means to examine for similarities or differences. Versus usually means against. In scientific writing, compared with is usually the best choice. Sometimes, however, versus is the best choice.

Roe vs Wade was a precedent-setting trial.

Graft-versus-host disease is a serious adverse event.

Compared to the same period last year, we're doing well.

Erin compared her findings with those of her colleagues.

complement, compliment To complement means to go with, to accompany, or free. To compliment means to praise or give laudatory comment. The latter is infrequent in pharma, biologic, device, and biotech writing.

The information in the brochure complements that in the annual report.

The two colors are complementary.

The hotel served a complementary breakfast.

The Human Resources director was unimpressed by the applicant's compliments.

comprise, is comprised of Do not use is comprised of; use comprise, which means include, contain, or embrace. An acceptable alternate is consists of, which means made up of, or composed of.

The NIH comprises several institutes. (Not the NIH is comprised of several institutes.)

The NIH consists of several institutes.

conjecture See assumption.

continual, continuous Continuous is the newer of these two terms, coming into usage in the seventeenth century. Since then, purists have striven to make a distinction between the two. Continual, they insist, means constantly recurring, and in that sense, is more likely than continuous to be used for repetition of something that can be interrupted. Continuous is generally understood to mean unceasing. Both terms, however, are often used in similar contexts.

Our department is continually interrupted by phone calls.

The new ventilation system supplies a continuous stream of cold air.

cord See chord.

criterion See benchmark.

current, currently Current is an adjective meaning belonging to the time actually passing. Currently is an adverb meaning at the present time.

At Site 452, this is the current practice.

We are currently writing the protocol for the trial.

database, data set, data Database is a file, typically electronic, that houses data which are retrievable and viewable. A data set is a body of data the database maintains, and a database may contain more than one data set. Data are discrete pieces of factual information. Note that data, while plural, is often used in the singular; this is because the singular datum is seldom used. Purists, however, still use the word with a plural verb.

We created a database to track the notebook numbering, distribution, retrieval, and microfilming.

The data are inconclusive.

decontaminate, inactivate, inhibit, sterilize Do not use these terms interchangeably. Decontaminate means to kill or remove bacteria or viruses (pathogens) from a substance. Inactivate means to stop a pathogen from replicating by damage to a key component. Inhibit means to reduce the ability of something to act as expected. Sterilize means to kill or remove all pathogens from something.

The plasma was decontaminated.

The device inactivates pathogens.

Viral growth was inhibited.

We have inactivated the bacteria, viruses, and leukocytes.

deduce, deduct, induce Deduce means infer or reason to a conclusion. Deduct means take away. Induce means to reach a conclusion through inductive reasoning (going from particular facts to a principle). Induce also means to bring about.

From these test results we may deduce that the condition had been present for some time.

Can you deduct my contributions from the balance?

Looking at the sequence of events that led to the out of specification result, we induced that the equipment suffered a short power failure.

Labor was induced.

defuse, diffuse To defuse means to remove the fuse from a device or to lesson, inactivate, or make less potent. To diffuse means to scatter or disperse.

The centrifuge diffused the particles.

The Institutional Review Board (IRB) defused a potentially difficult situation.

different from, different than Both usages are standard in American English, with different from being the most common. Purists maintain different from should be used only when the comparison is between two things. Different than typically finds a place when a clause follows.

Your dissolution results are different from mine.

The seminar was different than the one I attended last year.

definite, definitive Definite means clearly established and limited. Definitive means conclusive.

Their regulations are definite in their application.

The results of the study are conclusive.

diurnal See circadian.

dose/dosage A dose is a quantity to be administered at one time or the total quantity administered. Do not confuse it with dosage, the regulation or frequency of doses, or regimen, usually expressed in terms of a quantity per unit of time.

The subject received a dose of 10 to 15 mL.

The dosage was 10 to 15 mL.

due to See because.

effect See affect.

ensure See assure.

everyone, every one The same rules that apply to anyone and any one apply to everyone and every one.

explicit, implicit Explicit means stated outright. Implicit means implied, unstated.

He left explicit directions about the project.

Our trust in his judgment was implicit.

farther, further Farther and further are historically the same word, with further being the oldest form. When farther developed, however, further did not die out. Now the distinction made between the two is that farther refers to distance, and further to quantity or degree.

I can't travel any farther.

Jones carried the notion further.

fewer, less Fewer refers to items that can be counted. Less refers to a collective quantity that cannot be counted.

That department has fewer employees, but then, they have less to do.

There is less solution in that beaker.

following See after.

for See as.

former, latter Former refers to the first named of two things, latter to the second named.

His job responsibilities were to hire and fire employees. He loved doing the former, but hated the latter.

further See farther.

good, well There are many who will criticize using well in place of good and vice versa. However, you are safe no matter how you use well. And practically everybody agrees that both good and well after "feel" and "look" are equally acceptable; they serve as adjectives. Only good, however, can serve as an adjective to modify a noun. Yet good can serve as an adverb, too, but this form is usually reserved for speech, not writing.

You would do well to investigate the disparity further.

The office looks good.

Vilkov is a good manager because he treats his people well.

hardly see but.

he/him, she/her, I/me More and more frequently these pronouns are being used interchangeably. It happens because people believe that he, she, and I sound more polished. But in constructions such as prepositional phrases that call for the objective case, him, her, and me are correct. It's wrong to say, "She gave the report to William and I," because "to" is a preposition. Similarly, you shouldn't write, or say for that matter, "The new employee was delighted to go with Nancy and he." If you are in doubt, try the construction with just the pronoun. Surely no one would write "with he."

They presented the award to her and Johnson for their conjoint efforts.

Present your findings to Dr. Tavares and me.

That's for him and me to decide.

herself, himself, itself, myself, yourself These and other "self" pronouns are reflexive or intensive. They must refer to or intensify a noun or another pronoun in the sentence or in a previous sentence. Colloquially these pronouns are often used in place of personal pronouns, but this usage is inappropriate in professional writing. It is nonstandard to say "The attorney gave the report to the chairperson and myself." The correct pronoun in this case is "me." Similarly, it's incorrect to say "The boss and myself will handle the project together." Here, the pronoun should be "I."

She, herself, saw no harm in the fabrication.

The department heads disagreed among themselves.

Andrew Weil is wholly qualified to give the presentation.

hopefully Strictly speaking, hopefully means with hope, but it is often used colloquially to mean "with luck." Language mavens argue for its use only in the true meaning, but logically it is no different than other adverbs that modify complete sentences, such as "clearly."

I waited hopefully for a glimpse of the new president.

With luck, we'll find the right person for the job.

human, humane Human is a noun or adjective meaning homo sapiens. Humane is an adjective meaning kind.

What does it mean to be human?

For the study, we used non-human primates.

A humane solution is usually the best.

hypothesis See assumption.

I/me See he, him.

implicit See explicit.

imply, infer Imply means suggest. Infer means conclude.

The tone of his letter implies that he's more than a little concerned.

From the data, we can infer that there are no colony forming units.

in, into In indicates location or position. Into indicates movement or change. The colloquial use of into to mean "absorbed with" or "absorbed in" should be avoided.

Maureen is in Seattle.

Come into my office.

Lopez was totally absorbed in the project.

inactivate See decontaminate.

incidence, incident, incidents, prevalence Incidence is a noun meaning frequency of occurrence (the number of new cases occurring in a population of stated size during a stated period of time).

Incident is a noun meaning an occurrence. Incidents is the plural form of incident. Prevalence tells how common a condition is (the number of cases existing in a population of stated size at a particular time). Do not confuse incidence and prevalence.

The study monitored the incidence of breast cancer in Marin county.

Was this the first incident, or were there previous incidents?

There is a high prevalence of depression in Scandinavian countries.

induce See deduce.

infer See imply.

inhibit See decontaminate.

insure See assure.

inter, intra Inter, a prefix, means between. Intra, also a prefix, means within.

The obstruction was interventricular.

The origin was intranasal; that is, it was confined to the nasal cavity and had not spread.

irregardless, regardless People often use the colloquial irregardless when they mean regardless. Irregardless is thought to blend irrespective with regardless. Irregardless, though fairly widely used, has somewhat of a bad reputation.

Regardless of what you say, I stand by my decision.

is when, is where These mixed constructions are often incorrectly used to define. It's wrong to say "A job action is when employees refuse to work." Logic tells us a "job action" can't be time, and "when" signifies time. Is where is often misused as well. "Where" signifies "place" and should not be used to denote time, as in "The intermission is where I look for my colleagues."

That's the place where I used to work.

Friday is when we should make our proposal.

its, it's These two are confusing because, normally, to show possession, we use an apostrophe. However, pronouns have possession built in. "His" and "her" take no apostrophes. Yet a proper noun like John, used possessively becomes "John's." People logically reason the possessive form of it must be it's. But its denotes possession, and takes no apostrophe. It's is a contraction of "it is." You can avoid problems by writing "it is" when you mean it's. That way whenever you use its, write it without the apostrophe.

It's a good thing we know its name.

itself See herself.

lay, lie Lay is a verb that means to put or place; it is almost always followed by a direct object. Lie is a verb that means recline or be situated; it usually doesn't take an object.

If we lay out our plans systematically, the project will begin smoothly.

I lay awake all night worrying about the problems.

I lie down each afternoon.

lead, led Lead (rhymes with bed) is a noun naming a metal. Lead (rhymes with feed) is a verb meaning to guide. Led is the past tense of the verb lead and means guided.

Lead has been removed from modern day paint products.

The new vice president will lead us through the merger successfully.

Were we led down the wrong path?

lend, loan Lend is a verb meaning to give for temporary use. Loan is a noun meaning something furnished for the borrower's temporary use. Only careless writers use it to mean "lend."

Can you lend me those manuals?

She took a bank loan to finance her new car.

lessen, lesson Lessen is a verb meaning to diminish. Lesson is a noun meaning something to be learned.

The lesson was to lessen the workload of the overtaxed staff.

liable, likely These terms have been used interchangeably, but some grammarians insist that liable is the correct word to express responsibility or obligation, while likely is the word of choice to express probability.

Carmichael is liable for any incorrect data he submits.

She will likely complete the testing by Friday.

localize, locate Localize means to confine, restrict, attribute to a particular place, or have a characteristic resulting from an action. Locate means to find, specify, or place.

The infection localized in the sinus cavities.

We located the infection in the left pleural cavity.

lots, lots of Lots and lots of are substandard substitutes for a lot, many, and much.

We have a lot of material to review before the meeting.

Andrea was offered many opportunities to advance.

majority, most Majority means a number of items greater than half the total. Most is the term of choice when a quantitative expression is unnecessary.

The majority of patients in the study showed clear decline in symptoms.

Most doctors are licensed in but one state.

may See can.

maybe, may be Maybe is an adverb meaning perhaps. May be is a verb meaning might be.

It may be that the company will be ready to launch in May.

Maybe the investigation will determine the cause of the nonconformance.

method, technique Writers often use these words interchangeably to mean a procedure. Reserve method to mean a procedure, and use technique to mean the skill and application for carrying out a procedure.

We are writing the laboratory method.

The method required an unusual technique.

moral, morale Moral is an adjective meaning the ability to distinguish between right and wrong or a noun meaning lesson. Morale is a noun meaning moral or mental condition.

To encourage the administrative assistant to stay in her job or to pursue her own best interests became a moral dilemma for him.

Department morale was high because they had learned to work as a team, supporting and assisting one another.

morphology See anatomy.

most See majority.

myself See herself.

number See amount.

obsolete See archaic.

occur, recur Occur is means to happen. Recur means to happen again. Reoccur is a misspelling.

What has occurred will most likely recur.

option See alternative.

parental, parenteral Parental refers to mothers and fathers. Parenteral means introduced by a means other than the intestines.

The transdermal patch delivered the drug parenterally.

patience, patient, patients Patience is a noun meaning *t*he ability to wait and endure, the ability to be patient. Patient is an adjective meaning steadfast and uncomplaining. Patient is also a noun meaning person under treatment; patients is the plural of patient.

Arbitrators need patience to hear both sides of a dispute clearly.

He was a patient employee, not expecting immediate recognition.

The executive, used to giving directions, proved to be a terrible patient.

Nurse Gearhart had her favorite patients, and he was not one of them.

patient See patience or case.

people, persons In formal speech and writing, people refers to a general group. Persons refers to a specific collection of individuals.

The people who participate in the biostudies come from all walks of life.

Will the person or persons who saw the accident please see our safety officer?

Persons with myopia may opt for corrective surgery.

percent, percentage Percent means units per 100 units, and is represented by %. Percentage is a statement of a quantity or rate. Use the symbol with numbers, except at the beginning of a sentence.

Five percent of the complaints were related to packaging.

The difference between 25% and 50% is 25 percentage points.

A small percentage of those enrolled did not complete the trial.

plus Plus is often used to mean and or moreover. Some speculate the origin to be mathematical as in "Two plus two equals four," with plus meaning and. In recent times, it has moved into the conjunctive function, so you will often see it joining clauses.

Our firm has a solid future; moreover, we offer our executives excellent benefits.

Our firm has a solid future; plus, we offer our executives excellent benefits.

practicable, practical These two adjectives are not interchangeable. Practicable refers to something that seems possible, but has not been tested. Practical means useful and sensible.

At first glance the plan seemed practicable.

We seek a practical solution to allocating the additional duties our department has undertaken.

precede, proceed, proceeds Precede is a verb meaning to go before. Proceed is a verb meaning to continue. Proceeds is a noun meaning that which results from a transaction.

That which precedes the final thrust is also important to success.

Proceed carefully when you make decisions about people.

The proceeds from the sale were given to charity.

precision See accuracy.

prevalence See incidence.

principal, principle Principal is a noun meaning chief, official, or in finance, chief sum. Principal also functions as an adjective meaning foremost or major. Principle is a noun meaning rule or axiom.

He is the Principal Investigator.

How much of the principal remains?

His principal reason for admitting his error was his sense of principle.

procedure, protocol A protocol is a plan, usually for an experiment, study, or test. A procedure is the process for carrying out an activity, often a recurring one. Frequently, protocols define procedures as part of an overall plan.

All GLP development studies require QA-approved protocols.

Does the company have a facilities evacuation procedure?

proportional, proportionate More and more frequently, these two words are used interchangeably. Most standard dictionaries, however, make a distinction. Proportional should refer to a number of similar or related things; proportionate should be used when referring to two things in relation to each other.

A proportional distribution over time is the plan.

Your new desk and chair are proportionate.

protocol See procedure.

proved, proven Proven and proved are both acceptable forms of the past participle of "to prove." In scientific writing the preference is for proved.

We have proved that a retrovirus causes the syndrome.

quotation, quote Quotation is a noun. Quote is a verb. Writers with solid skills avoid using quote as a noun, although this usage is gaining acceptance.

May I quote you on that?

What's your favorite quotation?

recur See occur.

regardless See irregardless.

regime, regimen While both these words come from the Latin "regimen," regime is best reserved for reference to a ruling or guiding faction during a specific time. Use regimen when you mean a systematic process.

The patient dosing was by specific regimen.

respectfully, respectively Respectfully means with respect. Respectively means each in the order given.

She respectfully opposed the new plan.

We will address old business, the current issues, and possible problems, respectively.

sample See aliquot.

scarcely Scc but.

she/her See he/him.

since See as.

someone, some one The same rules that apply to "anyone, any one" apply to someone and some one.

standard See benchmark.

stationary, stationery Stationary is an adjective meaning not moving. Stationery is a noun meaning paper for writing letters.

Riding a stationary bicycle helps reduce stress.

Company letterhead is not to be used as personal stationery.

structure See anatomy.

subject See case.

take See bring.

technique See method.

than, then Than is a preposition meaning besides or a conjunction offering comparison. Then is an adverb meaning at that time or subsequently. "Then" is often erroneously used for "than."

No one, other than Cindy Brady, preferred working overtime.

Kristina Reynolds always finished her work earlier than the others.

He finished his tasks; then he assisted the others.

that, which, who These words, when used to begin subordinate clauses, sometimes give writers trouble. That and which add information about things and events to the sentence, while who and sometimes that add information about people. The information should be set off in commas or be unpunctuated, depending upon what you mean. If the meaning of the sentence is clear without the additional information, set the clause off in two commas if it comes in the middle of the sentence, or one if it comes at the beginning or end of the sentence. Use no commas if the information is necessary to clarify the main sentence.

Use the computer that's in the library.

You may use the computer, which is in the library, whenever you like.

The computer programmer who designed the organizational chart system is a valuable asset to the company.

Fredericks, who had no formal computer training, actually designed the Main menu software program for the entire company.

their, there, they're Their is the possessive form of "they." There indicates place. They're is a contraction of "they are." Avoid using the contraction in formal documents.

Their questions are right there on the roster, but they're not going to be addressed.

theory See assumption.

through, thorough Through is a preposition meaning in and out or by, or an adjective meaning finished. Thorough is an adjective meaning complete.

The mail manager passed through our office without delivering our mail.

I had planned to go through proper channels.

May I use the computer when you're through?

Patel expects a thorough evaluation of the project.

to, too, two To is a preposition meaning toward; it is also part of the infinitive form of a verb. Too is an adjective or adverb meaning also. Two is an adjective meaning one plus one.

I am going to the theater this evening.

Thomas has to complete the contract before week's end.

Monica Grimaldi indicated she'd like to work on the new product, too.

The director requested two sets of the documents.

toward, towards These prepositions are generally interchangeable; we've lost the distinction between the two. Toward, however, is the preferred word.

We are moving toward our goal.

utilize Utilize, as a verb, is generally accepted, but many writers prefer the simpler, more direct word use.

He wants the company to use all its resources.

versus See compare to.

wear, were, we're, where Wear is a verb meaning to have on. Were is a verb, the past tense of "to be". We're is a conjunction meaning we are. And where is an adverb or conjunction meaning in what place.

What you wear determines, to a large degree, the impression you make.

We were all set to vote, but then Diane introduced a variable we had not previously considered.

When we finish this job, I promise, we're all going to be handsomely rewarded.

Where can we find a competent benefits administrator who also has advanced computer skills?

well See good.

which, who See that.

who, whom, whoever, whomever The same principle works for these pronouns as for "he/she, her/him." Who and whoever are the nominative forms; whom and whomever are the objective forms.

I don't care who goes to the golf outing.

Give it to whoever wants it.

To whom shall we present the award?

The committee can choose whomever they wish.

who's, whose Who's is a contraction of who is. Whose is the possessive form of who.

He's the fellow who's going to head the department.

Whose manuscript is that?

your, you're Your, a pronoun, is the possessive form of you. You're is a contraction of you are.

When your report is complete, Lily would like to see it.

I believe you're exactly who the company needs.

yourself See herself.

14

Acronyms, Symbols, and Abbreviations

Users of English like to shorten and abbreviate whenever they can. Consider that "television," which means "to send a picture," has been shortened in England to "telly," but "TV" in the states. Truncating the language is common practice; it happens all the time. We take "taxis," not "taxicabs"; we say "24/7" when we mean "24 hours a day, seven days a week"; we say "post op" when we mean "after the operation," and we say "specs" for "specifications."

Other ways we shorten the language are by creating acronyms, abbreviating terms, and using symbols. Thus, we refer to the Food and Drug Administration as FDA, write "p.o." when we mean "by mouth," and record "7 J" when we mean "seven joules." To maintain control of this alphabet soup, it's best to follow set conventions when using acronyms, abbreviations, and symbols. Writing becomes easier to understand when consistency prevails. Regulatory agencies recognize that acronyms and abbreviations can be problematic because they can often mean more than one thing. That's why FDA posts an acronym list on its website. And in reports going to the government, companies include an acronyms and abbreviation list. The purpose is to eliminate the confusion or misunderstanding that can occur if a reader understands a term to mean one thing but the writer intends quite another.

Acronyms

In addition to acronyms your company may use internally, the industry is rampant with standard acronyms, including those that signal organizations, interest groups, and agencies as well as those reflective of the common vernacular.

The following are acronyms common to the pharmaceutical, medical device, and biologic industries. The ones relative to countries other than the United States identify the countries to which they belong. This list is far from complete since new acronyms develop rapidly. It's therefore a sound practice to make note of them as they appear.

AAALAC	American Association for Accreditation of Laboratory Animal Care
AAAS	American Association for the Advancement of Science
AABB	American Association of Blood Banks
AACR	American Association for Cancer Research
AACE	Association for the Advancement of Cost Engineering
AAFCS	American Association of Family and Consumer Sciences
AAMI	Association for the Advancement of Medical Instrumentation
AAPM	American Association of Physicists in Medicine
AAPS	American Association of Pharmaceutical Scientists
ABC	America's Blood Centers
ABIH	American Board of Industrial Hygiene
ABPI	Association of British Pharmaceutical Industries
ABSI	Association of the British Pharmaceutical Industry
ACE	adverse clinical event
ACIL	American Council of Independent Laboratories
ACM	Association for Computing Machinery
ACPE	American Council on Pharmaceutical Education
ACRPI	Association for Clinical Research in the Pharmaceutical Industry
ACS	American Chemical Society
ACT	Applied Clinical Trials
ADA	American Dental Association
ADE	adverse device effect
ADE	adverse drug event
ADER	adverse drug experience report
ADI	acceptable daily intake
ADME	absorbance, distribution, metabolism, excretion (refers to a pharmacokinetic study)
ADP	automated data processing
ADR	adverse drug reaction
ADROIT	Adverse Drug Reactions On Line Information Tracking
ADRS	adverse drug reporting system
AE	adverse event
AEM	Agency Español (Spanish Medicines Agency)
AEPAR	Associación Española de Professionales de Actividades del Registro (Spanish Society of Regulatory Affairs)
AERS	Adverse Event Reporting System
AESGP	Association Européenne des Spécialités Pharmaceutiques Grand Public (Association of the European Self-Medication Industry)
AFDO	Association of Food and Drug Officials

AFE	Association for Facilities Engineering
Afssaps	Agence Francaise de Sécurité Sanitaire des Produits de Santé (French Health Products Safety Agency)
AHCPR	Agency for Health Care Policy and Research
AIC	American Institute of Chemists
AICRC	Association of Independent Clinical Research Contractors
AIM	Active Implantable Medical Device Directive
AIM	active ingredient manufacturer
ALARP	as low as reasonably practicable
AMA	American Medical Association
AMT	American Medical Technologists
AMWA	American Medical Writers Association
ANADA	Abbreviated New Animal Drug Application
ANDA	Abbreviated New Drug Application (a simplified submission permitted for a duplicate of an already approved drug)
ANDI	Africa Nutrition Database Initiative
ANOVA	analysis of variance
ANSI	American National Standards Institute
ANVISA	Brazilian Agency of Sanitary Surveillance
APhA	American Pharmaceutical Association
APHIS	Animal and Plant Health Inspection Service
API	active pharmaceutical ingredient
AQL	acceptable quality level
AR	adverse reaction
ARC	American Red Cross
ARO	academic research organization
ASCP	American Society of Clinical Pathologists
ASH	American Society of Hematology
ASM	American Society for Microbiology
ASPI	Association of the British Pharmaceutical Industry
ASQ	American Society for Quality
ASTM	American Society for Testing and Materials
ATC	Anatomical Therapeutic Chemical (International system for classification of medicines)
ATF	Alcohol, Tobacco, and Firearms (Bureau of)
ATCC	American Type Culture Collection
AUC	area under the curve
BAN	British Approved Names (short names for substances)
BARQA	British Association for Research Quality Assurance

BCA	Blood Centers of America
Bcr	breakpoint cluster region
BCSP	Board of Certified Safety Professionals
BDP	biotechnology-derived products
BEUC	Bureau Européen des Unions de Consommateurs (European Consumers' Organization)
BfArM	Bundesinstitut für Arzneimittel und Medizinproducte (Federal Institute for Drugs and Medical)
BGH	bovine growth hormone
BHTA	British Healthcare Trades Association
BIMO	bioresearch monitoring
BIND	Biological Investigational New Drug
BIRA	British Institute of Regulatory Affairs
BLA	Biologics License Application
BMJ	British Medical Journal
BNF	British National Formulary
BP	British Pharmacopoeia
BPAC	Blood Products Advisory Committee
BPC	British Parmocopoeia Commission
BPCA	Best Pharmaceuticals for Children Act
BPI	bulk pharmaceutical ingredient
BPI	Bundesverband der Pharmazeutischen Industrie e. V. (German Federal Association of the Pharmaceutical Industry)
BRAS	Belgian Regulatory Affairs Society
BRMD	Bureau of Radiation and Medical Devices (Canada)
BSA	body surface area
BSA	bovine serum albumin
BSE	bovine spongiform encephalopathy
BSL	Biosafety Level
BVC	British Veterinary Codex
BWP	Biotechnology Working Party
CA	Competent Authority
CABs	Conformity Assessment Bodies
CAD	computer aided design
CAD	computer assisted drawing
CADREAC	Collaboration Agreement of Drug Regulatory Authorities in European Union Associated Countries
CANDA	Computer Assisted New Drug Application
CAPA	corrective and preventive action
CAPLA	Computer Assisted Product License Application

CAPRA	Canadian Association of Pharmaceutical Regulatory Affairs
CAPs	Conformity Assessment Procedures
CAS	chemical abstract system; universal numeric ID
CAS	corrective action system
CASE	computer-aided software engineering
CBER	Center for Biologics Evaluation and Research
CCDCP	Chinese Center for Disease Control and Prevention
CCRC	certified clinical research coordinator
CDC	Centers for Disease Control and Prevention
CDER	Center for Drug Evaluation and Research
CDP	clinical development plan
CDRH	Center for Devices and Radiological Health
CE Mark	Conformité Europée ne (the approval mark of the European community)
CEC	Commission of European Communities
CEECs	Central and Eastern European Countries (Bulgaria, Czech Republic, Estonia, Hungary, Latvia, Lithuania, Poland, Romania, Slovak Republic, Slovenia)
CEI	Commission Electrotechnique Internationale
CEN	European Committee for Standardization
CENELEC	European Committee for Electrotechnical Standardization
CENLECT/TC62	CENELEC Technical Committee for Medical Devices
CEP	Certificate of European Pharmacopoeia
CETF	Clinical Evaluation Task Force
CFI	Court of First Instance
CFPA	Center for Professional Advancement
CFR	Code of Federal Regulations
CFSAN	Center for Food Safety and Applied Nutrition
CFU	colony forming unit
cGLP	current Good Laboratory Practice
cGMP	current Good Manufacturing Practice
CH	clinical hold
CHPA	Consumer Health Products Association
CHPR	Canadian Health Protectorate Branch (Canadian equivalent of the FDA)
CIOMS	The Council for International Organization of Medical Sciences
CLIA	Clinical Laboratory Improvement Act (Regulations for Donor Centers)
CMC	Chemistry, Manufacturing and Controls
CME	Continuing Medical Education
CMIT	clinical malfunction incident tracking
CMS	concerned member state

CNS	central nervous system; the brain and the spinal cord
COA	Certificate of Analysis
COE	Council of Europe
COMP	Committee for Orphan Medicinal Products (within the EMEA)
COPE	Committee on Publication Ethics (UK)
CORE	comprehensive outsource risk evaluation
COREPER	Committee of Permanent Representatives from Member States (advisors to Council of Europe)
COSTART	coding symbols for a thesaurus of adverse reaction terms
COTS	commercial off-the-shelf
CPG	Compliance Policy Guidance (FDA)
CPM	Clinical Program Manager
CPMP	Committee for Proprietary Medicinal Products (within EMEA)
CPS	Chemistry, Pharmacy and Standards Subcommittee
CPSC	Consumer Product Safety Commission
CQA	customer quality assurance
CRA	Clinical Research Associate
CRADA	Cooperative Research and Development Agreement
CRC	Clinical Research Coordinator
CRF	Case Report Form
CRO	contract research organization
CRR	clinical research record
CRT	Case Report Tabulation or Clinical Report Tabulation
CSA	Canadian Standards Association
CSI	Consumer Safety Inspector
CSM	Committee on Safety Medicines
CSO	Consumer Safety Officer
CSR	Clinical Study Report
CSV	computer system validation
CSVM	computer system validation manager
CT	clinical trial
CTD	common technical document
CTFA	Cosmetic, Toiletry & Fragrance Association
CTX	clinical trial exemption
CV	curriculum vitae (similar to a résumé)
CVM	Center for Veterinary Medicine
CVMP	Committee for Veterinary Medicinal Products (within the EMEA)
DB	double-blind
DC	design control

DCF	data collection form
DDMAC	Division of Drug Marketing, Advertising, and Communications
DEA	Drug Enforcement Administration
DEFRA	Department for Environment, Food and Rural Affairs (UK)
DER	drug experience report
DESI	Drug Efficacy Study Implementation
DFI	Division of Field Investigations (a division of the US Department of Health and Human Services Public Health Service, Food and Drug Administration)
DG	Directorates-General (principal administrative agencies of the EC)
DGPharMed	Deutsche Gesellschaft für Pharmazeutische Medizin (German Society of Pharmaceutical Medicine)
DH	Department of Health (UK)
DHF	design history file
DHHS	Department of Health and Human Services
DHR	design history record
DHSS	Department of Health and Social Security (UK)
DIA	Drug Information Association
DIMDI	Deutsches Institut für Medizinsche Dokumentation und Information (German Institute for Medical Documentation and Information)
DIR	Device Incident Report
DIS	Draft International Standard
DMA	Danish Medicines Agency
DMF	Drug Master File
DMF	Device Master File
DMR	device malfunction report
DMR	device master record; tracks the manufacturing process of a medical device, much as a batch record tracks the manufacture of a drug
DNA	deoxyribonucleic acid
DNR	do not resuscitate
DPDD	Division of Pediatric Drug Development
DRA	Drug Regulatory Authority (Europe)
DRD	disposable requirements definition
DSMA	Division of Small Manufacturers Assistance
DSMB	Data and Safety Monitoring Board
DTC	direct to consumer
DTCI	direct to consumer information
DTH	delayed type hypersensitivity
EA	environmental assessment
EAB	Ethical Advisory Board

EBSA	European Biosafety Association
EBMT	European Bone Marrow Transplant registry
EBSA	European Biosafety Association
EC	European Commission
EC	European Community
ECG	electrocardiogram
ECJ	European Court of Justice
ECPHIN	European Community Pharmaceutical Information Network
ECRI	Emergency Case Research Institute
eCTD	electronic common technical document
EDC	electronic data control or electronic data capture
EDCTP	European and Developing Countries Clinical Trials Partnership (15 EU member states and Norway)
EDL	essential drug list (WHO)
EDMA	European Diagnostic Manufacturers Association
EDQM	European Directorate for the Quality of Medicines
EEA	European Economic Area
EEC	European Economic Community
EFB	European Federation of Biotechnology
EFGCP	European Federation for Good Clinical Practices
EFOIA	Electronic Freedom of Information Act
EFTA	European Free Trade Association
EGA	European Generic Medicines Association
EIR	Establishment Inspection Report
ELA	Establishment License Application
ELDU	extra-label drug use
ELM	European Laboratory Medicine (formerly European Confederation for Laboratory Medicine [ECLM])
EMC	electromagnetic compatibility
eMC	Electronic Medicines Compendium (website)
EMEA	European Medicines Evaluation Agency; represents 15 member nations
EN	European normalization (standardization)
EN46000	European Quality System Standard for Medical Devices
EOF	National Organization for Medicines (Greece's health authority)
EORTC	European Organisation for the Research and Treatment of Cancer
EOTC	European Organisation for Conformity Assessment
EP	European Pharmacopoeia
EPA	Environmental Protection Agency
EPAR	European Public Assessment Report

EPI	Expanded Programme Immunisation (WHO)
EPO	European Patent Office
EPROM	erasable programmable read only memory
ERK	electronic record keeping
ESCP	European Society of Clinical Pharmacy
ESD	electrostatic discharge
ESOP	European Society of Pharmacovigilance
ESRA	European Society of Regulatory Affairs
ETOMEP	European Technical Office for Medicinal Products
EU	European Union
EUCOMED	European Medical Technology Industry Association (formerly European Confederation of Medical Devices Association)
EUDMED	European Union Database on Medical Devices
EUDRA	European Union Drug Regulatory Authorities
EudraLex	Compilation of EU pharmaceutical legislation and guidelines (from the European Commission Pharmaceuticals web server)
EudraNet	European Union Drug Regulatory Network Authorities
EudraTrack	Database system of marketing applications within Mutual Recognition procedure.
EudraWatch	Tool for managing pharmacovigilance reports between regulators and industry
FACS	Fellow of the American College of Surgeons
FAI	further action indicated
FC	feline conjunctivitis
FCA	False Claims Act
FCC	Federal Communications Commission
FDA	Food and Drug Administration
FDAMA	Food and Drug Administration Modernization Act
FDC	Food, Drug, and Cosmetic
FDLI	Food and Drug Law Institute
FEDESA	Féderation Européenne de la Santé Animale (European Federation of Animal Health)
FERQAS	Federation of European Quality Assurance Societies
FFDCA	Federal Food, Drug, and Cosmetic Act
FFEA	Functional Failures and Effects Analysis
FMEA	failure modes effects analysis
FMECA	failure modes and effects criticality analysis
FOB	functional observational battery
FOI	Freedom of Information
FOIA	Freedom of Information Act

FONSI	finding of no significant impact
FP	finished product
FPL	final printed labeling
FR	Federal Register
FSA	Food Standards Agency (UK)
FSIS	Food Safety Inspection Service (USDA)
FTA	fault tree analysis
FTC	Federal Trade Commission
GATT	General Agreement on Tariffs and Trade
GC	gas chromatography
GCP	Good Clinical Practice
GGP	Good Guidance Practice
GHTF	Global Harmonisation Task Force
GLP	Good Laboratory Practice
GMC	General Medical Council (UK)
GMDN	Global Medical Device Nomenclature
GMO	genetically modified organism
GMP	Good Manufacturing Practice
GPI	General Pharmaceutical Inspectorate (Belgium Health Authority)
GRAS	generally recognized as safe (food ingredients)
GRASE	generally recognized as safe and effective
GRP	Good Regulatory Practices
GRP	Good Review Practice
GUI	graphical user interface
GxP	A generic term referring to various government regulations, such as GLP, GMP, and GCP
HA	hazard analysis
HACCP	hazard analysis and critical control points (inspection technique)
HCFA	Health Care Financing Administration
HEVRA	Heads of the European Veterinary Regulatory Authorities (for medicinal products)
HHS	Health and Human Services (Department of)
HIMA	Health Industry Manufacturers Association
HIPAA	Health Insurance Portability and Protection Act
HPB	Health Protection Branch (Canada's equivalent to the FDA)
HPLC	high performance liquid chromatography; also referred to as high pressure liquid chromatography
IACET	International Association for Continuing Education and Training
IACUC	Institutional Animal Care and Use Committee

IAG	Interagency Agreement
IAHC	International Animal Health Code
IB	Investigator's Brochure
IBD	international birth date
IBW	ideal body weight
IC	informed consent
ICD	informed consent document
ICF	informed consent form
ICH	International Conference on Harmonisation
IC	Informed Consent
ID	identification
IDA	Interchange of Data between Administrations
IDE	Investigational Device Exemption
IDMC	Independent Data Monitoring Committee
IDS	information disclosure statement
IDSMB	Independent Data and Safety Monitoring Board
IEC	Independent Ethics Committee
IEC	Institutional Ethics Committee or Independent Ethics Committee
IEC/TC 62	IEC Technical Committee for Electrical Equipment in Medical Practice
IEEE	Institute of Electrical and Electronics Engineers
IFAH	International Federation Animal Health Industry
IFCC	International Federation of Clinical Chemistry and Laboratory Medicine
IFPMA	International Federation of Pharmaceutical Manufacturers' Associations
IGPA	International Generic Pharmaceutical Alliance
IMB	Irish Medicines Board
IMCA	Icelandic Medicines Control Agency
INAD	Investigational New Animal Drug
INADA	Investigational New Animal Drug Application
INCI	International Nomenclature of Cosmetic Ingredients
IND	Investigational New Drug
INDA	Investigational New Drug Application
INFARMED	Instituto Nacional da Farmácia e do Medicamento (Portugal's health authority: National Institute of Pharmacy and Medicines)
INHEM	Instituto Nacional de Higiene, Epidemiologia y Microbiologia (Cuba)
INN	international non-proprietary name
IP	Intellectual Property
IPEC	International Pharmaceutical Excipients Council (IPEC Americas, IPEC Europe, and IPEC Japan)
IPR	intellectual property rights

IPRO	Independent Pharmaceutical Research Organization
IQ/OQ/PQ	installation qualification/operation qualification/performance qualification
IRB	Institutional Review Board
IRS	identical, related, or similar
ISBT	International Society of Blood Transfusion
ISCB	International Society for Clinical Biostatistics
ISI	International Sensitivity Index
ISO	International Organization for Standardization
ISO 9000	generic international standards for quality managements, quality assurance, and implementation of a quality system
ISO 9001	specific generic international conformance standard to which a firm, involved with the design, development, production, installation, and servicing, may comply with to gain compliance certification
ISO 9002	specific generic international conformance standard to which a firm, involved with production, installation, and servicing, may comply with to gain compliance certification
ISO 9003	specific generic international conformance standard to which a firm, involved with final inspection and test activity, may comply with to gain compliance certification
ISO 9004	specific generic international conformance standard for Quality Management and Quality System Elements Guideline
ISO 10011	specific generic international conformance standard for Quality Assurance Auditing Guideline
ISO 10013	specific generic internal conformance standard for Quality Manuals
ISO 13485	specific generic international conformance standard that combines ISO 9000 and EN46000
ISO 14000	specific generic international conformance standard for environmental control systems
ISO 117025	ISO accreditation for laboratories; similar to GLP compliance
ISOQOL	Internal Society of Quality of Life
ISPE	International Society for Pharmacoepidemiology
ISPOR	International Society of Pharacoeconomic and Outcomes Research
ISS	Integrated Safety Summary
ISS	Instituto Duperiore Di Sanità (Italian National Institute of Health)
ISTH	International Society for Thrombosis and Haemostasis
IU	international unit
IVD	in vitro diagnostic
IVDD	in vitro Device Directive
IVRS	Interactive Voice Recognition Systems
IWP	Immunological Working Party (within EMEA)
JAMA	Journal of the American Medical Association

JCAH	Joint Commission for the Accreditation of Hospitals
JCAHCA	Joint Commission on Accreditation of Health Care Organizations
JPMA	Japan Pharmaceutical Manufacturers Association
JRC	Joint Research Center (EU)
LC	liquid chromatography
LC/TMS	Liquid chromatography/tandem mass spectrometry
LDA	limiting dilution assay
LOA	letter of access or letter of agreement
LOCF	last observation carried forward
LOD	limit of detection
LOD	loss on drying
LOQ	limit of quantitation
LSD	least significant difference
LTE	less than effective
MA	marketing authorization
MAA	marketing authorization applicant
MAH	marketing authorization holder
MAPPS	Manual of Policies and Procedures (CDER's official policies and procedures manual)
MCN	manufacturer's control number
MDD	Medical Device Directive
MDEG	Medical Devices Expert Group
MDR	Medical Device Record
MDR	medical device report (ing)
MDUFMA	Medical Device User Fee and Modernization Act
MEB	Medicines Evaluation Board (Netherland's health authority)
MECI/SME	Manufacturing Engineering Certification Institute/The Society of Manufacturing Engineers
MedDRA	Medical Dictionary for Regulatory Activities
MEGRA	MittelEuropäische Gesllschaft für Regulatory Affairs (Mid-European group of regulatory professionals)
MEP	Member of the European Parliament
MHW	Ministry of Health and Welfare (Japan)
MHPR	Ministry of Health People's Republic (China)
MHRA	Medicines and Healthcare Products Regulatory Agency (UK)
MHW	Ministry of Health and Welfare (Japan's equivalent to the FDA)
Mil STD	Military standard; a statistical sampling plan
MOA	mechanism of action

MOH	Ministry of Health (Numerous countries including Brazil, Chile, Columbia, Costa Rica, Germany, Indonesia, Israel, Italy, Jordan, Kenya, Luxembourg, Malaysia, Malta, Morocco, Slovak Republic, Spain, Turkey, and the United Arab Emirates.)
MOU	Memorandum of Understanding
MOSS	Market-Oriented, Sector Selective (US/Japan agreement)
MPA	Medical Products Agency (Swedish national authority)
MPI	Ministry of Public Health (Belgium)
MPI	multivariate prognostic index
MRA	Mutual Recognition Agreement
MRP	mutual recognition procedure
MRFG	Mutual Recognition Facilitation Group
MRI	Magnetic Resonance Imaging
MS	Member state (European union)
MS	mass spectroscopy
MSDS	Material Safety Data Sheet
MSOG	Market Surveillance Operations Group
MTD	maximum tolerated dose
MVaP	Master Validation Plan
MVeP	Master Verification Plan
NA	not available or not applicable or not approvable
NADA	New Animal Drug Application
NAF	notice of adverse findings
NAFTA	North American Free Trade Agreement
NAI	no action indicated. (This is a designation that results from a satisfactory regulatory inspection of a facility.)
NAM	National Agency for Medicines (Finland)
NAPM	National Association of Pharmaceutical Manufacturers
NAT	Nucleic Acid Amplification Testing
NB	notified body
NCCLS	National Committee for Clinical Laboratory Standards
NCE	new chemical entity
NCI	National Cancer Institute
NCS	not clinically significant
NCTR	National Center for Toxicological Research
NDA	Non-Disclosure Agreement
NDA	New Drug Application
NDE	New Drug Evaluation
NDS	New Drug Study

NEMA	National Electrical Manufacturers Association
NF	National Formulary
NGO	non-governmental organization
NHS	National Health Service (UK)
NIA	National Institute on Aging
NIBSC	National Institute for Biological Standards and Control (UK)
NICE	National Institute for Clinical Excellence (UK)
NIH	National Institutes of Health
NIP	National Institute for Pharmacy (Hungary)
NIST	National Institute of Standards & Technology
NLEA	Nutrition Labeling and Education Act (1990)
NLM	National Library of Medicine
NMDC	National Medical Device Coalition
NMDP	National Marrow Donor Program
NME	new molecular entity
NMR	Nuclear Magnetic Resonance
NOEL	no observed effect level
NOM	National Organization for Medicine (Greece)
NoMA	Norwegian Medicines Agency
NPF	Nurse Prescribers' Formulary (EU)
NPHF	Nederlandse Public Health Federatie (National Health Federation of the Nederlands)
NRC	National Research Council
NRC	Nuclear Regulatory Commission
NSE	not substantially equivalent (510[k])
NSF	National Science Foundation
NSR	nonsignificant risk
OAI	official action indicated. (A designation which is the result of an FDA inspection that finds a facility wanting.)
OBD	optimal biologic dose
OC	Office of the Commissioner
OC	Office of Compliance (FDA for devices)
OCA	Office of Consumer Affairs
OCI	Office of Criminal Investigation
ODAC	Oncologic Drugs Advisory Committee
ODB	observational database
ODE	Office of Device Evaluation (of FDA)
OEA	Office of External Affairs (of FDA)

OECD	Organisation for Economic Co-Operation and Development (30 member countries)
OELPS	Office of Establishment Licensing and Product Surveillance (of FDA)
OEM	original equipment manufacturer
OHA	Office of Health Affairs
OHRP	Office of Human Research Protections
OIE	Office International des Epizooties (animal reporting agreement; 162 member nations)
OJEC	Official Journal of the European Communities (EC legislation)
OLA	Office of Legislative Affairs
OMCL	Official Medicines Control Laboratories (EU)
ONADE	Office of New Animal Drug Evaluation
OO	Office of Operations (of FDA)
OOS	out of specification. An OOS results from quality control testing and indicates that a product does not meet specification.
OPA	Office of Public Affairs
OPE	Office of Planning and Evaluation
ORA	Office of Regulatory Affairs
ORCA	Organization of Regulatory and Clinical Associates
OS	operating system
OSHA	Occupational Safety and Health Administration
OSR	Office of Standards and Regulations (of FDA)
OST	Office of Science and Technology (of FDA)
OTC	over-the-counter (therapeutic drugs)
PAHO	Pan-American Health Organization (WHO/UN)
PAI	Pre-Approval Inspection
PAMP	Post Approval Monitoring Program
PC	personal computer
PC	protocol change (amendment)
PC	process control
PCT	photochemical treatment
PD	pharmacodynamics
PDA	Parenteral Drug Association
pdf	portable document format
PDMA	Prescription Drug Marketing Act
PDP	Product Development Protocol (for medical devices)
PDR	Physician's Desk Reference
PDUFA	Prescription Drug User Fee Act (1992)
PECAA	Protocol to the European Agreement on Conformity Assessment and Acceptance of Products

PEFRAS	Pan European Federation of Regulatory Affairs Societies
PEG	Paediatric Expert Group (EMEA)
PEI	Paul-Ehrlick-Institute (Germany's health authority for sera, vaccines, blood products)
PERF	Pan European Regulatory Forum
PERI	Pharmaceutical Research and Manufacturers of America Education and Research Institute
PES	programmable electronic system
PFU	plaque forming units
PHARE	Poland and Hungary Assistance for Economic Restructuring (10 candidate countries)
PhRMA	Pharmaceutical Research and Manufacturers of America
PHS	Public Health Service
PI	Principal Investigator
PI	package insert (approved product labeling)
PIC	Pharmaceutical Inspection Convention
PIL	patient information leaflet
PhVWP	Pharmacovigilance Working Party
PK	pharmacokinetics
PL	package leaflet
PLA	Product License Application (for biologics)
PLA/ELA	Product License Application/Establishment License Application
PMA	Premarket Approval
PMAA	Premarket Approval Application
PMS	postmarketing surveillance
POM	prescription-only medicine
POMA	Pharmaceutical Outsourcing Management Association
PPA	Poison Prevention Act
PPI	patient package insert
PQA	Preproduction Quality Assurance
PQG	Pharmaceutical Quality Group
PRO	patient reported outcome
PSUR	periodic safety update report
PTCC	Pharmacology/Toxicology Coordinating Committee (CDER)
PTO	Patent and Trademark Office
PVC	polyvinylchloride
QA	quality assurance
QAU	quality assurance unit
QC	quality control

QMS	quality management system
QNS	quantity not sufficient
QoL	quality of life
QRD	quality review of records
QRE	Quality and Reliability Engineering
QS	Quality System
QSIT	Quality Systems Inspection Technique
QSR	Quality Systems Regulations
QWP	Quality Working Party (EMEA)
QU	quality unit
R & D	research and development
RA	risk analysis
RAB	Registration Accreditation Board, the US ISO 9000 accreditation body
RAC	Reviewer Affairs Committee (CDER)
RAC	Regulatory Affairs Certification
RAC EU	European Regulatory Affairs Certification
RAPS	Regulatory Affairs Professionals Society
RRC	Regulatory Research Center (RAPS)
RCT	randomized clinical trial
RDE	remote data entry
RF	radio frequency
RFDD	Regional Food and Drug Director
RHIA	Registered Health Information Administrators
RIA	radioimmunoassay
RL	regulatory letter
RMA	return material authorization
RMF	risk management file
RNA	ribonucleic acid
RPN	risk priority number (a component of risk analysis)
RPSGB	Royal Pharmaceutical Society for Great Britain
RTF	Refuse to File (FDA decision to refuse to file an application)
SADR	suspected adverse drug reaction
SAE	serious adverse event
SAM	State Agency of Medicines (Estonia)
SAS	statistical analysis system
SBA	summary basis of approval
SBU	strategic business unit
SC	Study Coordinator
SCD	sterile connection device

SCMP	software configuration management plan
SD	standard deviation
S/D	solvent/detergent
SDD	software design description
SDS	system design specification
SDSOP	software development standard operating procedure
SE	substantially equivalent (510[k])
SE	standard error
SEA	Single European Act
SEC	Security and Exchange Commission
SEI	Software Engineering Institute
SEM	standard error of the mean
SERC	Statistics and Epidemiology Research Corporation
SHA	software hazard analysis
SI	Systeme International
SIAR	Societa Italianà Attività Regolatorie (Italian Society Regulatory Activity)
SLAM	Scientists/Lab Managers
SMDA	Safe Medical Devices Act (1990)
SME	significant medical event
SNDA	Supplemental New Drug Application
SNF	Skilled Nursing Facility
SNITEM	Syndicat National de L'Industrie des Technologies Médicales (French National Association for Medical Technology Industries)
SOE	Statement of Experience (clinical trial investigator)
SOP	standard operating procedure
SPC	Statistical Process Control
SPC	summary of product characteristics
SPSS	statistical package for the social sciences
SQAP	software quality assurance plan
SRD	systems requirements definition
SRS	software requirements specification
SSE	Summary of Safety and Effectiveness
STS	Society of Thoracic Surgeons
SUD	single use device
SUPAC	scale up and post approval changes
SUSAR	suspected unexpected serious adverse reactions
Swissmedic	Swiss Agency for Therapeutic Products
SVVP	Software verification and validation plan
SWP	Safety Working Party (EMEA)

TABU	official Finland journal that publishes marketing authorizations
TGA	Therapeutic Goods Administration (Australia)
TK	toxicokinetics
TLC	thin layer chromatography; chromatography through a thin layer of cellulose or similar inert material supported on a glass or plastic plate
TSE	transmissible spongiform encephalopathy
TRIPS	Trade-Related Aspects of Intellectual Property Rights (WTO Agreement)
TÜV	German Test Agency and Notified Body. (This organization does the assessment for a CE Mark.)
UADE	unanticipated adverse device effect
UDS	unconditional stimulus
UDS	unscheduled DNA synthesis
UL	Underwriters Laboratories, Inc.
UN	United Nations
USC	US Code
USCA	US Code Annotated
USDA	US Department of Agriculture
USP	US Pharmacopeia
USS	undiluted sampling system
UV	ultraviolet A, B, and C light with wavelengths ranging from 200 to 400 nm
UVA	ultraviolet A light with wavelengths ranging from 320 to 400 nm
UVB	ultraviolet B light with wavelengths ranging from 280 to 320 nm
UVR	ultraviolet radiation using a xenon-arc solar simulator
V & V	verification and validation
VA	Veterans Affairs (US Department)
VAERS	Vaccine Adverse Event Reporting System
VAI	voluntary action indicated (a designation that is the result of a regulatory inspection that finds minor and correctable violations)
VEG	Vaccine Expert Group (EU)
VMD	Veterinary Medicines Director
VPN	virtual private network
WHA	World Health Assembly
WHO	World Health Organization
WHOART	World Health Organization Adverse Reaction Terms (terminology)
WL	warning letter
WNL	within normal limits
WSMI	World Self-Medication Industry (54 member countries)
WTO	World Trade Organization

Symbols and Abbreviations

To present information succinctly, writers often use symbols and abbreviations. These are useful devices, but when these conventions are inconsistent, readability suffers and so does the professional appearance of documents. That's why many companies develop style guides for their own documents. Documents where authors use an x, ×, or • interchangeably to indicate multiplication impairs consistency. It's better to use a symbol than a letter; and it's best to determine which one is preferable for the company's documents. A style guide that addresses such an issue will help prevent a sentence like this "The total AUC was 10.67 pg/mL•min" from being followed by a sentence like this "The AUC was 8.42 pg/mL × min" further on in the document.

The following are commonly used symbols and abbreviations.

+	plus; add; positive
–	minus; subtract; negative
±	plus or minus; positive or negative
× or •	times, multiplied by
÷ or /	divided by
= or : :	equals; as identical to; congruent with
	not identical; not equal to
%	percent
≤	less than or equal to
≥	greater than or equal to
:	ratio
∝	varies as (proportional to)
Σ	summation of
™	trademark
©	copyright mark
®	registered mark
°	degree
,	solid angle
α	alpha
β	beta
μ	micron (mu) (also called micrometer)
μg	microgram
μm	micrometer (micron)
	wavelength (lambda)
	velocity; specific volume (volume divided by mass)
	gamma
	lambda; wavelength; coefficient of linear expansion
A	ampere
Å	angstrom (unit of wavelength measure)
abs	absolute; absorbance
ac	alternating current
ad lib.	as desired
amor	amorphous
amt	amount
anhyd	anhydrous
aq	aqueous
at no	atomic number
at wt	atomic weight
atm	atmosphere
Bé	Baumé degree
b.i.d.	twice a day
bp	boiling point; base pair
Btu	British thermal unit
Bx	biopsy
C	Celsius
ca.	approximately
Cal	large calorie (kilogram calorie or kilocalorie)
cal	small calorie (gram calorie)
cc	cubic centimeter
gal	gallon
gr	grain
Gy	gray; absorbed dose
h	hour
h	Planck's constant
Hb	hemoglobin

cd	candela; luminous intensity
cg	centigram
Ci	curie(s); activity
c.i.	confidence intervals
cm	centimeter
cm^2	square centimeter
cm^3	cubic centimeter
conc	concentration
conc'd.	concentrated
cos	cosine
cP	candlepower
cu	cubic
cu mm	cubic millimeter
D	density diopter
"D"	Rh factor
d	diameter
d	day
dB	decibel
dc	direct current
df	degrees of freedom
dg	decigram
dil.	dilute
dr	dram
ds	double strand, double stranded
dse	disease
Dx	diagnosis
dyn	dyne
e.g.,	for example
esu	electrostatic unit
et al.	and others
etc.	and so forth
et seq.	and the following
eV	electronvolt
expt.	experiment
ext.	extension
F	Fahrenheit
F	frictional loss
f	frequency
FH	family history
fl oz	fluid ounce
ft	foot
ft^2	square foot
ft^3	cubic foot
ft·c	foot-candle
ft·lb	foot-pound
g	acceleration due to gravity
g	gram
G	gauss
g·cal	gram-calorie
max.	maximum (abbreviation in tables)
mCi	millicurie
meq	milliequivalent
MeV	million (or mega) electron volt
MHz	million (or mega-) hertz
MeV	million (or mega-) electronvolt

hp	horsepower
hp·h	horsepower-hour
hr	hour
h.s.	bedtime (hour of sleep)
Hx	history
hyg	hygroscopic
Hz	hertz (formerly cycles per second, cps)
I	electric current
ibid.	in the same place
i.d.	identification
i.e.	that is
Ig	immunoglobulin
in	inch
in^2	square inch
in^3	cubic inch
insol.	insoluble
i.p.	intraperitoneal
iso	isotropic
i.v.	intravenous; within a vein or veins
J	joule; energy, work, quantity of heat
J	flux (density)
JK	Kelvin; unit of thermodynamic temperature
k	kilo; thousand
kc	kilocycle
kcal	kilogram-calorie
kd	daltons/kilodaltons
kg	kilogram; unit of mass
kg/m^3	kilogram per cubic meter
lft	left
kW	kilowatt
kWh	kilowatt-hour
L	liter
l	lumin
lx	lux
lb	pound
lb/ft^3	pound per cubic foot
ln	natural logarithm
log	logarithm
MW	molecular weight; mass
m	meter
m^2	square meter
m^3	cubic meter
m^3/kg	cubic meter per kilogram
q2h	every two hours
q.i.d.	four times a day
q.n.s.	quantity not sufficient
q.o.d.	every other day
q.s.	quantity sufficient; as much as needed
qt	quart
q.v.	which see
mg	milligram
min	minute
min.	minimum (abbreviation in tables)
mks	meter-kilogram-second (system of units)
mL	milliliter
mm	millimeter
mm^2	square millimeter
mm^3	cubic millimeter
mm	millimeter of mercury
mol	mole
mp	melting point
mph	miles per hour
N	newtons
N	normality (as 1 *N*)
n	index of refraction
n.a.	not applicable
Na	sodium
NaCl	saline
n.d.	not done
ng	nanogram
nm	nanometer
noc.	night (nocturnal)
nr	number
oz	ounce
P	poise
p or *P*	probability (p-value)
Pa	pascal
pfu	plaque forming unit
pH	relative hydrogen ion concentration expressed on a logarithmic scale; used in expressing both acidity and alkalinity
pk	pharmacokinetic
p.o.	per os (by mouth)
post op	postoperation
ppb	parts per billion
ppm	parts per million
ppt	precipitate
psi	pounds per square inch
p. sol.	partly soluble
pt	patient
px	prophylactic
Q1	first quarter of the year; also Q2, Q3, Q4
q	every
qd	every day
qh	every hour
sol'n.	solution
sp	specific
sp gr	specific gravity
sp ht	specific heat
sq	square
sq rt	square root
sr	steradian (solid angle)

R	roentgen (international unit for x-rays)
r	radius
rad	radian (plane angle)
rd	rad; absorbed dose
Rh	rhesus factor
R/O	rule out
rpm	revolutions per minute
rps	revolutions per second
rt	right
RT	room temperature
rx	prescription: a written formula for a remedy or a medicinal preparation
rxn	reaction
sat'd.	saturated
s	second
sd	standard deviation
SI	International System of Units
sin	sine
ss	single stranded
Sx	symptom
T	temperature
t	time
t	mass; metric ton
tan	tangent
t.i.d.	three times a day
t.i.w.	three times a week
tx	therapeutic
txn	transfusion
U	unit
US	United States
V	volt; volume
vs	versus
W	watt
Wh	watt-hour
wk	week
wt	weight

REFERENCES

Aaron, Jane E. 2001. *The Little Brown Compact Handbook,* Custom Fourth Edition. Reading, MA: Addison Wesley Educational Publishers.

Alvarez, Joseph A. 1980. *The Elements of Technical Writing.* New York: Harcourt Brace Jovanovich.

American Medical Association. 1998. *American Medical Association Manual of Style: A Guide for Authors and Editors,* 9th ed. Baltimore: Williams & Wilkins.

American National Standard for Bibliographic References. 1976. (ANSI, Z39.29). New York: American National Standards Institute.

Baker, C. L. 1989. *English Syntax.* Cambridge, MA: MIT Press.

Baker, Sheridan. 1977. *The Practical Stylist,* 4th ed. New York: Harper and Row.

Baugh, Albert C. and Thomas Cable. 1978. *A History of the English Language,* 3rd ed. Englewood Cliffs, NJ: Prentice-Hall.

Bernstein, Theodore M. 1971. *Miss Thistlebottom's Hobgoblins: The Careful Writer's Guide to the Taboos, Bugbears and Outmoded Rules of English Usage.* New York: Simon & Schuster.

British Medical Journal. http://bmj.com Accessed November 3, 2003.

CDER Acronym List. http://www.fda.gov/cder/handbook/acronym.htm. Accessed July 2003.

CDRH Abbreviations and Acronyms. http://www.fda.gov/cdrh/ost/reports/fy95/abbreviations.html. Accessed July 2003.

Celce-Murcia, Marianne and Diane Larsen-Freeman. 1983. *The Grammar Book: An ESL/EFL Teacher's Course.* New York: Newbury House Publishers.

The Chicago Manual of Style, 14th ed. 1993. Chicago: University of Chicago Press.

Cianfrani, Charles A, Joseph J. Tsiakals, and John E. West. 2001. *ISO 9001:2000 Explained,* 2nd ed. Milwaukee, WI: ASQC Quality Press.

Code of Federal Regulations. 1998. Title 21, Part 211, *Good manufacturing practices for finished pharmaceuticals.* Washington, DC: U.S. Government Printing Office.

Corbett, Edward P. J. 1971. *Classical Rhetoric for the Modern Student.* New York: Oxford University Press.

Corbett, Edward P. J. and Sheryl L. Finkle. 1992, *The Little English Handbook,* 6th ed. New York: HarperCollins Publishers.

Cran, William, Robert McCrum, and Robert MacNeil. 1987. *The Story of English.* New York: Penguin Books.

Crystal, David. 1987. *The Cambridge Encyclopedia of Language.* New York: Cambridge University Press.

DeSain, C. V. and C. V. Sutton. 1996. *Documentation Practices: A Complete Guide to Document Development and Management for GMP and ISO9000 Compliant Industries.* Cleveland, OH: Advanstar Communications.

Dicker, Susan J. 1996. *Languages in America: A Pluralist View.* Clevedon, UK: Multilingual Matters.

Dornan, Edward A. and Charles W. Dawe. 1984. *The Brief English Handbook.* Boston: Little, Brown and Company.

Drug Enforcement Administration (DEA). Washington, D.C.: U.S. Government, Department of Justice.

Drug GMP Reports. Arlington, VA: Washington Business Information.

Dumont, Raymond A. and John M. Lannon. 1987. *Business Communications,* 2nd ed. Boston: Little, Brown and Company.

Einsohn, Amy. 2000. *The Copyeditor's Handbook: A Guide for Book Publishing and Corporate Communications.* Berkeley: University of California Press.

EU Regulatory Affairs Acronyms & Definitions. 2003. Rockville, MD: Regulatory Affairs Professional Society.

fda.gov/cder/guidance/index.htm. Guidance for Industry. Investigating Out of Specification (OOS) Test Results for Pharmaceutical Production.

Food and Drug Administration, Center for Devices and Radiological Health. 1993. *Write It Right: Recommendations for Developing User Instruction Manuals for Medical Devices Used in Home Health Care.*

Food and Drug Administration. *FDA/CVM-Related Acronyms and Abbreviations.* http://www.fda.gov/cvm/index/other/acronym.htm. Accessed August 2003.

Food and Drug Administration. 1988. *Guidance for Industry. Guideline for the Monitoring of Clinical Investigations.*

Fowler, H. Ramsey. 1980. *The Little, Brown Handbook.* Boston: Little, Brown and Company.

GAMP Forum. 2001. *GAMP 4 The Good Automatic Manufacturing Practice (GAMP) Guide for Validation of Automated Systems in Pharmaceutical Manufacture.* Amsterdam: GAMP.

The Genesis Report/RX. 1998. New products could double the dermatology market by 2007. 7(5):9. Montclair, NJ: Genesis.

The Gold Sheet. Chevy Chase, MD: F-D-C-Reports.

Gough, Janet. 2001. *Hosting a Compliance Inspection.* Bethesda, MD: Parenteral Drug Association (PDA) and Godalming, Surrey, UK: Davis Horwood International.

Government and Regulatory Bodies. http://www.pharmweb.net/pwmirror/pwk/pharmwebk.html. Accessed August 2003.

Grimaldi, Monica and Janet Gough. 2001. *The External Quality Audit.* Bethesda, MD: Parenteral Drug Association (PDA) and Godalming, Surrey, UK: Davis Horwood International.

Grimaldi, Monica and Janet Gough. 2001. *The Internal Quality Audit.* Bethesda, MD: Parenteral Drug Association (PDA) and Godalming, Surrey, UK: Davis Horwood International.

Harnack, Gordon. 1999. *Mastering and Managing the FDA Maze, Medical Device Overview.* Milwaukee, WI: ASQ Quality Press.

Harty, Kevin J. 1980. *Strategies for Business and Technical Writing.* New York: Brace Jovanovich.

Heinson, J. 1972. *Article and Noun in English.* The Hague: Mouton.

Hornby, A. S. 1986. *Oxford Advanced Learner's Dictionary of Current English,* 4th ed. A. P. Cowie, Chief Ed. Oxford: Oxford University Press, 1986.

http://eduserv.hscer.washington.edu/bioethics/topics/consent.html#ques1, Accessed October 10, 2004.

Kollin, Martha. 1982. *Understanding English Grammar.* New York: Macmillan Publishing.

Lang, Thomas A. and Michelle Secic. 1997. *How To Report Statistics In Medicine.* Philadelphia, PA: American College of Physicians, 1997.

Lannon, John M. 1985. *Technical Writing,* 3rd ed. Boston: Little, Brown and Company.

Lederer R. and Dowis R. 1999. *Sleeping Dogs Don't Lay: Practical Advice for the Grammatically Challenged.* New York: St. Martin's Press.

Leich, G. 1971. *Meaning and the English Verb.* London: Longman.

Leki, Illona. 1992. *Understand ESL Writers: A Guide for Teachers.* Portsmith, NH: Boynton/Cook Publishers.

Liu, Margaret B. and Kate Davis. 2001. *Lessons From A Horse Named Jim. Durham.* Durham, NC: Duke Clinical Research Institute.

Locke, Joanne. 2003. The plain language movement. *AMWA Journal,* 18(1).

Lowrey, R. G, Charles Moorman, and Robert Barnes. 1960. *Mechanics of English.* New York: Appleton-Century-Crofts.

Marius, Richard. 1985. *A Writer's Companion.* New York: Alfred A. Knopf.

McArthur, Tom. 1986. *Longman Lexicon of Contemporary English.* Singapore: Longman Singapore Publishers.

Melnick, Arnold D.O. 2002. KYOS – five easy questions to erase your writer's block. The Writer Side: Melnick on Writing. *AMWA Journal,* 17(1).

Merriam Webster's Collegiate Dictionary, 10th ed. 1993. Springfield, MA: Merriam-Webster Publishers.

Nettleton, David and Janet Gough. 2004. *Achieving and Maintaining Compliance with 21 CFR Part 11 and 45 CFR Parts 160, 162, and 164.* Boca Raton, FL: CRC Press.

Nettleton, David and Janet Gough. 2003. *Commercial Off-the-Shelf (COTS) Software Validation for 21 CFR Part 11 Compliance.* Bethesda, MD: Parenteral Drug Association (PDA) and Godalming, Surrey, UK: Davis Horwood International.

Orwell, George. *Politics and the English Language.* http://www.resort.com/~prime8/Orwell/patee.html

The Pink Sheet. Chevy Chase, MD: F-D-C Reports, Inc.

Pyles, Thomas and John Algeo. 1993. *The Origins and Development of the English Language.* 4th edition. Fort Worth, TX: Harcourt Brace College Publishers.

Raimes, Ann. 1992. *Exploring Through Writing: A Process Approach to ESL Writing,* 2nd ed. New York: St. Martins Press.

Richek, Margaret Ann. 1993. *The World of Words: Vocabulary for College Students,* 9th ed. Boston: Houghton Mifflin Company.

Rollins School of Public Health. RSPH Public Health InfoLinks: Public Health Sites by Country. http://www.sph.emory.edu/PHIL/country.html. Accessed August 2003.

Schwager, Edith. 1991. *Medical English Usage and Abusage.* Phoenix, AZ: Oryx Press.

Stedman's Medical Dictionary, 27th ed. 2000. Baltimore, MD: Lippincot Williams & Wilkins.

Strunk, William Jr. and E. B. White. 1979. *The Elements of Style,* 3rd ed. New York: Macmillan Publishing.

Style Manual Committee Council of Biology Editors. 1994. *Scientific Style and Format. The CBE Manual for Authors, Editors, and Publishers,* 6th ed. Cambridge, UK: Cambridge University Press.

Swan, Michael. 1996. *Practical English Usage.* New York: Oxford University Press.

The Tan Sheet. Chevy Chase, MD: F-D-C Reports, Inc.

Trautman, Kimberly A. 1997. *The FDA and Worldwide Quality System Requirements Guidebook for Medical Devices.* Milwaukee, Wisconsin: ASQC Quality Press.

United States Pharmacopeia and the National Formulary 26.

Webster's Dictionary of English Usage. 1989. Springfield, MA: Merriam-Webster Publishers.

Willis-Pence-Gonzalez, A. 1976. *Grammar and Composition Skills: Generating Sentences and Paragraphs,* 3rd ed. New York: Holt, Rinehart and Winston.

The World's Major Languages. 1990. Bernard Comrie, Ed. New York: Oxford University Press.

www.fda.gov/for/warning letters. Accessed July 2003.
www.plainlanguagelanguage.gov. Accessed November 3, 2003
Zinsser, William. 2001. *On Writing Well*. New York: Harper Resource.

Index

D

E

F

Q

R

S